Reinforced
concrete
design

Other titles of interest to civil engineers:

Reinforced concrete design

to Eurocode 2

SIXTH EDITION

BILL MOSLEY

FORMERLY NANYANG TECHNOLOGICAL UNIVERSITY, SINGAPORE
AND DEPARTMENT OF CIVIL ENGINEERING
UNIVERSITY OF LIVERPOOL

JOHN BUNGEY

DEPARTMENT OF ENGINEERING
UNIVERSITY OF LIVERPOOL

RAY HULSE

FORMERLY FACULTY OF ENGINEERING AND COMPUTING
COVENTRY UNIVERSITY

palgrave
macmillan

First published 2007 by
PALGRAVE MACMILLAN
Houndmills, Basingstoke, Hampshire RG21 6XS and
175 Fifth Avenue, New York, N.Y. 10010
Companies and representatives throughout the world

PALGRAVE MACMILLAN is the global academic imprint of the Palgrave
Macmillan division of St. Martin's Press, LLC and of Palgrave Macmillan Ltd.
Macmillan is a registered trademark in the United States, United Kingdom
and other countries. Palgrave is a registered trademark in the European
Union and other countries.

ISBN-13 978–0–230–50071–6
ISBN-10 0–230–50071–4

This book is printed on paper suitable for recycling and made from fully
managed and sustained forest sources. Logging, pulping and manufacturing
processes are expected to conform to the environmental regulations of the
country of origin.

A catalogue record for this book is available from the British Library.

A catalog record for this book is available from the Library of Congress.

10   9   8   7   6   5   4
16   15   14   13   12   11   10   09   08

Printed in China

Contents

Preface

The purpose of this book is to provide a straightforward introduction to the principles and methods of design for concrete structures. It is directed primarily at students and young engineers who require an understanding of the basic theory and a concise guide to design procedures. Although the detailed design methods are generally according to European Standards (*Eurocodes*), much of the theory and practice is of a fundamental nature and should, therefore, be useful to engineers in countries outside Europe.

The search for harmonisation of Technical Standards across the European Community (EC) has led to the development of a series of these *Structural Eurocodes* which are the technical documents intended for adoption throughout all the member states. The use of these common standards is intended to lower trade barriers and enable companies to compete on a more equitable basis throughout the EC. *Eurocode 2* (EC2) deals with the design of concrete structures, which has most recently been covered in the UK by British Standard BS8110. BS8110 is scheduled for withdrawal in 2008. *Eurocode 2*, which will consist of 4 parts, also adopts the limit state principles established in British Standards. This book refers primarily to part 1, dealing with general rules for buildings. *Eurocode 2* must be used in conjunction with other European Standards including Eurocode 0 (Basis of Design) that deals with analysis and Eurocode 1 (Actions) that covers loadings on structures. Other relevant Standards are Eurocode 7 (Geotechnical Design) and Eurocode 8 (Seismic Design).

Several UK bodies have also produced a range of supporting documents giving commentary and background explanation for some of the requirements of the code. Further supporting documentation includes, for each separate country, the *National Annex* which includes information specific to the individual member states and is supported in the UK by the British Standards publication PD 6687:2006 which provides background information. Additionally, the British Cement Association has produced *The Concise Eurocode for the Design of Concrete Buildings* which contains material that has been distilled from EC2 but is presented in a way that makes it more user-friendly than the main Eurocode and contains only that information which is essential for the design of more everyday concrete structures. The Institution of Structural Engineers has also produced a new edition of their Design Manual. These latter two documents also contain information not included in EC2 such as design charts and design methods drawn from previous British Standards. In this text, reference is made to both EC2 and the Concise Code.

The presentation of EC2 is oriented towards computer solution of equations, encompasses higher concrete strengths and is quite different from that of BS8110. However the essential feature of EC2 is that the principles of design embodied in the document are almost identical to the principles inherent in the use of BS8110. Hence, although there are some differences in details, engineers who are used to designing to the existing British Standard should have no difficulty in grasping the essential features of this new code. New grades of reinforcing steel have been recently been introduced and design is now based on concrete cylinder strength, with both of these changes incorporated in this edition.

Changes in terminology, arising partly from language differences, have resulted in the introduction of a few terms that are unfamiliar to engineers who have worked with BS8110. The most obvious of these is the use of *actions* to describe the loading on structures and the use of the terms *permanent* and *variable actions* to describe dead and imposed loads. Notwithstanding this, UK influence in drafting the document has been very strong and terminology is broadly the same as in existing British Standards. Throughout this text, terminology has been kept generally in line with commonly accepted UK practice and hence, for example, *loads* and *actions* are used interchangeably. Other 'new' terminology is identified at appropriate points in the text.

The subject matter in this book has been arranged so that chapters 1 to 5 deal mostly with theory and analysis while the subsequent chapters cover the design and detailing of various types of member and structure. In order to include topics that are usually in an undergraduate course, there is a section on earth-retaining structures and also chapters on prestressed concrete and composite construction. A new section on seismic design has also been added.

Important equations that have been derived within the text are highlighted by an asterisk adjacent to the equation number and in the Appendix a summary of key equations is given. Where it has been necessary to include material that is not directly provided by the Eurocodes, this has been based on currently accepted UK good practice.

In preparing this new edition [which replaces *Reinforced Concrete Design to EC2* (1996) by the same authors], the principal aim has been to retain the structure and features of the well-established book *Reinforced Concrete Design* by Mosley, Bungey and Hulse (Palgrave) which is based on British Standards. By comparing the books it is possible to see the essential differences between Eurocode 2 and existing British Standards and to contrast the different outcomes when structures are designed to either code.

It should be emphasised that Codes of Practice are always liable to be revised, and readers should ensure that they are using the latest edition of any relevant standard.

Finally, the authors would like to thank Mrs Mary Davison for her hard work, patience and assistance with the preparation of the manuscript.

Acknowledgements

Permission to reproduce EC2 Figures 5.2, 5.3, 6.7, 8.2, 8.3, 8.7, 8.9, 9.4 and 9.9 and Tables A1.1 (EN 1990), 7.4, 8.2 and 8.3 from BS EN 1992-1-1: 2004 is granted by BSI. British Standards can be obtained from BSI Customer Services, 389 Chiswick High Road, London W4 4AL (tel. +44 (0)20 8996 9001, email: cservices@bsi-global.com). We would also like to acknowledge and thank ARUP for permission to reproduce the photographs shown in chapters 2 to 8, and 12.

The photograph of The Tower, East Side Plaza, Portsmouth (cover and chapter 1) is reproduced by courtesy of Stephenson RC Frame Contractor, Oakwood House, Guildford Road, Bucks Green, Horsham, West Sussex.

Dedicated to all our families for their encouragement and patience whilst writing this text

Notation

Notation is generally in accordance with EC2 and the principal symbols are listed below. Other symbols are defined in the text where necessary. The symbols ε for strain and f for stress have been adopted throughout, with the general system of subscripts such that the first subscript refers to the material, c – concrete, s – steel, and the second subscript refers to the type of stress, c – compression, t – tension.

E	modulus of elasticity
F	load (action)
G	permanent load
I	second moment of area
K	prestress loss factor
M	moment or bending moment
N	axial load
Q	variable load
T	torsional moment
V	shear force

a	deflection
b	breadth or width
d	effective depth of tension reinforcement
d'	depth to compression reinforcement
e	eccentricity
h	overall depth of section in plane of bending
i	radius of gyration
k	coefficient
l	length or span
n	ultimate load per unit area
$1/r$	curvature of a beam
s	spacing of shear reinforcement *or* depth of stress block
t	thickness
u	punching shear perimeter
x	neutral axis depth
z	lever arm

A_c	concrete cross-sectional area
A_p	cross-sectional area of prestressing tendons
A_s	cross-sectional area of tension reinforcement
A_s'	cross-sectional area of compression reinforcement
$A_{s,req}$	cross-sectional area of tension reinforcement required at the ultimate limit state

$A_{s,prov}$	cross-sectional area of tension reinforcement provided at the ultimate limit state
A_{sw}	cross-sectional area of shear reinforcement in the form of links or bent-up bars
E_{cm}	secant modulus of elasticity of concrete
E_s	modulus of elasticity of reinforcing or prestressing steel
G_k	characteristic permanent load
I_c	second moment of area of concrete
M_{bal}	moment on a column corresponding to the balanced condition
M_{Ed}	design value of moment
M_u	ultimate moment of resistance
N_{bal}	axial load on a column corresponding to the balanced condition
N_{Ed}	design value of axial force
P_0	initial prestress force
Q_k	characteristic variable load
T_{Ed}	design value of torsional moment
V_{Ed}	design value of shear force
W_k	characteristic wind load
b_w	minimum width of section
f_{ck}	characteristic cylinder strength of concrete
f_{cm}	mean cylinder strength of concrete
f_{ctm}	mean tensile strength of concrete
f_{pk}	characteristic yield strength of prestressing steel
f_{yk}	characteristic yield strength of reinforcement
g_k	characteristic permanent load per unit area
k_1	average compressive stress in the concrete for a rectangular parabolic stress block
k_2	a factor that relates the depth to the centroid of the rectangular parabolic stress block and the depth to the neutral axis
l_a	lever-arm factor $= z/d$
l_0	effective height of column or wall
q_k	characteristic variable load per unit area
α	coefficient of thermal expansion
α_e	modular ratio
ψ	action combination factor
γ_c	partial safety factor for concrete strength
γ_f	partial safety factor for loads (actions), F
γ_G	partial safety factor for permanent loads, G
γ_Q	partial safety factor for variable loads, Q
γ_s	partial safety factor for steel strength
δ	moment redistribution factor
ε	strain
σ	stress
ϕ	bar diameter

Notation for composite construction, Chapter 12

A_a	Area of a structural steel section
A_v	Shear area of a structural steel section
b	Width of the steel flange

b_{eff}	Effective width of the concrete flange
d	Clear depth of steel web *or* diameter of the shank of the shear stud
E_a	Modulus of elasticity of steel
$E_{c,eff}$	Effective modulus of elasticity of concrete
E_{cm}	Secant modulus of elasticity of concrete
f_{ctm}	Mean value of the axial tensile strength of concrete
f_y	Nominal value of the yield strength of the structural steel
f_u	Specified ultimate tensile strength
h	Overall depth; thickness
h_a	Depth of structural steel section
h_f	Thickness of the concrete flange
h_p	Overall depth of the profiled steel sheeting excluding embossments
h_{sc}	Overall nominal height of a shear stud connector
I_a	Second moment of area of the structural steel section
I_{transf}	Second moment of area of the transformed concrete area and the structural steel area
k_1	Reduction factor for resistance of headed stud with profiled steel sheeting parallel with the beam
k_t	Reduction factor for resistance of headed stud with profiled steel sheeting transverse to the beam
L	Length, span
M_c	Moment of resistance of the composite section
n	Modular ratio *or* number of shear connectors
n_f	Number of shear connectors for full shear connection
P_{Rd}	Design value of the shear resistance of a single connector
R_{cf}	Resistance of the concrete flange
R_{cx}	Resistance of the concrete above the neutral axis
R_s	Resistance of the steel section
R_{sf}	Resistance of the steel flange
R_{sx}	Resistance of the steel flange above the neutral axis
R_v	Resistance of the clear web depth
R_w	Resistance of the overall web depth $= R_s = 2R_{sf}$
R_{wx}	Resistance of the web above the neutral axis
t_f	Thickness of the steel flange
t_w	Thickness of the steel web
$W_{pl,y}$	Plastic section modulus of a steel structural section
$\bar{x}$	Distance to the centroid of a section
z	Lever arm
δ	Deflection at mid span
ε	Constant equal to $\sqrt{235/f_y}$ where f_y is in N/mm^2
γ	factor of safety
v_{Ed}	Longitudinal shear stress in the concrete flange
η	Degree of shear connection

CHAPTER 1

Properties of reinforced concrete

CHAPTER INTRODUCTION

Reinforced concrete is a strong durable building material that can be formed into many varied shapes and sizes ranging from a simple rectangular column, to a slender curved dome or shell. Its utility and versatility are achieved by combining the best features of concrete and steel. Consider some of the widely differing properties of these two materials that are listed below.

	Concrete	Steel
strength in tension	poor	good
strength in compression	good	good, but slender bars will buckle
strength in shear	fair	good
durability	good	corrodes if unprotected
fire resistance	good	poor – suffers rapid loss of strength at high temperatures

It can be seen from this list that the materials are more or less complementary. Thus, when they are combined, the steel is able to provide the tensile strength and probably some of the shear strength while the concrete, strong in compression, protects the steel to give durability and fire resistance. This chapter can present only a brief introduction to the basic properties of concrete and its steel reinforcement. For a more comprehensive study, it is recommended that reference should be made to the specialised texts listed in Further Reading at the end of the book.

1.1 ▌ Composite action

The tensile strength of concrete is only about 10 per cent of the compressive strength. Because of this, nearly all reinforced concrete structures are designed on the assumption that the concrete does not resist any tensile forces. Reinforcement is designed to carry these tensile forces, which are transferred by bond between the interface of the two materials. If this bond is not adequate, the reinforcing bars will just slip within the concrete and there will not be a composite action. Thus members should be detailed so that the concrete can be well compacted around the reinforcement during construction. In addition, bars are normally ribbed so that there is an extra mechanical grip.

In the analysis and design of the composite reinforced concrete section, it is assumed that there is a perfect bond, so that the strain in the reinforcement is identical to the strain in the adjacent concrete. This ensures that there is what is known as 'compatibility of strains' across the cross-section of the member.

The coefficients of thermal expansion for steel and for concrete are of the order of 10×10^{-6} per °C and $7-12 \times 10^{-6}$ per °C respectively. These values are sufficiently close that problems with bond seldom arise from differential expansion between the two materials over normal temperature ranges.

Figure 1.1 illustrates the behaviour of a simply supported beam subjected to bending and shows the position of steel reinforcement to resist the tensile forces, while the compression forces in the top of the beam are carried by the concrete.

Figure 1.1
Composite action

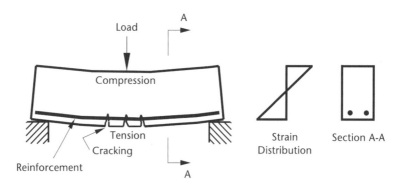

Wherever tension occurs it is likely that cracking of the concrete will take place. This cracking, however, does not detract from the safety of the structure provided there is good reinforcement bonding to ensure that the cracks are restrained from opening so that the embedded steel continues to be protected from corrosion.

When the compressive or shearing forces exceed the strength of the concrete, then steel reinforcement must again be provided, but in these cases it is only required to supplement the load-carrying capacity of the concrete. For example, compression reinforcement is generally required in a column, where it takes the form of vertical bars spaced near the perimeter. To prevent these bars buckling, steel binders are used to assist the restraint provided by the surrounding concrete.

1.2 | Stress–strain relations

The loads on a structure cause distortion of its members with resulting stresses and strains in the concrete and the steel reinforcement. To carry out the analysis and design of a member it is necessary to have a knowledge of the relationship between these stresses and strains. This knowledge is particularly important when dealing with reinforced concrete which is a composite material; for in this case the analysis of the stresses on a cross-section of a member must consider the equilibrium of the forces in the concrete and steel, and also the compatibility of the strains across the cross-section.

1.2.1 Concrete

Concrete is a very variable material, having a wide range of strengths and stress–strain curves. A typical curve for concrete in compression is shown in figure 1.2. As the load is applied, the ratio between the stresses and strains is approximately linear at first and the concrete behaves almost as an elastic material with virtually full recovery of displacement if the load is removed. Eventually, the curve is no longer linear and the concrete behaves more and more as a plastic material. If the load were removed during the plastic range the recovery would no longer be complete and a permanent deformation would remain. The ultimate strain for most structural concretes tends to be a constant value of approximately 0.0035, although this is likely to reduce for concretes with cube strengths above about 60 N/mm^2. BS EN1992 'Design of Concrete Structures' – commonly known as Eurocode 2 (or EC2) recommends values for use in such cases. The precise shape of the stress–strain curve is very dependent on the length of time the load is applied, a factor which will be further discussed in section 1.4 on creep. Figure 1.2 is typical for a short-term loading.

Concrete generally increases its strength with age. This characteristic is illustrated by the graph in figure 1.3 which shows how the increase is rapid at first, becoming more gradual later. The precise relationship will depend upon the type of cement used. That shown is for the typical variation of an adequately cured concrete made with commonly used class 42.5 Portland Cement. Some codes of practice allow the concrete strength

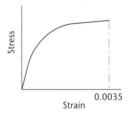

Figure 1.2
Stress–strain curve for concrete in compression

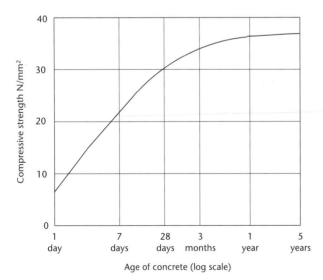

Figure 1.3
Increase of concrete strength with age. Typical curve for a concrete made with a class 42.5 Portland cement with a 28 day compressive strength of 30 N/mm^2

used in design to be varied according to the age of the concrete when it supports the design load. European Codes, however, do not permit the use of strengths greater than the 28-day value in calculations, but the modulus of elasticity may be modified to account for age as shown later.

In the United Kingdom, compressive stress has traditionally been measured and expressed in terms of 150 mm cube crushing strength at an age of 28 days. Most other countries use 150 mm diameter cylinders which are 300 mm long. For normal strength concretes, the cylinder strength is, on average, about $0.8 \times$ the cube strength. All design calculations to EC2 are based on the characteristic cylinder strength f_{ck} as defined in section 2.2.1. Cube strengths may however be used for compliance purposes, with the characteristic strength identified as $f_{ck, cube}$.

Concretes will normally be specified in terms of these 28-day characteristic strengths, for example strength class C35/45 concrete has a characteristic cylinder strength of 35 N/mm^2 and a characteristic cube strength of 45 N/mm^2. It will be noted that there is some 'rounding off' in these values, which are usually quoted in multiples of 5 N/mm^2 for cube strength. Concretes made with lightweight aggregates are identified by the prefix LC.

Modulus of elasticity of concrete

It is seen from the stress–strain curve for concrete that although elastic behaviour may be assumed for stresses below about one-third of the ultimate compressive strength, this relationship is not truly linear. Consequently it is necessary to define precisely what value is to be taken as the modulus of elasticity.

$$E = \frac{\text{stress}}{\text{strain}}$$

A number of alternative definitions exist, but the most commonly adopted is $E = E_{cm}$ where E_{cm} is known as the *secant* or *static modulus*. This is measured for a particular concrete by means of a static test in which a cylinder is loaded to just above one-third of the corresponding mean control cube stress $f_{cm, cube}$, or 0.4 mean cylinder strength, and then cycled back to zero stress. This removes the effect of initial 'bedding-in' and minor stress redistributions in the concrete under load. The load is reapplied and the behaviour will then be almost linear; the average slope of the line up to the specified stress is taken as the value for E_{cm}. The test is described in detail in BS 1881 and the result is generally known as the *secant modulus of elasticity*.

The *dynamic modulus of elasticity*, E_d, is sometimes referred to since this is much easier to measure in the laboratory and there is a fairly well-defined relationship between E_{cm} and E_d. The standard test is based on determining the resonant frequency of a prism specimen and is also described in BS 1881. It is also possible to obtain a good estimate of E_d from ultrasonic measuring techniques, which may sometimes be used on site to assess the concrete in an actual structure. The standard test for E_d is on an unstressed specimen. It can be seen from figure 1.4 that the value obtained represents the slope of the tangent at zero stress and E_d is therefore higher than E_{cm}. The relationship between the two moduli is often taken as

Secant modulus $E_{cm} = (1.25E_d - 19)$ kN/mm^2

This equation is sufficiently accurate for normal design purposes.

The actual value of E for a concrete depends on many factors related to the mix, but a general relationship is considered to exist between the modulus of elasticity and the compressive strength.

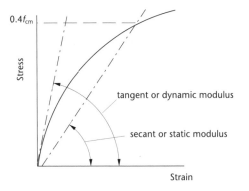

Figure 1.4
Moduli of elasticity of concrete

Typical values of E_{cm} for various concrete classes using gravel aggregates which are suitable for design are shown in table 1.1. For limestone aggregates these values should be reduced by a factor of 0.9, or for basalt increased by a factor of 1.2. The magnitude of the modulus of elasticity is required when investigating the deflection and cracking of a structure. When considering short-term effects, member stiffness will be based on the static modulus E_{cm} defined above. If long-term effects are being considered, it can be shown that the effect of creep can be represented by modifying the value of E_{cm} to an effective value $E_{c, eff}$, and this is discussed in section 6.3.2.

The elastic modulus at an age other than 28 days may be estimated from this table by using the anticipated strength value at that age. If a typical value of Poisson's ratio is needed, this should be taken as 0.2 for regions which are not subject to tension cracking.

1.2.2 Steel

Figure 1.5 shows typical stress–strain curves for (a) hot rolled high yield steel, and (b) cold-worked high yield steel. Mild steel behaves as an elastic material, with the strain proportional to the stress up to the yield, at which point there is a sudden increase in strain with no change in stress. After the yield point, this becomes a plastic material and the strain increases rapidly up to the ultimate value. High yield steel, which is most

Table 1.1 Short-term modulus of elasticity of normal-weight gravel concrete

28 day characteristic strength (N/mm²) $f_{ck}/f_{ck, cube}$ (cylinder/cube)	Static (secant) modulus E_{cm} (kN/mm²) Mean
20/25	30
25/30	31
30/37	33
35/45	34
40/50	35
45/55	36
50/60	37
60/75	39
70/85	41
80/95	42
90/105	44

Figure 1.5
Stress–strain curves for high
yield reinforcing steel

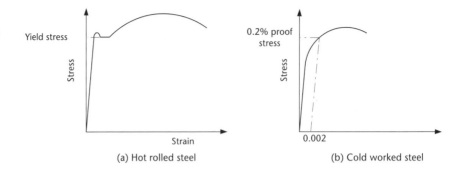

(a) Hot rolled steel   (b) Cold worked steel

commonly used for reinforcement, may behave in a similar manner or may, on the other hand, not have such a definite yield point but may show a more gradual change from elastic to plastic behaviour and reduced ductility depending on the manufacturing process. All materials have a similar slope of the elastic region with elastic modulus $E_s = 200$ kN/mm^2 approximately.

The specified strength used in design is based on either the yield stress or a specified proof stress. A 0.2 per cent proof stress is defined in figure 1.5 by the broken line drawn parallel to the linear part of the stress–strain curve.

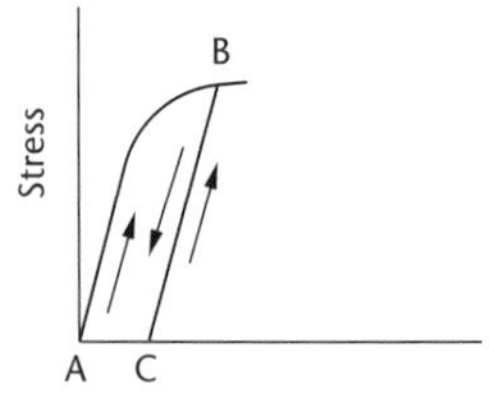

Removal of the load within the plastic range would result in the stress–strain diagram following a line approximately parallel to the loading portion – see line BC in figure 1.6. The steel will be left with a permanent strain AC, which is known as 'slip'. If the steel is again loaded, the stress–strain diagram will follow the unloading curve until it almost reaches the original stress at B and then it will curve in the direction of the first loading. Thus, the proportional limit for the second loading is higher than for the initial loading. This action is referred to as 'strain hardening' or 'work hardening'.

Figure 1.6
Strain hardening

The load deformation of the steel is also dependent on the length of time the load is applied. Under a constant stress the strains will gradually increase – this phenomenon is known as 'creep' or 'relaxation'. The amount of creep that takes place over a period of time depends on the grade of steel and the magnitude of the stress. Creep of the steel is of little significance in normal reinforced concrete work, but it is an important factor in prestressed concrete where the prestressing steel is very highly stressed.

1.3   Shrinkage and thermal movement

As concrete hardens there is a reduction in volume. This shrinkage is liable to cause cracking of the concrete, but it also has the beneficial effect of strengthening the bond between the concrete and the steel reinforcement. Shrinkage begins to take place as soon as the concrete is mixed, and is caused initially by the absorption of the water by the concrete and the aggregate. Further shrinkage is caused by evaporation of the water which rises to the concrete surface. During the setting process the hydration of the cement causes a great deal of heat to be generated, and as the concrete cools, further shrinkage takes place as a result of thermal contraction. Even after the concrete has hardened, shrinkage continues as drying out persists over many months, and any subsequent wetting and drying can also cause swelling and shrinkage. Thermal shrinkage may be reduced by restricting the temperature rise during hydration, which may be achieved by the following procedures:

1. Use a mix design with a low cement content or suitable cement replacement (e.g. Pulverised Fuel Ash or Ground Granulated Blast Furnace Slag).

2. Avoid rapid hardening and finely ground cement if possible.

3. Keep aggregates and mixing water cool.

4. Use steel shuttering and cool with a water spray.

5. Strike the shuttering early to allow the heat of hydration to dissipate.

A low water–cement ratio will help to reduce drying shrinkage by keeping to a minimum the volume of moisture that can be lost.

If the change in volume of the concrete is allowed to take place freely and without restraint, there will be no stress change within the concrete. Restraint of the shrinkage, on the other hand, will cause tensile strains and stresses. The restraint may be caused externally by fixity with adjoining members or friction against an earth surface, and internally by the action of the steel reinforcement. For a long wall or floor slab, the restraint from adjoining concrete may be reduced by constructing successive bays instead of alternate bays. This allows the free end of every bay to contract before the next bay is cast.

Day-to-day thermal expansion of the concrete can be greater than the movements caused by shrinkage. Thermal stresses and strains may be controlled by the correct positioning of movement or expansion joints in a structure. For example, the joints should be placed at an abrupt change in cross-section and they should, in general, pass completely through the structure in one plane.

When the tensile stresses caused by shrinkage or thermal movement exceed the strength of the concrete, cracking will occur. To control the crack widths, steel reinforcement must be provided close to the concrete surface; the codes of practice specify minimum quantities of reinforcement in a member for this purpose.

Calculation of stresses induced by shrinkage

(a) Shrinkage restrained by the reinforcement

The shrinkage stresses caused by reinforcement in an otherwise unrestrained member may be calculated quite simply. The member shown in figure 1.7 has a free shrinkage strain of ε_{cs} if made of plain concrete, but this overall movement is reduced by the inclusion of reinforcement, giving a compressive strain ε_{sc} in the steel and causing an effective tensile strain ε_{ct} the concrete.

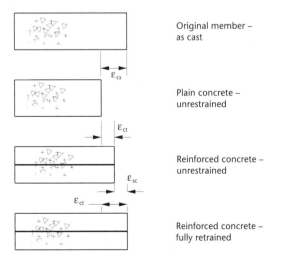

Original member – as cast

Plain concrete – unrestrained

Reinforced concrete – unrestrained

Reinforced concrete – fully retrained

Figure 1.7
Shrinkage strains

Thus

$$\varepsilon_{cs} = \varepsilon_{ct} + \varepsilon_{sc} = \frac{f_{ct}}{E_{cm}} + \frac{f_{sc}}{E_s} \tag{1.1}$$

where f_{ct} is the tensile stress in concrete area A_c and f_{sc} is the compressive stress in steel area A_s

Equating forces in the concrete and steel for equilibrium gives

$$A_c f_{ct} = A_s f_{sc} \tag{1.2}$$

therefore

$$f_{ct} = \frac{A_s}{A_c} f_{sc}$$

Substituting for f_{ct} in equation 1.1

$$\varepsilon_{cs} = f_{sc}\left(\frac{A_s}{A_c E_{cm}} + \frac{1}{E_s}\right)$$

Thus if $\alpha_e = \dfrac{E_s}{E_{cm}}$

$$\varepsilon_{cs} = f_{sc}\left(\frac{\alpha_e A_s}{A_c E_s} + \frac{1}{E_s}\right)$$
$$= \frac{f_{sc}}{E_s}\left(\frac{\alpha_e A_s}{A_c} + 1\right)$$

Therefore steel stress

$$f_{sc} = \frac{\varepsilon_{cs} E_s}{1 + \dfrac{\alpha_e A_s}{A_c}} \tag{1.3}$$

EXAMPLE 1.1

Calculation of shrinkage stresses in concrete that is restrained by reinforcement only

A member contains 1.0 per cent reinforcement, and the free shrinkage strain ε_{cs} of the concrete is 200×10^{-6}. For steel, $E_s = 200$ kN/mm^2 and for concrete $E_{cm} = 15$ kN/mm^2. Hence from equation 1.3:

$$\text{stress in reinforcement } f_{sc} = \frac{\varepsilon_{cs} E_s}{1 + \alpha_e \dfrac{A_s}{A_c}}$$

$$= \frac{200 \times 10^{-6} \times 200 \times 10^3}{1 + \dfrac{200}{15} \times 0.01}$$

$$= 35.3 \text{ N/mm}^2 \text{ compression}$$

$$\text{stress in concrete } f_{ct} = \frac{A_s}{A_c} f_{sc}$$

$$= 0.01 \times 35.3$$

$$= 0.35 \text{ N/mm}^2 \text{ tension}$$

The stresses produced in members free from external restraint are generally small as in example 1.1, and can be easily withstood both by the steel and the concrete.

(b) Shrinkage fully restrained

If the member is fully restrained, then the steel cannot be in compression since $\varepsilon_{sc} = 0$ and hence $f_{sc} = 0$ (figure 1.7). In this case the tensile strain induced in the concrete ε_{ct} must be equal to the free shrinkage strain ε_{cs}, and the corresponding stress will probably be high enough to cause cracking in immature concrete.

EXAMPLE 1.2

Calculation of fully restrained shrinkage stresses

If the member in example 1.1 were fully restrained, the stress in the concrete would be given by

$$f_{ct} = \varepsilon_{ct} E_{cm}$$

where

$$\varepsilon_{ct} = \varepsilon_{cs} = 200 \times 10^{-6}$$

then

$$f_{ct} = 200 \times 10^{-6} \times 15 \times 10^3$$
$$= 3.0 \text{N/mm}^2$$

When cracking occurs, the uncracked lengths of concrete try to contract so that the embedded steel between cracks is in compression while the steel across the cracks is in tension. This feature is accompanied by localised bond breakdown, adjacent to each crack. The equilibrium of the concrete and reinforcement is shown in figure 1.8 and calculations may be developed to relate crack widths and spacings to properties of the cross-section; this is examined in more detail in chapter 6, which deals with serviceability requirements.

$A_s f_{st}$

$A_c f_{ct}$

$A_s f_{sc}$

Figure 1.8
Shrinkage forces adjacent to a crack

Thermal movement

As the coefficients of thermal expansion of steel and concrete ($\alpha_{T,s}$ and $\alpha_{T,c}$) are similar, differential movement between the steel and concrete will only be very small and is unlikely to cause cracking.

The differential thermal strain due to a temperature change T may be calculated as

$$T(\alpha_{T,c} - \alpha_{T,s})$$

and should be added to the shrinkage strain ε_{cs} if significant.

The overall thermal contraction of concrete is, however, frequently effective in producing the first crack in a restrained member, since the required temperature changes could easily occur overnight in a newly cast member, even with good control of the heat generated during the hydration processes.

EXAMPLE 1.3

Thermal shrinkage

Find the fall in temperature required to cause cracking in a restrained member if ultimate tensile strength of the concrete $f_{ct,\,eff} = 2$ N/mm^2, $E_{cm} = 16$ kN/mm^2 and $\alpha_{T,\,c} = \alpha_{T,\,s} = 10 \times 10^{-6}$ per °C.

Ultimate tensile strain of concrete

$$\varepsilon_{ult} = \frac{f_{ct,\,eff}}{E_{cm}}$$

$$= \frac{2}{16 \times 10^3} = 125 \times 10^{-6}$$

Minimum temperature drop to cause cracking

$$= \frac{\varepsilon_{ult}}{\alpha_{T,\,c}} = \frac{125}{10} = 12.5°C$$

It should be noted that full restraint, as assumed in this example, is unlikely to occur in practice; thus the temperature change required to cause cracking is increased. A maximum 'restraint factor' of 0.5 is often used, with lower values where external restraint is likely to be small. The temperature drop required would then be given by the theoretical minimum divided by the 'restraint factor'. i.e. $12.5/0.5 = 25°C$ in this example

1.4 ▎Creep

Creep is the continuous deformation of a member under sustained load. It is a phenomenon associated with many materials, but it is particularly evident with concrete. The precise behaviour of a particular concrete depends on the aggregates and the mix design as well as the ambient humidity, member cross-section, and age at first loading, but the general pattern is illustrated by considering a member subjected to axial compression. For such a member, a typical variation of deformation with time is shown by the curve in figure 1.9.

The characteristics of creep are

1. The final deformation of the member can be three to four times the short-term elastic deformation.

2. The deformation is roughly proportional to the intensity of loading and to the inverse of the concrete strength.

3. If the load is removed, only the instantaneous elastic deformation will recover – the plastic deformation will not.

4. There is a redistribution of load between the concrete and any steel present.

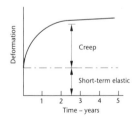

Figure 1.9
Typical increase of deformation with time for concrete

The redistribution of load is caused by the changes in compressive strains being transferred to the reinforcing steel. Thus the compressive stresses in the steel are increased so that the steel takes a larger proportion of the load.

The effects of creep are particularly important in beams, where the increased deflections may cause the opening of cracks, damage to finishes, and the non-alignment of mechanical equipment. Redistribution of stress between concrete and steel occurs primarily in the uncracked compressive areas and has little effect on the tension reinforcement other than reducing shrinkage stresses in some instances. The provision of reinforcement in the compressive zone of a flexural member, however, often helps to restrain the deflections due to creep.

1.5 Durability

Concrete structures, properly designed and constructed, are long lasting and should require little maintenance. The durability of the concrete is influenced by

1. the exposure conditions;
2. the cement type;
3. the concrete quality;
4. the cover to the reinforcement;
5. the width of any cracks.

Concrete can be exposed to a wide range of conditions such as the soil, sea water, de-icing salts, stored chemicals or the atmosphere. The severity of the exposure governs the type of concrete mix required and the minimum cover to the reinforcing steel. Whatever the exposure, the concrete mix should be made from impervious and chemically inert aggregates. A dense, well-compacted concrete with a low water–cement ratio is all important and for some soil conditions it is advisable to use a sulfate-resisting cement. Air entrainment is usually specified where it is necessary to cater for repeated freezing and thawing.

Adequate cover is essential to prevent corrosive agents reaching the reinforcement through cracks and pervious concrete. The thickness of cover required depends on the severity of the exposure and the quality of the concrete (as shown in table 6.2). The cover is also necessary to protect the reinforcement against a rapid rise in temperature and subsequent loss of strength during a fire. Part 1.2 of EC2 provides guidance on this and other aspects of fire design. Durability requirements with related design calculations to check and control crack widths and depths are described in more detail in chapter 6.

1.6 Specification of materials

1.6.1 Concrete

The selection of the type of concrete is frequently governed by the strength required, which in turn depends on the intensity of loading and the form and size of the structural members. For example, in the lower columns of a multi-storey building a higher-strength concrete may be chosen in preference to greatly increasing the size of the column section with a resultant loss in clear floor space.

As indicated in section 1.2.1, the concrete strength is assessed by measuring the crushing strength of cubes or cylinders of concrete made from the mix. These are usually cured, and tested after 28 days according to standard procedures. Concrete of a given strength is identified by its 'class' – a Class 25/30 concrete has a characteristic cylinder crushing strength (f_{ck}) of 25 N/mm^2 and cube strength of 30 N/mm^2. Table 1.2 shows a list of commonly used classes and also the lowest class normally appropriate for various types of construction.

Exposure conditions and durability can also affect the choice of the mix design and the class of concrete. A structure subject to corrosive conditions in a chemical plant, for example, would require a denser and higher class of concrete than, say, the interior members of a school or office block. Although Class 42.5 Portland cement would be used in most structures, other cement types can also be used to advantage. Blast-furnace or sulfate-resisting cement may be used to resist chemical attack, low-heat cements in massive sections to reduce the heat of hydration, or rapid-hardening cement when a high early strength is required. In some circumstances it may be useful to replace some of the cement by materials such as Pulverised Fuel Ash or Ground Granulated Blast Furnace Slag which have slowly developing cementitious properties. These will reduce the heat of hydration and may also lead to a smaller pore structure and increased durability. Generally, natural aggregates found locally are preferred; however, manufactured lightweight material may be used when self-weight is important, or a special dense aggregate when radiation shielding is required.

The concrete mix may either be classified as 'designed' or 'designated'. A 'designed concrete' is one where the strength class, cement type, and limits to composition, including water–cement ratio and cement content, are specified. With a 'designated concrete' the producer must provide a material to satisfy the designated strength class and consistence (workability) using a particular aggregate size. 'Designated concretes' are identified as RC30 (for example) based on cube strength up to RC50 according to the application involved. 'Designed concretes' are needed in situations where 'designated concretes' cannot be used on the basis of durability requirements (e.g.

Table 1.2 Strength classes of concrete

Class	f_{ck} (N/mm^2)	Normal lowest class for use as specified
C16/20	16	Plain concrete
C20/25	20	Reinforced concrete
C25/30	25	
C28/35	28	Prestressed concrete/Reinforced concrete subject to chlorides
C30/37	30	Reinforced concrete in foundations
C32/40	32	
C35/45	35	
C40/50	40	
C45/55	45	
C50/60	50	
C55/67	55	
C60/75	60	
C70/85	70	
C80/95	80	
C90/105	90	

chloride-induced corrosion). Detailed requirements for mix specification and compliance are given by BS EN206 'Concrete – Performance, Production, Placing and Compliance Criteria' and BS8500 'Concrete – Complementary British Standard to BS EN206'

1.6.2   Reinforcing steel

Table 1.3 lists the characteristic design strengths of some of the more common types of reinforcement currently used in the UK. Grade 500 (500N/mm^2 characteristic strength) has replaced Grade 250 and Grade 460 reinforcing steel throughout Europe. The nominal size of a bar is the diameter of an equivalent circular area.

Grade 250 bars are hot-rolled mild-steel bars which usually have a smooth surface so that the bond with the concrete is by adhesion only. This type of bar can be more readily bent, so they have in the past been used where small radius bends are necessary, such as links in narrow beams or columns, but plain bars are not now recognised in the European Union and they are no longer available for general use in the UK.

High-yield bars are manufactured with a ribbed surface or in the form of a twisted square. Square twisted bars have inferior bond characteristics and have been used in the past, although they are now obsolete. Deformed bars have a mechanical bond with the concrete, thus enhancing ultimate bond stresses as described in section 5.2. The bending of high-yield bars through a small radius is liable to cause tension cracking of the steel, and to avoid this the radius of the bend should not be less than two times the nominal bar size for small bars ($\leq$16 mm) or 3½ times for larger bars (see figure 5.11). The ductility of reinforcing steel is also classified for design purposes. Ribbed high yield bars may be classified as:

Class A – which is normally associated with small diameter ($\leq$12 mm) cold-worked bars used in mesh and fabric. This is the lowest ductility category and will include limits on moment redistribution which can be applied (see section 4.7) and higher quantities for fire resistance.
Class B – which is most commonly used for reinforcing bars.
Class C – high ductility which may be used in earthquake design or similar situations.

Floor slabs, walls, shells and roads may be reinforced with a welded fabric of reinforcement, supplied in rolls and having a square or rectangular mesh. This can give large economies in the detailing of the reinforcement and also in site labour costs of handling and fixing. Prefabricated reinforcement bar assemblies are also becoming increasingly popular for similar reasons. Welded fabric mesh made of ribbed wire greater than 6 mm diameter may be of any of the ductility classes listed above.

Table 1.3   Strength of reinforcement

Designation	Normal sizes (mm)	Specified characteristic strength f_{yk} (N/mm^2)
Hot-rolled high yield (BS4449)	All sizes	500
Cold-worked high yield (BS4449)	Up to and including 12	500

*Note that BS4449 will be replaced by BS EN10080 in due course.

The cross-sectional areas and perimeters of various sizes of bars, and the cross-sectional area per unit width of slabs are listed in the Appendix. Reinforcing bars in a member should either be straight or bent to standard shapes. These shapes must be fully dimensioned and listed in a schedule of the reinforcement which is used on site for the bending and fixing of the bars. Standard bar shapes and a method of scheduling are specified in BS8666. The bar types as previously described are commonly identified by the following codes: H for high yield steel, irrespective of ductility class or HA, HB, HC where a specific ductility class is required; this notation is generally used throughout this book.

CHAPTER 2

Limit state design

CHAPTER INTRODUCTION

Limit state design of an engineering structure must ensure that (1) under the worst loadings the structure is safe, and (2) during normal working conditions the deformation of the members does not detract from the appearance, durability or performance of the structure. Despite the difficulty in assessing the precise loading and variations in the strength of the concrete and steel, these requirements have to be met. Three basic methods using factors of safety to achieve safe, workable structures have been developed over many years; they are

1. The permissible stress method in which ultimate strengths of the materials are divided by a factor of safety to provide design stresses which are usually within the elastic range.
2. The load factor method in which the working loads are multiplied by a factor of safety.
3. The limit state method which multiplies the working loads by partial factors of safety and also divides the materials' ultimate strengths by further partial factors of safety.

The permissible stress method has proved to be a simple and useful method but it does have some serious inconsistencies and is generally no longer in use. Because it is based on an elastic stress distribution, it is not really applicable to a semi-plastic material such as concrete, nor is it suitable when the deformations are not proportional to the load, as in slender columns. It has also been found to be unsafe when dealing with the stability of structures subject to overturning forces (see example 2.2). ➡

15

→

In the load factor method the ultimate strength of the materials should be used in the calculations. As this method does not apply factors of safety to the material stresses, it cannot directly take account of the variability of the materials, and also it cannot be used to calculate the deflections or cracking at working loads. Again, this is a design method that has now been effectively superseded by modern limit state design methods.

The limit state method of design, now widely adopted across Europe and many other parts of the world, overcomes many of the disadvantages of the previous two methods. It does so by applying partial factors of safety, both to the loads and to the material strengths, and the magnitude of the factors may be varied so that they may be used either with the plastic conditions in the ultimate state or with the more elastic stress range at working loads. This flexibility is particularly important if full benefits are to be obtained from development of improved concrete and steel properties.

2.1 Limit states

The purpose of design is to achieve acceptable probabilities that a structure will not become unfit for its intended use – that is, that it will not reach a limit state. Thus, any way in which a structure may cease to be fit for use will constitute a limit state and the design aim is to avoid any such condition being reached during the expected life of the structure.

The two principal types of limit state are the ultimate limit state and the serviceability limit state.

(a) Ultimate limit state

This requires that the structure must be able to withstand, with an adequate factor of safety against collapse, the loads for which it is designed to ensure the safety of the building occupants and/or the safety of the structure itself. The possibility of buckling or overturning must also be taken into account, as must the possibility of accidental damage as caused, for example, by an internal explosion.

(b) Serviceability limit states

Generally the most important serviceability limit states are:

1. Deflection – the appearance or efficiency of any part of the structure must not be adversely affected by deflections nor should the comfort of the building users be adversely affected.

2. Cracking – local damage due to cracking and spalling must not affect the appearance, efficiency or durability of the structure.

3. Durability – this must be considered in terms of the proposed life of the structure and its conditions of exposure.

Other limit states that may be reached include:

4. Excessive vibration – which may cause discomfort or alarm as well as damage.

5. Fatigue – must be considered if cyclic loading is likely.

6. Fire resistance – this must be considered in terms of resistance to collapse, flame penetration and heat transfer.

7. Special circumstances – any special requirements of the structure which are not covered by any of the more common limit states, such as earthquake resistance, must be taken into account.

The relative importance of each limit state will vary according to the nature of the structure. The usual procedure is to decide which is the crucial limit state for a particular structure and base the design on this, although durability and fire resistance requirements may well influence initial member sizing and concrete class selection. Checks must also be made to ensure that all other relevant limit states are satisfied by the results produced. Except in special cases, such as water-retaining structures, the ultimate limit state is generally critical for reinforced concrete although subsequent serviceability checks may affect some of the details of the design. Prestressed concrete design, however, is generally based on serviceability conditions with checks on the ultimate limit state.

In assessing a particular limit state for a structure it is necessary to consider all the possible variable parameters such as the loads, material strengths and all constructional tolerances.

2.2 Characteristic material strengths and characteristic loads

2.2.1 Characteristic material strengths

The strengths of materials upon which a design is based are, normally, those strengths below which results are unlikely to fall. These are called 'characteristic' strengths. It is assumed that for a given material, the distribution of strength will be approximately 'normal', so that a frequency distribution curve of a large number of sample results would be of the form shown in figure 2.1. The characteristic strength is taken as that value below which it is unlikely that more than 5 per cent of the results will fall.

This is given by

$$f_k = f_m - 1.64s$$

where f_k = characteristic strength, f_m = mean strength and s = standard deviation.

The relationship between characteristic and mean values accounts for variations in results of test specimens and will, therefore, reflect the method and control of manufacture, quality of constituents, and nature of the material.

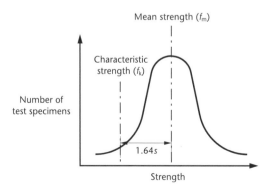

Figure 2.1
Normal frequency distribution of strengths

2.2.2 Characteristic actions

In Eurocode terminology the set of applied forces (or loads) for which a structure is to be designed are called 'actions' although the terms 'actions' and 'loads' tend to be used interchangeably in some of the Eurocodes. 'Actions' can also have a wider meaning including the effect of imposed deformations caused by, for example, settlement of foundations. In this text we will standardise on the term 'actions' as much as possible. Ideally it should be possible to assess actions statistically in the same way that material characteristic strengths can be determined statistically, in which case

characteristic action = mean action $\pm$ 1.64 standard deviations

In most cases it is the maximum value of the actions on a structural member that is critical and the upper, positive value given by this expression is used; but the lower, minimum value may apply when considering stability or the behaviour of continuous members.

These characteristic values represent the limits within which at least 90 per cent of values will lie in practice. It is to be expected that not more than 5 per cent of cases will exceed the upper limit and not more than 5 per cent will fall below the lower limit. They are design values that take into account the accuracy with which the structural loading can be predicted.

Usually, however, there is insufficient statistical data to allow actions to be treated in this way, and in this case the standard loadings, such as those given in BS EN 1991, Eurocode 1 – Actions on Structures, should be used as representing characteristic values.

2.3 ▌ Partial factors of safety

Other possible variations such as constructional tolerances are allowed for by partial factors of safety applied to the strength of the materials and to the actions. It should theoretically be possible to derive values for these from a mathematical assessment of the probability of reaching each limit state. Lack of adequate data, however, makes this unrealistic and, in practice, the values adopted are based on experience and simplified calculations.

2.3.1 Partial factors of safety for materials (γ_m)

$$\text{Design strength} = \frac{\text{characteristic strength } (f_k)}{\text{partial factor of safety } (\gamma_m)}$$

The following factors are considered when selecting a suitable value for γ_m:

1. The strength of the material in an actual member. This strength will differ from that measured in a carefully prepared test specimen and it is particularly true for concrete where placing, compaction and curing are so important to the strength. Steel, on the other hand, is a relatively consistent material requiring a small partial factor of safety.

2. The severity of the limit state being considered. Thus, higher values are taken for the ultimate limit state than for the serviceability limit state.

Recommended values for γ_m are given in table 2.1 The values in the first two columns should be used when the structure is being designed for *persistent design situations* (anticipated normal usage) or *transient design situations* (temporary

Table 2.1   Partial factors of safety applied to materials (γ_m)

Limit state	Persistent and transient		Accidental	
	Concrete	Reinforcing and Prestressing Steel	Concrete	Reinforcing and Prestressing Steel
Ultimate				
Flexure	1.50	1.15	1.20	1.00
Shear	1.50	1.15	1.20	1.00
Bond	1.50	1.15	1.20	1.00
Serviceability	1.00	1.00		

situations such as may occur during construction). The values in the last two columns should be used when the structure is being designed for exceptional *accidental design situations* such as the effects of fire or explosion.

2.3.2   Partial factors of safety for actions (γ_f)

Errors and inaccuracies may be due to a number of causes:

1. design assumptions and inaccuracy of calculation;
2. possible unusual increases in the magnitude of the actions;
3. unforeseen stress redistributions;
4. constructional inaccuracies.

These cannot be ignored, and are taken into account by applying a partial factor of safety (γ_f) on the characteristic actions, so that

design value of action = characteristic action × partial factor of safety (γ_f)

The value of this factor should also take into account the importance of the limit state under consideration and reflects to some extent the accuracy with which different types of actions can be predicted, and the probability of particular combinations of actions occurring. It should be noted that design errors and constructional inaccuracies have similar effects and are thus sensibly grouped together. These factors will account adequately for normal conditions although gross errors in design or construction obviously cannot be catered for.

Recommended values of partial factors of safety are given in tables 2.2 and 2.3 according to the different categorisations of actions shown in the tables. Actions are categorised as either *permanent* (G_k), such as the self-weight of the structure, or *variable* (Q_k), such as the temporary imposed loading arising from the traffic of people, wind and snow loading, and the like. *Variable actions* are also categorised as *leading* (the predominant variable action on the structure such as an imposed crowd load – $Q_{k,1}$) and *accompanying* (secondary variable action(s) such as the effect of wind loading, $Q_{k,i}$, where the subscript 'i' indicates the i'th action).

The terms *favourable* and *unfavourable* refer to the effect of the action(s) on the design situation under consideration. For example, if a beam, continuous over several spans, is to be designed for the largest sagging bending moment it will have to sustain any action that has the effect of **increasing** the bending moment will be considered unfavourable whilst any action that **reduces** the bending moment will be considered to be favourable.

Table 2.2 Partial safety factors at the ultimate limit state

Persistent or transient design situation	Permanent actions (G_k)		Leading variable action (Q_{k,1})		Accompanying variable actions (Q_{k,i})	
	Unfavourable	Favourable	Unfavourable	Favourable	Unfavourable	Favourable
(a) For checking the static equilibrium of a building structure	1.10	0.90	1.50	0	1.50	0
(b) For the design of structural members (excluding geotechnical actions)	1.35	1.00	1.50	0	1.50	0
(c) As an alternative to (a) and (b) above to design for both situations with one set of calculations	1.35	1.15	1.50	0	1.50	0

Table 2.3 Partial safety factors at the serviceability limit state

Design Situation	Permanent actions	Variable actions
All	1.0	1.0

Example 2.1 shows how the partial safety factors at the ultimate limit state from tables 2.1 and 2.2 are used to design the cross-sectional area of a steel cable supporting permanent and variable actions.

EXAMPLE 2.1

Simple design of a cable at the ultimate limit state

Determine the cross-sectional area of steel required for a cable which supports a total characteristic permanent action of 3.0 kN and a characteristic variable action of 2.0 kN as shown in figure 2.2.

Figure 2.2
Cable design

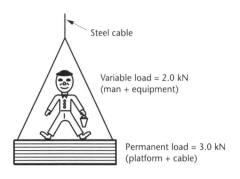

Steel cable

Variable load = 2.0 kN
(man + equipment)

Permanent load = 3.0 kN
(platform + cable)

The characteristic yield stress of the steel is 500 N/mm². Carry out the calculation using limit state design with the following factors of safety:

$\gamma_G = 1.35$ for the permanent action,

$\gamma_Q = 1.5$ for the variable action, and

$\gamma_m = 1.15$ for the steel strength.

$$\text{Design value} = \gamma_G \times \text{permanent action} + \gamma_Q \times \text{variable action}$$

$$= 1.35 \times 3.0 + 1.5 \times 2.0$$

$$= 7.05\,\text{kN}$$

$$\text{Design stress} = \frac{\text{characteristic yield stress}}{\gamma_m}$$

$$= \frac{500}{1.15}$$

$$= 434\,\text{N/mm}^2$$

$$\text{Required cross-sectional area} = \frac{\text{design value}}{\text{design stress}}$$

$$= \frac{7.05 \times 10^3}{434}$$

$$= 16.2\,\text{mm}^2$$

For convenience, the partial factors of safety in the example are the same as those recommended in EC2. Probably, in a practical design, higher factors of safety would be preferred for a single supporting cable, in view of the consequences of a failure.

Example 2.2 shows the design of a foundation to resist uplift at the ultimate limit state using the partial factors of safety from table 2.2. It demonstrates the benefits of using the limit state approach instead of the potentially unsafe overall factor of safety design used in part (b).

EXAMPLE 2.2

Design of a foundation to resist uplift

Figure 2.3 shows a beam supported on foundations at A and B. The loads supported by the beam are its own uniformly distributed permanent weight of 20 kN/m and a 170 kN variable load concentrated at end C.

Determine the weight of foundation required at A in order to resist uplift:

(a) by applying a factor of safety of 2.0 to the reaction calculated for the working loads.

(b) by using an ultimate limit state approach with partial factors of safety of $\gamma_G = 1.10$ or 0.9 for the permanent action and $\gamma_Q = 1.5$ for the variable action.

Investigate the effect on these designs of a 7 per cent increase in the variable action.

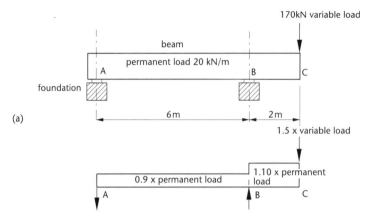

Figure 2.3
Uplift calculation example

(b) Loading arrangement for uplift at A at the ultimate limit state

(a) Factor of safety on uplift = 2.0

Taking moments about B

$$\text{Uplift } R_A = \frac{(170 \times 2 - 20 \times 8 \times 2)}{6.0} = 3.33 \text{ kN}$$

Weight of foundation required $= 3.33 \times$ safety factor

$$= 3.33 \times 2.0 = 6.7 \text{ kN}$$

With a 7 per cent increase in the variable action

$$\text{Uplift } R_A = \frac{(1.07 \times 170 \times 2 - 20 \times 8 \times 2)}{6.0} = 7.3 \text{ kN}$$

Thus with a slight increase in the variable action there is a significant increase in the uplift and the structure becomes unsafe.

(b) Limit state method – ultimate load pattern

As this example includes a cantilever and also involves the requirement for static equilibrium at A, partial factors of safety of 1.10 and 0.9 were chosen for the permanent actions as given in the first row of values in table 2.2

The arrangement of the loads for the maximum uplift at A is shown in figure 2.3b.

Design permanent action over BC $= \gamma_G \times 20 \times 2 = 1.10 \times 20 \times 2 = 44 \text{ kN}$

Design permanent action over AB $= \gamma_G \times 20 \times 6 = 0.9 \times 20 \times 6 = 108 \text{ kN}$

Design variable action $= \gamma_Q \times 170 = 1.5 \times 170 = 255 \text{ kN}$

Taking moments about B for the ultimate actions

$$\text{Uplift } R_A = \frac{(255 \times 2 + 44 \times 1 - 108 \times 3)}{6.0} = 38 \text{ kN}$$

Therefore weight of foundation required = 38 kN.

A 7 per cent increase in the variable action will not endanger the structure, since the actual uplift will only be 7.3 kN as calculated previously. In fact in this case it would require an increase of 61 per cent in the variable load before the uplift would exceed the weight of a 38 kN foundation.

Parts (a) and (b) of example 2.2 illustrate how the limit state method of design can ensure a safer result when the stability or strength of a structure is sensitive to a small numerical difference between the effects of two opposing actions of a similar magnitude.

2.4   Combination of actions

Permanent and variable actions will occur in different combinations, all of which must be taken into account in determining the most critical design situation for any structure. For example, the self-weight of the structure may be considered in combination with the weight of furnishings and people, with or without the effect of wind acting on the building (which may also act in more than one direction)

In cases where actions are to be combined it is recommended that, in determining suitable design values, each characteristic action is not only multiplied by the partial factors of safety, as discussed above, but also by a further factor given the symbol Ψ. This factor is generally taken as 1.0 other than where described below:

(i)   **Combination values of variable actions**
       Where more than one variable action is to be considered (i.e a combination) then the variable actions should be multiplied by a value of Ψ (denoted as Ψ_0) as given in table 2.4. This ensures that the probability of a combination of actions being exceeded is approximately the same as that for a single action. As can be seen in the table this is also dependent on the type of structure being designed. Combination values are used for designing for (i) the ultimate limit state and (ii) irreversible serviceability limit states such as irreversible cracking due to temporary but excessive overloading of the structure.

(ii)   **Frequent values of variable actions**
       Frequent combinations of actions are used in the consideration of (i) ultimate limit states involving accidental actions and (ii) reversible limit states such as the serviceability limit states of cracking and deflection where the actions causing these effects are of a short transitory nature. In these cases the variable actions are multiplied by a value of Ψ (denoted as Ψ_1) as given in table 2.4. The values of Ψ_1 give an estimation of the proportion of the total variable action that is likely to be associated with this particular combination of actions.

(iii) **Quasi-permanent values of variable action**
       EC2 requires that, in certain situations, the effects of 'quasi-permanent' actions should be considered. Quasi-permanent (meaning 'almost' permanent) actions are those that may be sustained over a long period but are not necessarily as permanent as, say, the self-weight of the structure. An example of such a loading would be the effect of snow on the roofs of buildings at high altitudes where the weight of the snow may have to be sustained over weeks or months.

Quasi-permanent combinations of actions are used in the consideration of (i) ultimate limit states involving accidental actions and (ii) serviceability limit states attributable to, for example, the long-term effects of creep and where the actions causing these effects, whilst variable, are of a more long-term, sustained nature. In these cases the variable actions are multiplied by a value of Ψ (denoted as Ψ_2) as given in table 2.4. The values of Ψ_2 give an estimation of the proportion of the total variable action that is likely to be associated with this particular combination of actions.

Table 2.4 Values of Ψ for different load combinations

Action	Combination Ψ_0	Frequent Ψ_1	Quasi-permanent Ψ_2
Imposed load in buildings, category (see EN 1991-1-1)			
Category A: domestic, residential areas	0.7	0.5	0.3
Category B: office areas	0.7	0.5	0.3
Category C: congregation areas	0.7	0.7	0.6
Category D: shopping areas	0.7	0.7	0.6
Category E: storage areas	1.0	0.9	0.8
Category F: traffic area, vehicle weight < 30kN	0.7	0.7	0.6
Category G: traffic area, 30 kN < vehicle weight < 160 kN	0.7	0.5	0.3
Category H: roofs	0.7	0	0
Snow loads on buildings (see EN 1991-1-3)			
For sites located at altitude $H > 1000$ m above sea level	0.7	0.5	0.2
For sites located at altitude $H \leq 1000$ m above sea level	0.5	0.2	0
Wind loads on buildings (see EN 1991-1-4)	0.5	0.2	0

Figure 2.4
Wind and imposed load acting
on an office building – stability
check

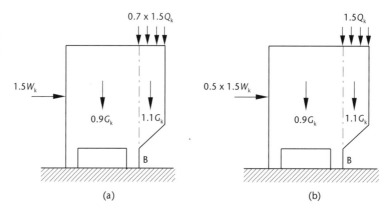

Figure 2.4 illustrates how the factors in table 2.2 and 2.4 can be applied when considering the stability of the office building shown for overturning about point B. Figure 2.4(a) treats the wind load (W_k) as the leading variable action and the live load (Q_k) on the roof as the accompanying variable action. Figure 2.4(b) considers the live load as the leading variable action and the wind as the accompanying variable action.

2.4.1 Design values of actions at the ultimate limit state

In general terms, for *persistent and transient design situations* the design value can be taken as:

Design value (E_d) = (factored permanent actions) combined with (factored single leading variable action) combined with (factored remaining accompanying variable actions)

The 'factors' will, in all cases, be the appropriate partial factor of safety (γ_f) taken together with the appropriate value of Ψ as given in table 2.4.

The design value can be expressed formalistically as:

$$E_d = \left| \sum_{j \geq 1} \gamma_{G,j} G_{k,j} \right| + \gamma_{Q,1} Q_{k,1} + \left| \sum_{i>1} \gamma_{Q,i} \Psi_{0,i} Q_{k,i} \right| \qquad (2.1)$$

Note that the $+$ sign in this expression is not algebraic: it simply means 'combined with'. The $\sum$ symbol indicates the combined effect of all the similar action effects. e.g. $\left| \sum_{j \geq 1} \gamma_{G,j} G_{k,j} \right|$ indicates the combined effects of all factored permanent actions, summed from the first to the 'j'th action, where there are a total of j permanent actions acting on the structure. Two other similar equations are given in EC2, the least favourable of which can alternatively be used to give the design value. However, equation (2.1) will normally apply for most standard situations.

For *accidental design situations* the design value of actions can be expressed in a similar way with the permanent and variable actions being combined with the effect of the accidental design situation such as fire or impact. As previously indicated, such accidental design situations will be based on the *frequent* or *quasi-permanent* values of actions with the load combinations calculated using the appropriate Ψ value(s) from table 2.4

2.4.2 Typical common design values of actions at the ultimate limit state

For the routine design of the members within reinforced concrete structures the standard design loading cases will often consist of combinations of the permanent action with a single variable action and possibly with wind. If the single variable action is considered to be the *leading* variable action then wind loading will be the *accompanying* variable action. The reverse may, however, be true and both scenarios must be considered. In such cases the factors given in table 2.5 can be used to determine the design value of the actions. The value of 1.35 for unfavourable permanent actions is conservative, and used throughout this book for simplicity. Alternative equations indicated in 2.4.1 may, in some cases, give greater economy.

Table 2.5 Combination of actions and load factors at the ultimate limit state

Persistent or transient design situation	Permanent actions (G_k)		Variable action $(Q_{k,1})$		Wind
	Unfavourable	Favourable	Unfavourable	Favourable	
Permanent + Variable	1.35	1.00[1]	1.50	0	–
Permanent + Wind	1.35	1.00	–	–	1.50
Permanent + Variable + Wind	1.35	1.00	1.50	0	$\Psi_0 \times 1.50$ $= 0.5^{(2)} \times 1.50$ $= 0.75$
Either of these two cases may be critical – both should be considered	1.35	1.00	$\Psi_0{}^{(3)} \times 1.50$	0	1.50

(1) For continuous beams with cantilevers, the partial safety factor for the favourable effect of the permanent action should be taken as 1.0 for the span adjacent to the cantilever (see figure 7.21).

(2) Based on the 'combination' figure in table 2.4 for wind

(3) Ψ_0 to be selected from table 2.4 depending on category of building (most typical value = 0.7)

(4) The partial safety factor for earth pressures may be taken as 1.30 when unfavourable and 0.0 when favourable

2.4.3 Design values of actions at the serviceability limit state

The design values of actions at the serviceability limit state can be expressed in a similar way to equation 2.1, but taking account of the different combinations of actions to be used in the three different situations discussed above. In the case of serviceability the partial factor of safety, γ_f will be taken as equal to 1.0 in all cases.

(i) **Combination values of variable actions**

$$E_d = \left| \sum_{j \geq 1} G_{k,j} \right| + Q_{k,1} + \left| \sum_{i > 1} \Psi_{0,i} Q_{k,i} \right| \tag{2.2}$$

(ii) **Frequent values of variable actions**

$$E_d = \left| \sum_{j \geq 1} G_{k,j} \right| + \Psi_{1,1} Q_{k,1} + \left| \sum_{i > 1} \Psi_{2,i} Q_{k,i} \right| \tag{2.3}$$

(iii) **Quasi-permanent values of variable actions**

$$E_d = \left| \sum_{j \geq 1} G_{k,j} \right| + \left| \sum_{i \geq 1} \Psi_{2,i} Q_{k,i} \right| \tag{2.4}$$

Note that, as before, the + signs in these expressions are not necessarily algebraic: they simply mean 'combined with'. The terms in the expressions have the following meanings:

$\left| \sum_{j \geq 1} G_{k,j} \right| =$ the combined effect of all the characteristic permanent actions where the subscript 'j' indicates that there could be between one and 'j' permanent actions on the structure

$\Psi Q_{k,1} =$ the single leading characteristic variable action multiplied by the factor Ψ, where Ψ takes the value of 1, Ψ_1 or Ψ_2 as appropriate from table 2.4. The subscript '1' indicates that this relates to the single leading variable action on the structure.

$\left| \sum_{i > 1} \Psi Q_{k,i} \right| =$ the combined effect of all the 'accompanying' characteristic variable actions each multiplied by the factor Ψ, where Ψ takes the value of Ψ_0 or Ψ_2 as appropriate from table 2.4. The subscript 'i' indicates that there could be up to 'i' variable actions on the structure in addition to the single leading variable action

---

EXAMPLE 2.3

Combination of actions at the serviceability limit state

A simply supported reinforced concrete beam forms part of a building within a shopping complex. It is to be designed for a characteristic permanent action of 20 kN/m (its own self-weight and that of the supported structure) together with a characteristic, single leading variable action of 10 kN/m and an accompanying variable action of 2 kN/m (both representing the imposed loading on the beam). Calculate each of the serviceability limit state design values as given by equations (2.2) to (2.4).

From table 2.4 the building is classified as category D. Hence, $\Psi_0 = 0.7$, $\Psi_1 = 0.7$ and $\Psi_2 = 0.6$.

Combination value

$$E_{\mathrm{d}} = \left| \sum_{j \geq 1} G_{\mathrm{k},j} \right| + Q_{\mathrm{k},1} + \left| \sum_{i > 1} \Psi_{0,i}\, Q_{\mathrm{k},i} \right| = 20 + 10 + (0.7 \times 2) = 31.4\,\mathrm{kN/m}$$

Frequent value

$$E_{\mathrm{d}} = \left| \sum_{j \geq 1} G_{\mathrm{k},j} \right| + \Psi_{1,1}\, Q_{\mathrm{k},1} + \left| \sum_{i > 1} \Psi_{2,i}\, Q_{\mathrm{k},i} \right| = 20 + (0.7 \times 10) + (0.6 \times 2)$$

$$= 28.2\,\mathrm{kN/m}$$

Quasi-permanent value

$$E_{\mathrm{d}} = \left| \sum_{j \geq 1} G_{\mathrm{k},j} \right| + \left| \sum_{i \geq 1} \Psi_{2,i} Q_{\mathrm{k},i} \right| = 20 + (0.6 \times 10) + (0.6 \times 2) = 27.2\,\mathrm{kN/m}$$

2.5 Global factor of safety

The use of partial factors of safety on materials and actions offers considerable flexibility, which may be used to allow for special conditions such as very high standards of construction and control or, at the other extreme, where structural failure would be particularly disastrous.

The global factor of safety against a particular type of failure may be obtained by multiplying the appropriate partial factors of safety. For instance, a beam failure caused by yielding of tensile reinforcement would have a factor of

$$\gamma_{\mathrm{m}} \times \gamma_{\mathrm{f}} = 1.15 \times 1.35 = 1.55 \quad \text{for permanent loads only}$$

or

$$1.15 \times 1.5 = 1.72 \quad \text{for variable loads only}$$

Thus practical cases will have a value between these two figures, depending on the relative loading proportions, and this can be compared with the value of 1.8 which was the order of magnitude used by the load factor method prior to the introduction of limit state design.

Similarly, failure by crushing of the concrete in the compression zone has a factor of $1.5 \times 1.5 = 2.25$ due to variable actions only, which reflects the fact that such failure is generally without warning and may be very serious. Thus the basic values of partial factors chosen are such that under normal circumstances the global factor of safety is similar to that used in earlier design methods.

CHAPTER 3

Analysis of the structure at the ultimate limit state

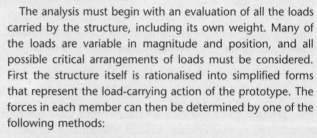

CHAPTER INTRODUCTION

A reinforced concrete structure is a combination of beams, columns, slabs and walls, rigidly connected together to form a monolithic frame. Each individual member must be capable of resisting the forces acting on it, so that the determination of these forces is an essential part of the design process. The full analysis of a rigid concrete frame is rarely simple; but simplified calculations of adequate precision can often be made if the basic action of the structure is understood.

The analysis must begin with an evaluation of all the loads carried by the structure, including its own weight. Many of the loads are variable in magnitude and position, and all possible critical arrangements of loads must be considered. First the structure itself is rationalised into simplified forms that represent the load-carrying action of the prototype. The forces in each member can then be determined by one of the following methods:

1. applying moment and shear coefficients
2. manual calculations
3. computer methods

Tabulated coefficients are suitable for use only with simple, regular structures such as equal-span continuous beams carrying uniform loads. Manual calculations are possible for the vast majority of structures, but may be tedious for large or complicated ones. The computer can be an invaluable help in the analysis of even quite small frames, and for some calculations it is almost indispensable. However, the amount of output from a computer analysis is sometimes almost overwhelming; and then the results are most readily interpreted when they are presented diagrammatically. ➡️

> →
> Since the design of a reinforced concrete member is generally based on the ultimate limit state, the analysis is usually performed for loadings corresponding to that state. Prestressed concrete members, however, are normally designed for serviceability loadings, as discussed in chapter 11.

3.1 | Actions

The actions (loads) on a structure are divided into two types: permanent actions, and variable (or imposed) actions. Permanent actions are those which are normally constant during the structure's life. Variable actions, on the other hand, are transient and not constant in magnitude, as for example those due to wind or to human occupants. Recommendations for the loadings on structures are given in the European Standards, some of which are EN 1991-1-1 General actions, EN 1991-1-3 Snow loads, EN 1991-1-4 Wind actions, EN 1991-1-7 Accidental actions from impact and explosions, and EN 1991-2 Traffic loads on bridges.

A table of values for some useful permanent loads and variable loads is given in the appendix.

3.1.1 Permanent actions

Permanent actions include the weight of the structure itself and all architectural components such as exterior cladding, partitions and ceilings. Equipment and static machinery, when permanent fixtures, are also often considered as part of the permanent action. Once the sizes of all the structural members, and the details of the architectural requirements and permanent fixtures have been established, the permanent actions can be calculated quite accurately; but, first of all, preliminary design calculations are generally required to estimate the probable sizes and self-weights of the structural concrete elements.

For most reinforced concretes, a typical value for the self-weight is 25 kN per cubic metre, but a higher density should be taken for heavily reinforced or dense concretes. In the case of a building, the weights of any permanent partitions should be calculated from the architects' drawings. A minimum partition loading equivalent to 1.0 kN per square metre is often specified as a variable action, but this is only adequate for lightweight partitions.

Permanent actions are generally calculated on a slightly conservative basis, so that a member will not need redesigning because of a small change in its dimensions. Over-estimation, however, should be done with care, since the permanent action can often actually reduce some of the forces in parts of the structure as will be seen in the case of the hogging moments in the continuous beam in figure 3.1.

3.1.2 Variable actions

These actions are more difficult to determine accurately. For many of them, it is only possible to make conservative estimates based on standard codes of practice or past experience. Examples of variable actions on buildings are: the weights of its occupants, furniture, or machinery; the pressures of wind, the weight of snow, and of retained earth or water; and the forces caused by thermal expansion or shrinkage of the concrete.

A large building is unlikely to be carrying its full variable action simultaneously on all its floors. For this reason EN 1991-1-1: 2002 (Actions on Structures) clause 6.2.2(2) allows a reduction in the total variable floor actions when the columns, walls or foundations are designed, for a building more than two storeys high. Similarly from the same code, clause 6.3.1.2(10), the variable action may be reduced when designing a beam span which supports a large floor area.

Although the wind load is a variable action, it is kept in a separate category when its partial factors of safety are specified, and when the load combinations on the structure are being considered.

3.2 | Load combinations and patterns

3.2.1 Load combinations and patterns for the ultimate limit state

Various combinations of the characteristic values of permanent G_k, variable actions Q_k, wind actions W_k, and their partial factors of safety must be considered for the loading of the structure. The partial factors of safety specified in the code are discussed in chapter 2, and for the ultimate limit state the following loading combinations from tables 2.2, 2.4 and 2.5 are commonly used.

1. Permanent and variable actions

$$1.35G_k + 1.5Q_k$$

2. Permanent and wind actions

$$1.35G_k + 1.5W_k$$

The variable load can usually cover all or any part of the structure and, therefore, should be arranged to cause the most severe stresses. So, for a three-span continuous beam, load combination 1 would have the loading arrangement shown in figure 3.1, in order to cause the maximum sagging moment in the outer spans and the maximum possible hogging moment in the centre span. A study of the deflected shape of the beam would confirm this to be the case.

Load combination 2, permanent + wind load is used to check the stability of a structure. A load combination of permanent + variable + wind load could have the arrangements shown in figure 2.4 and described in section 2.4 of Chapter 2.

Figure 3.1
Three-span beam

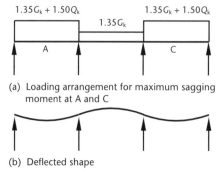

1.35G_k + 1.50Q_k 1.35G_k 1.35G_k + 1.50Q_k

A   C

(a) Loading arrangement for maximum sagging moment at A and C

(b) Deflected shape

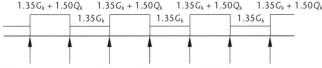

(i) Loading arrangements for maximum moments in the spans

(ii) Loading arrangements for maximum support moment at A

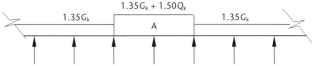

(iii) Loading for design moments at the supports according to EC2

Note that when there is a cantilever span the minimum load on the span adjacent to the cantilever should be $1.0G_k$ for loading pattern (i)

Figure 3.2
Multi-span beam loading patterns

Figure 3.2 shows the patterns of vertical loading on a multi-span continuous beam to cause (i) maximum design sagging moments in alternate spans and maximum possible hogging moments in adjacent spans, (ii) maximum design hogging moments at support A, and (iii) the design hogging moment at support A as specified by the EC2 code for simplicity. Thus there is a similar loading pattern for the design hogging moment at each internal support of a continuous beam. It should be noted that the UK National Annex permits a simpler alternative to load case (iii) where a single load case may be considered of all spans loaded with the maximum loading of $(1.35G_k + 1.50Q_k)$.

3.3   Analysis of beams

To design a structure it is necessary to know the bending moments, torsional moments, shearing forces and axial forces in each member. An elastic analysis is generally used to determine the distribution of these forces within the structure; but because – to some extent – reinforced concrete is a plastic material, a limited redistribution of the elastic moments is sometimes allowed. A plastic yield-line theory may be used to calculate the moments in concrete slabs. The properties of the materials, such as Young's modulus, which are used in the structural analysis should be those associated with their characteristic strengths. The stiffnesses of the members can be calculated on the basis of any one of the following:

1 the entire concrete cross-section (ignoring the reinforcement);

2. the concrete cross-section plus the transformed area of reinforcement based on the modular ratio;

3. the compression area only of the concrete cross-section, plus the transformed area of reinforcement based on the modular ratio.

The concrete cross-section described in (1) is the simpler to calculate and would normally be chosen.

A structure should be analysed for each of the critical loading conditions which produce the maximum stresses at any particular section. This procedure will be illustrated in the examples for a continuous beam and a building frame. For these structures it is conventional to draw the bending-moment diagram on the tension side of the members.

Sign Conventions

1. For the moment-distribution analysis anti-clockwise support moments are positive as, for example, in table 3.1 for the fixed end moments (FEM).
2. For subsequently calculating the moments along the span of a member, moments causing sagging are positive, while moments causing hogging are negative, as illustrated in figure 3.4.

3.3.1 Non-continuous beams

One-span, simply supported beams or slabs are statically determinate and the analysis for bending moments and shearing forces is readily performed manually. For the ultimate limit state we need only consider the maximum load of $1.35G_k + 1.5Q_k$ on the span.

EXAMPLE 3.1

Analysis of a non-continuous beam

The one-span simply supported beam shown in figure 3.3a carries a distributed permanent action including self-weight of 25 kN/m, a permanent concentrated action of 40 kN at mid-span, and a distributed variable action of 10 kN/m.

Figure 3.3
Analysis of one-span beam

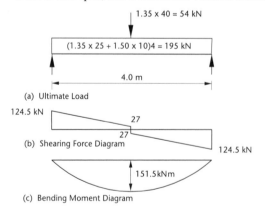

(a) Ultimate Load

(b) Shearing Force Diagram

(c) Bending Moment Diagram

Figure 3.3 shows the values of ultimate load required in the calculations of the shearing forces and bending moments.

$$\text{Maximum shear force} = \frac{54}{2} + \frac{195}{2} = 124.5\,\text{kN}$$

$$\text{Maximum bending moment} = \frac{54 \times 4}{4} + \frac{195 \times 4}{8} = 151.5\,\text{kN m}$$

The analysis is completed by drawing the shearing-force and bending-moment diagrams which would later be used in the design and detailing of the shear and bending reinforcement.

3.3.2 Continuous beams

The methods of analysis for continuous beams may also be applied to continuous slabs which span in one direction. A continuous beam is considered to have no fixity with the supports so that the beam is free to rotate. This assumption is not strictly true for beams framing into columns and for that type of continuous beam it is more accurate to analyse them as part of a frame, as described in section 3.4.

A continuous beam should be analysed for the loading arrangements which give the maximum stresses at each section, as described in section 3.2.1 and illustrated in figures 3.1 and 3.2. The analysis to calculate the bending moments can be carried out manually by moment distribution or equivalent methods, but tabulated shear and moment coefficients may be adequate for continuous beams having approximately equal spans and uniformly distributed loads.

For a beam or slab set monolithically into its supports, the design moment at the support can be taken as the moment at the face of the support.

Continuous beams – the general case

Having determined the moments at the supports by, say, moment distribution, it is necessary to calculate the moments in the spans and also the shear forces on the beam. For a uniformly distributed load, the equations for the shears and the maximum span moments can be derived from the following analysis.

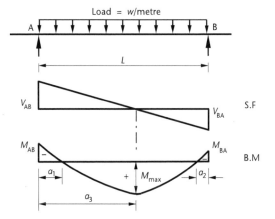

Figure 3.4
Shears and moments in a beam

Using the sign convention of figure 3.4 and taking moments about support B:

$$V_{AB}L - \frac{wL^2}{2} + M_{AB} - M_{BA} = 0$$

therefore

$$V_{AB} = \frac{wL}{2} - \frac{(M_{AB} - M_{BA})}{L} \tag{3.1}$$

and

$$V_{BA} = wL - V_{AB} \tag{3.2}$$

Maximum span moment M_{max} occurs at zero shear, and distance to zero shear

$$a_3 = \frac{V_{AB}}{w} \qquad (3.3)$$

therefore

$$M_{max} = \frac{V_{AB}^2}{2w} + M_{AB} \qquad (3.4)$$

The points of contraflexure occur at $M = 0$, that is

$$V_{AB}x - \frac{wx^2}{2} + M_{AB} = 0$$

where x the distance from support A. Taking the roots of this equation gives

$$x = \frac{V_{AB} \pm \sqrt{\left(V_{AB}^2 + 2wM_{AB}\right)}}{w}$$

so that

$$a_1 = \frac{V_{AB} - \sqrt{\left(V_{AB}^2 + 2wM_{AB}\right)}}{w} \qquad (3.5)$$

and

$$a_2 = L - \frac{V_{AB} + \sqrt{\left(V_{AB}^2 + 2wM_{AB}\right)}}{w} \qquad (3.6)$$

A similar analysis can be applied to beams that do not support a uniformly distributed load. In manual calculations it is usually not considered necessary to calculate the distances a_1, a_2 and a_3 which locate the points of contraflexure and maximum moment – a sketch of the bending moment is often adequate – but if a computer is performing the calculations these distances may as well be determined also.

At the face of the support, width s

$$M'_{AB} = M_{AB} - \left(V_{AB} - \frac{ws}{4}\right)\frac{s}{2}$$

EXAMPLE 3.2

Analysis of a continuous beam

The continuous beam shown in figure 3.5 has a constant cross-section and supports a uniformly distributed permanent action including its self-weight of $G_k = 25$ kN/m and a variable action $Q_k = 10$ kN/m.

The critical loading patterns for the ultimate limit state are shown in figure 3.5 where the 'stars' indicate the region of maximum moments, sagging or possible hogging. Table 3.1 is the moment distribution carried out for the first loading arrangement: similar calculations would be required for each of the remaining load cases. It should be noted that the reduced stiffness of $\frac{3}{4}\frac{I}{L}$ has been used for the end spans.

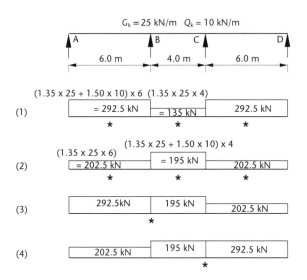

Figure 3.5
Continuous beam loading patterns

Table 3.1 Moment distribution for the first loading case

	A	B		C	D
Stiffness (k)	$\dfrac{3}{4}\cdot\dfrac{I}{L}$ $=\dfrac{3}{4}\cdot\dfrac{1}{6}=0.125$	$\dfrac{I}{L}$ $=\dfrac{1}{4}=0.25$		$\dfrac{3}{4}\cdot\dfrac{I}{L}$ $=0.125$	
Distr. factors	$\dfrac{0.125}{0.125+0.25}$ $=1/3$	2/3	$\dfrac{0.25}{0.125+0.25}$ 2/3	1/3	
Load (kN)	292		135		292
F.E.M.	0	$-\dfrac{292\times6}{8}$	$\pm\dfrac{135\times4}{12}$	$+\dfrac{292\times6}{8}$	0
	0	− 219.4	+ 45.0	− 45.0	+219.4	0
Balance		+ 58.1	+116.3	− 116.3	− 58.1	
Carry over			− 58.1	+ 58.1		
Balance		+ 19.4	+ 38.7	− 38.7	− 19.4	
Carry over			− 19.4	+ 19.4		
Balance		+ 6.5	+ 12.9	− 12.9	− 6.5	
Carry over			− 6.5	+ 6.5		
Balance		+ 2.2	+ 4.3	− 4.3	− 2.2	
Carry over			− 2.2	+ 2.2		
Balance		+ 0.7	+ 1.5	− 1.5	− 0.7	
M (kN m)	0	− 132.5	+132.5	− 132.5	+132.5	0

The shearing forces, the maximum span bending moments, and their positions along the beam, can be calculated using the formulae previously derived. Thus for the first loading arrangement and span AB, using the sign convention of figure 3.4:

$$\text{Shear } V_{AB} = \frac{\text{load}}{2} - \frac{(M_{AB} - M_{BA})}{L}$$
$$= \frac{292.5}{2} - \frac{132.5}{6.0} = 124.2 \text{ kN}$$
$$V_{BA} = \text{load} - V_{AB}$$
$$= 292.5 - 124.2 = 168.3 \text{ kN}$$

$$\text{Maximum moment, span AB} = \frac{V_{AB}^2}{2w} + M_{AB}$$

where $w = 292.5/6.0 = 48.75$ kN/m. Therefore:

$$M_{max} = \frac{124.2^2}{2 \times 48.75} + 0 = 158.2 \text{ kNm}$$
$$\text{Distance from A, } a_3 = \frac{V_{AB}}{w}$$
$$= \frac{124.2}{48.75} = 2.55 \text{ m}$$

The bending-moment diagrams for each of the loading arrangements are shown in figure 3.6, and the corresponding shearing-force diagrams are shown in figure 3.7. The individual bending-moment diagrams are combined in figure 3.8a to give the bending-moment design envelope. Similarly, figure 3.8b is the shearing-force design envelope. Such envelope diagrams are used in the detailed design of the beams, as described in chapter 7.

In this example, simple supports with no fixity have been assumed for the end supports at A and D. Even so, the sections at A and D should be designed for a hogging moment due to a partial fixity equal to 25 per cent of the maximum moment in the span, that is $158/4 = 39.5$ kNm.

Figure 3.6
Bending-moment diagrams (kN m)

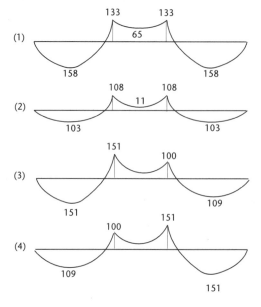

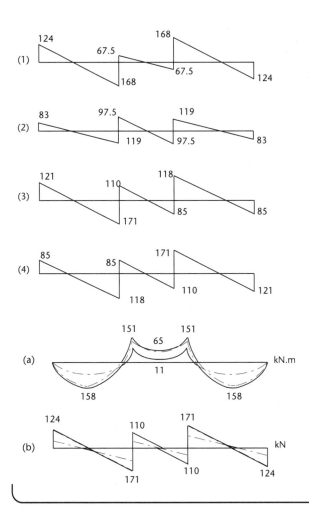

Figure 3.7
Shearing-force diagrams (kN)

Figure 3.8
Bending-moment and
shearing-force envelopes

Continuous beams with approximately equal spans and uniform loading

The ultimate bending moments and shearing forces in continuous beams of three or more approximately equal spans without cantilevers can be obtained using relevant coefficients provided that the spans differ by no more than 15 per cent of the longest span, that the loading is uniform, and that the characteristic variable action does not exceed the characteristic permanent action. The values of these coefficients are shown in diagrammatic form in figure 3.9 for beams (equivalent simplified values for slabs are given in chapter 8).

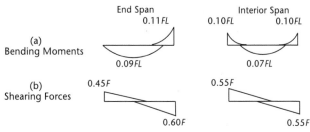

F = Total ultimate load on span = $(1.35G_k + 1.50Q_k)$ kN
L = Effective span

Figure 3.9
Bending-moment and
shearing-force coefficients
for beams

The possibility of hogging moments in any of the spans should not be ignored, even if it is not indicated by these coefficients. For example, a beam of three equal spans may have a hogging moment in the centre span if Q_k exceeds $0.45G_k$.

3.4 Analysis of frames

In situ reinforced concrete structures behave as rigid frames, and should be analysed as such. They can be analysed as a complete space frame or be divided into a series of plane frames. Bridge deck-type structures can be analysed as an equivalent grillage, whilst some form of finite-element analysis can be utilised in solving complicated shear wall buildings. All these methods lend themselves to solution by computer, but many frames can be simplified for a satisfactory solution by hand calculations.

The general procedure for a building is to analyse the slabs as continuous members supported by the beams or structural walls. The slabs can be either one-way spanning or two-way spanning. The columns and main beams are considered as a series of rigid plane frames which can be divided into two types: (1) braced frames supporting vertical loads only, (2) frames supporting vertical and lateral loads.

Type one frames are in buildings where none of the lateral loads such as wind are transmitted to the columns and beams but are resisted by much more stiffer elements such as shear walls, lift shafts or stairwells. Type two frames are designed to resist the lateral loads, which cause bending, shearing and axial loads in the beams and columns. For both types of frames the axial forces in the columns can generally be calculated as if the beams and slabs were simply supported.

3.4.1 Braced frames supporting vertical loads only

A building frame can be analysed as a complete frame, or it can be simplified into a series of substitute frames for the vertical loading analysis. The frame shown in figure 3.10, for example, can be divided into any of the subframes shown in figure 3.11.

The substitute frame 1 in figure 3.11 consists of one complete floor beam with its connecting columns (which are assumed rigidly fixed at their remote ends). An analysis of this frame will give the bending moments and shearing forces in the beams and columns for the floor level considered.

Substitute frame 2 is a single span combined with its connecting columns and two adjacent spans, all fixed at their remote ends. This frame may be used to determine the bending moments and shearing forces in the central beam. Provided that the central span is greater than the two adjacent spans, the bending moments in the columns can also be found with this frame.

Substitute frame 3 can be used to find the moments in the columns only. It consists of a single junction, with the remote ends of the members fixed. This type of subframe would be used when beams have been analysed as continuous over simple supports.

In frames 2 and 3, the assumption of fixed ends to the outer beams over-estimates their stiffnesses. These values are, therefore, halved to allow for the flexibility resulting from continuity.

The various critical loading patterns to produce maximum stresses have to be considered. In general these loading patterns for the ultimate limit state are as shown in figure 3.2, except when there is also a cantilever span which may have a beneficial minimum loading condition ($1.0G_k$) – see figure 7.21.

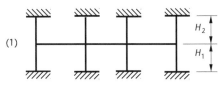

(1)

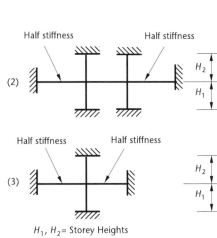

Half stiffness   Half stiffness

(2)

H_1, H_2= Storey Heights

Figure 3.10
Building frame

Figure 3.11
Substitute frames

When considering the critical loading arrangements for a column, it is sometimes necessary to include the case of maximum moment and minimum possible axial load, in order to investigate the possibility of tension failure caused by the bending.

EXAMPLE 3.3

Analysis of a substitute frame

The substitute frame shown in figure 3.12 is part of the complete frame in figure 3.10. The characteristic actions carried by the beams are permanent actions (including self-weight) $G_k = 25$ kN/m, and variable action, $Q_k = 10$ kN/m, uniformly distributed along the beam. The analysis of the subframe will be carried out by moment distribution: thus the member stiffnesses and their relevant distribution factors are first required.

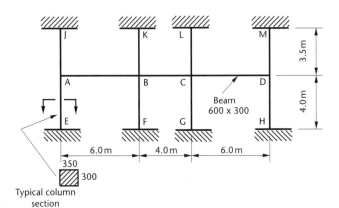

Figure 3.12
Substitute frame

Stiffnesses, k

Beam

$$I = \frac{0.3 \times 0.6^3}{12} = 5.4 \times 10^{-3} \, \text{m}^4$$

Spans AB and CD

$$k_{AB} = k_{CD} = \frac{5.4 \times 10^{-3}}{6.0} = 0.9 \times 10^{-3}$$

Span BC

$$k_{BC} = \frac{5.4 \times 10^{-3}}{4.0} = 1.35 \times 10^{-3}$$

Columns

$$I = \frac{0.3 \times 0.35^3}{12} = 1.07 \times 10^{-3} \, \text{m}^4$$

Upper

$$k_U = \frac{1.07 \times 10^{-3}}{3.5} = 0.31 \times 10^{-3}$$

Lower

$$k_L = \frac{1.07 \times 10^{-3}}{4.0} = 0.27 \times 10^{-3}$$

$$k_U + k_L = (0.31 + 0.27)10^{-3} = 0.58 \times 10^{-3}$$

Distribution factors

Joints A and D

$$\sum k = 0.9 + 0.58 = 1.48$$

$$\text{D.F.}_{AB} = \text{D.F.}_{DC} = \frac{0.9}{1.48} = 0.61$$

$$\text{D.F.}_{\cdot\text{cols}} = \frac{0.58}{1.48} = 0.39$$

Joints B and C

$$\sum k = 0.9 + 1.35 + 0.58 = 2.83$$

$$\text{D.F.}_{BA} = \text{D.F.}_{CD} = \frac{0.9}{2.83} = 0.32$$

$$\text{D.F.}_{BC} = \text{D.F.}_{CB} = \frac{1.35}{2.83} = 0.48$$

$$\text{D.F.}_{\cdot\text{cols}} = \frac{0.58}{2.83} = 0.20$$

The critical loading patterns for the ultimate limit state are identical to those for the continuous beam in example 3.2, and they are illustrated in figure 3.5. The moment distribution for the first loading arrangement is shown in table 3.2. In the table, the distribution for each upper and lower column have been combined, since this simplifies the layout for the calculations.

Table 3.2 Moment distribution for the first loading case

	A Cols.(∑M) 0.39	A AB 0.61	Load	B BA 0.32	B Cols.(∑M) 0.20	B BC 0.48	Load	C CB 0.48	C Cols.(∑M) 0.20	C CD 0.32	Load	D DC 0.61	D Cols.(∑M) 0.39
D.F.s	0.39	0.61		0.32	0.20	0.48		0.48	0.20	0.32		0.61	0.39
Load kN			292				135				292		
F.E.M.		+ 146		− 146		+ 45.0		− 45.0		+ 146		− 146	
Bal.	56.9	− 89.1		+ 32.3	+ 20.2	+ 48.5		+ 48.5	20.2	− 32.3		+ 89.1	56.9
C.O.		+ 16.2		− 44.6		− 24.2		+ 24.2		+ 44.6		− 16.2	
Bal.	6.3	− 9.9		+ 22.0	+ 13.8	+ 33.0		− 33.0	13.8	− 22.0		+ 9.9	6.3
C.O.		+ 11.0		− 5.0		− 16.5		+ 16.5		+ 5.0		− 11.0	
Bal.	4.3	− 6.7		+ 6.9	+ 4.3	+ 10.3		− 10.3	4.3	− 6.9		+ 6.7	4.3
C.O.		+ 3.4		− 3.4		− 5.2		+ 5.2		+ 3.4		− 3.4	
Bal.	1.3	− 2.1		+ 2.8	+ 1.7	+ 4.1		− 4.1	1.7	− 2.8		+ 2.1	1.3
M (kN m)	+ 68.8	+ 68.8		135.0	+ 40.0	+ 95.0		95.0	40.0	+ 135.0		− 68.8	+ 68.8

The shearing forces and the maximum span moments can be calculated from the formulae of section 3.3.2. For the first loading arrangement and span AB:

$$\text{Shear } V_{AB} = \frac{\text{load}}{2} - \frac{(M_{AB} - M_{BA})}{L}$$

$$= \frac{292.5}{2} - \frac{(-68.8 + 135.0)}{6.0} = 135\,\text{kN}$$

$$V_{BA} = \text{load} - V_{AB}$$

$$= 292.5 - 135 = 157\,\text{kN}$$

$$\text{Maximum moment, span AB} = \frac{V_{AB}^{\,2}}{2w} + M_{AB}$$

$$= \frac{135^2}{2 \times 48.75} - 68.8$$

$$= 118\,\text{kN m}$$

$$\text{Distance from A, } a_3 = \frac{V_{AB}}{w} = \frac{135}{48.75} = 2.8\,\text{m}$$

Figure 3.13 shows the bending moments in the beams for each loading pattern; figure 3.14 shows the shearing forces. These diagrams have been combined in figure 3.15 to give design envelopes for bending moments and shearing forces.

A comparison of the design envelopes of figure 3.15 and figure 3.8 will emphasise the advantages of considering the concrete beam as part of a frame, not as a continuous beam as in example 3.2. Not only is the analysis of a subframe more precise, but many moments and shears in the beam are smaller in magnitude.

The moment in each column is given by

$$M_{\text{col}} = \sum M_{\text{col}} \times \frac{k_{\text{col}}}{\sum k_{\text{col}}}$$

Figure 3.13
Beam bending-moment
diagrams (kNm)

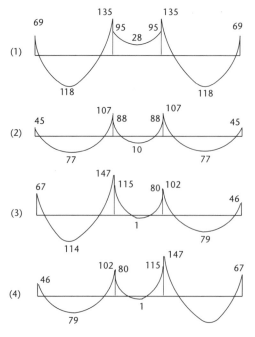

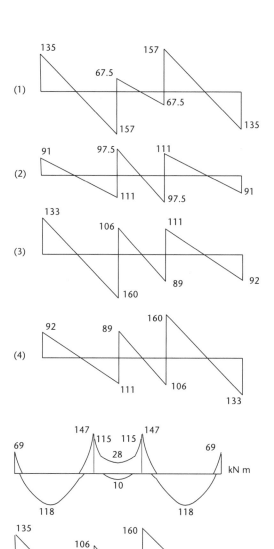

Figure 3.14
Beam shearing-force
diagrams (kN)

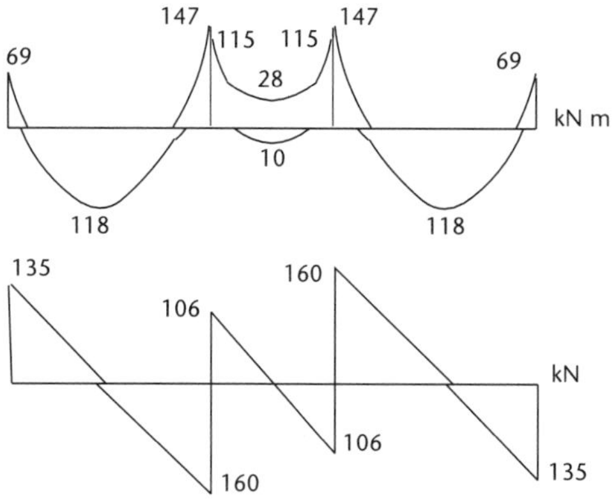

Figure 3.15
Bending-moment and
shearing-force envelopes

Thus, for the first loading arrangement and taking $\sum M_{\text{col}}$ table 3.2 gives

$$\text{Column moment } M_{\text{AJ}} = 68.8 \times \frac{0.31}{0.58} = 37 \, \text{kN m}$$

$$M_{\text{AE}} = 68.8 \times \frac{0.27}{0.58} = 32 \, \text{kN m}$$

$$M_{\text{BK}} = 40 \times \frac{0.31}{0.58} = 21 \, \text{kN m}$$

$$M_{\text{BF}} = 40 \times \frac{0.27}{0.58} = 19 \, \text{kN m}$$

This loading arrangement gives the maximum column moments, as plotted in figure 3.16.

Figure 3.16
Column bending
moments (kN m)

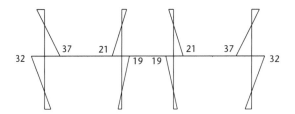

EXAMPLE 3.4

Analysis of a substitute frame for a column

The substitute frame for this example, shown in figure 3.17, is taken from the building frame in figure 3.10. The loading to cause maximum column moments is shown in the figure for $G_k = 25\,\text{kN/m}$ and $Q_k = 10\,\text{kN/m}$.

Figure 3.17
Substitute frame

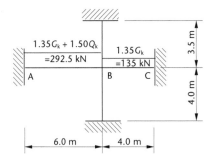

The stiffnesses of these members are identical to those calculated in example 3.3, except that for this type of frame the beam stiffnesses are halved. Thus

$$k_{AB} = \frac{1}{2} \times 0.9 \times 10^{-3} = 0.45 \times 10^{-3}$$

$$k_{BC} = \frac{1}{2} \times 1.35 \times 10^{-3} = 0.675 \times 10^{-3}$$

$$\text{upper column } k_U = 0.31 \times 10^{-3}$$

$$\text{lower column } k_L = 0.27 \times 10^{-3}$$

$$\sum k = (0.45 + 0.675 + 0.31 + 0.27) \times 10^{-3} = 1.705 \times 10^{-3}$$

$$\text{fixed-end moment } M_{BA} = 292.5 \times \frac{6}{12} = 146\,\text{kN m}$$

$$\text{fixed-end moment } M_{BC} = 135 \times \frac{4}{12} = 45\,\text{kN m}$$

Column moments are

$$\text{upper column } M_U = (146 - 45) \times \frac{0.31}{1.705} = 18\,\text{kN m}$$

$$\text{lower column } M_L = (146 - 45) \times \frac{0.27}{1.705} = 16\,\text{kN m}$$

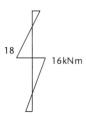

18

16kNm

Figure 3.18
Column moments

The column moments are illustrated in figure 3.18. They should be compared with the corresponding moments for the internal column in figure 3.16.

In examples 3.3 and 3.4 the second moment of area of the beam was calculated as $bh^3/12$ a rectangular section for simplicity, but where an *in situ* slab forms a flange to the beam, the second moment of area may be calculated for the T-section or L-section.

3.4.2 Lateral loads on frames

Lateral loads on a structure may be caused by wind pressures, by retained earth or by seismic forces. A horizontal force should also be applied at each level of a structure resulting from a notional inclination of the vertical members representing imperfections. The value of this depends on building height and number of columns (EC2 clause 5.2), but will typically be less than 1% of the vertical load at that level for a braced structure. This should be added to any wind loads at the ultimate limit state

An unbraced frame subjected to wind forces must be analysed for all the vertical loading combinations described in section 3.2.1. The vertical-loading analysis can be carried out by the methods described previously. The analysis for the lateral loads should be kept separate. The forces may be calculated by an elastic computer analysis or by a simplified approximate method. For preliminary design calculations, and also only for medium-size regular structures, a simplified analysis may well be adequate.

A suitable approximate analysis is the cantilever method. It assumes that:

1. points of contraflexure are located at the mid-points of all columns and beams; and
2. the direct axial loads in the columns are in proportion to their distances from the centre of gravity of the frame. It is also usual to assume that all the columns in a storey are of equal cross-sectional area.

It should be emphasised that these approximate methods may give quite inaccurate results for irregular or high-rise structures. Application of this method is probably best illustrated by an example, as follows.

EXAMPLE 3.5

Simplified analysis for lateral loads – cantilever method

Figure 3.19 shows a building frame subjected to a characteristic wind action of 3.0 kN per metre height of the frame. This action is assumed to be transferred to the frame as a concentrated load at each floor level as indicated in the figure.

By inspection, there is tension in the two columns to the left and compression in the columns to the right; and by assumption 2 the axial forces in columns are proportional to their distances from the centre line of the frame.

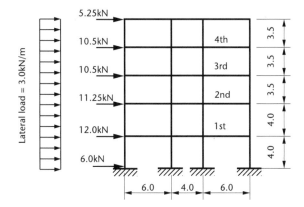

Figure 3.19
Frame with lateral load

Figure 3.20
Subframes at the roof
and 4th floor

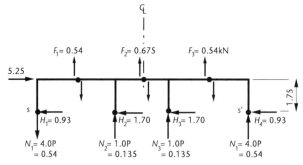

(a) Roof

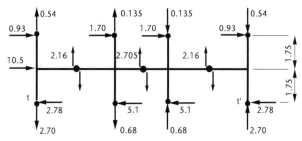

(b) 4th Floor

Thus

Axial force in exterior column : axial force in interior column $= 4.0P : 1.0P$

The analysis of the frame continues by considering a section through the top-storey columns: the removal of the frame below this section gives the remainder shown in figure 3.20a. The forces in this subframe are calculated as follows.

(a) Axial forces in the columns

Taking moments about point s, $\sum M_s = 0$, therefore
$$5.25 \times 1.75 + P \times 6.0 - P \times 10.0 - 4P \times 16.0 = 0$$
and therefore
$$P = 0.135 \, \text{kN}$$
thus
$$N_1 = -N_4 = 4.0P = 0.54 \, \text{kN}$$
$$N_2 = -N_3 = 1.0P = 0.135 \, \text{kN}$$

(b) Vertical shearing forces F in the beams

For each part of the subframe, $\sum F = 0$, therefore
$$F_1 = N_1 = 0.54 \, \text{kN}$$
$$F_2 = N_1 + N_2 = 0.675 \, \text{kN}$$

(c) Horizontal shearing forces H in the columns

Taking moments about the points of contraflexure of each beam, $\sum M = 0$, therefore
$$H_1 \times 1.75 - N_1 \times 3.0 = 0$$
$$H_1 = 0.93 \, \text{kN}$$

and

$$(H_1 + H_2)1.75 - N_1 \times 8.0 - N_2 \times 2.0 = 0$$
$$H_2 = 1.70\,\text{kN}$$

The calculations of the equivalent forces for the fourth floor (figure 3.20b) follow a similar procedure, as follows.

(d) Axial forces in the columns

For the frame above section tt', $\sum M_t = 0$, therefore

$$5.25(3 \times 1.75) + 10.5 \times 1.75 + P \times 6.0 - P \times 10.0 - 4P \times 16.0 = 0$$
$$P = 0.675\,\text{kN}$$

therefore

$$N_1 = 4.0P = 2.70\ \text{kN}$$
$$N_2 = 1.0P = 0.68\,\text{kN}$$

(e) Beam shears

$$F_1 = 2.70 - 0.54$$
$$= 2.16\,\text{kN}$$
$$F_2 = 2.70 + 0.68 - 0.54 - 0.135$$
$$= 2.705\,\text{kN}$$

(f) Column shears

$$H_1 \times 1.75 + 0.93 \times 1.75 - (2.70 - 0.54)3.0 = 0$$
$$H_1 = 2.78\,\text{kN}$$
$$H_2 = \frac{1}{2}(10.5 + 5.25) - 2.78$$
$$= 5.1\,\text{kN}$$

Values calculated for sections taken below the remaining floors are

third floor	$N_1 = 7.03\,\text{kN}$	$N_2 = 1.76\,\text{kN}$
	$F_1 = 4.33\,\text{kN}$	$F_2 = 5.41\,\text{kN}$
	$H_1 = 4.64\,\text{kN}$	$H_2 = 8.49\,\text{kN}$
second floor	$N_1 = 14.14\,\text{kN}$	$N_2 = 3.53\,\text{kN}$
	$F_1 = 7.11\,\text{kN}$	$F_2 = 8.88\,\text{kN}$
	$H_1 = 6.61\,\text{kN}$	$H_2 = 12.14\,\text{kN}$
first floor	$N_1 = 24.37\,\text{kN}$	$N_2 = 6.09\,\text{kN}$
	$F_1 = 10.23\,\text{kN}$	$F_2 = 12.79\,\text{kN}$
	$H_1 = 8.74\,\text{kN}$	$H_2 = 16.01\,\text{kN}$

The bending moments in the beams and columns at their connections can be calculated from these results by the following formulae

beams $M_B = F \times \frac{1}{2}\text{beam span}$

columns $M_C = H \times \frac{1}{2}\text{storey height}$

Figure 3.21
Moments (kN m) and
reactions (kN)

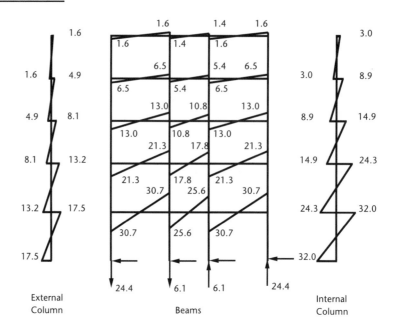

so that the roof's external connection

$$M_B = 0.54 \times \frac{1}{2} \times 60 = 1.6 \, \text{kN m}$$

$$M_C = 0.93 \times \frac{1}{2} \times 3.5 = 1.6 \, \text{kN m}$$

As a check at each joint, $\sum M_B = \sum M_C$.

The bending moments due to characteristic wind loads in all the columns and beams of this structure are shown in figure 3.21.

3.5 | Shear wall structures resisting horizontal loads

A reinforced concrete structure with shear walls is shown in figure 3.22 . Shear walls are very effective in resisting horizontal loads such as F_z in the figure which act in the direction of the plane of the walls. As the walls are relatively thin they offer little resistance to loads which are perpendicular to their plane.

The floor slabs which are supported by the walls also act as rigid diaphragms which transfer and distribute the horizontal forces into the shear walls. The shear walls act as vertical cantilevers transferring the horizontal loads to the structural foundations.

3.5.1 Symmetrical arrangement of walls

With a symmetrical arrangement of walls as shown in figure 3.23 the horizontal load is distributed in proportion to the the relative stiffness k_i of each wall. The relative

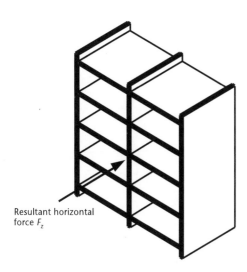

Figure 3.22
Shear wall structure

Resultant horizontal
force F_z

stiffnesses are given by the second moment of area of each wall about its major axis such that

$$k_i \approx h \times b^3$$

where h is the thickness of the wall and b is the length of the wall.
   The force P_i distributed into each wall is then given by

$$P_i = F \times \frac{k_i}{\sum k}$$

EXAMPLE 3.6

Symmetrical arrangement of shear walls

A structure with a symmetrical arrangement of shear walls is shown in figure 3.23.
Calculate the proportion of the 100 kN horizontal load carried by each of the walls.

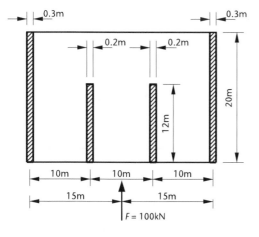

Figure 3.23
Symmetrical arrangement
of shear walls

Relative stiffnesses:

Walls A $\qquad k_A = 0.3 \times 20^3 = 2400$

Walls B $\qquad k_B = 0.2 \times 8^3 = 346$

$$\sum k = 2(2400 + 346) = 5492$$

Force in each wall:

$$P_A = \frac{k_A}{\sum k} \times F = \frac{2400}{5492} \times 100 = 43.7\,\text{kN}$$

$$P_B = \frac{k_B}{\sum k} \times F = \frac{346}{5492} \times 100 = 6.3\,\text{kN}$$

Check $2(43.7 + 6.3) = 100\,\text{kN} = F$

3.5.2 Unsymmetrical arrangement of walls

With an unsymmetrical arrangement of shear walls as shown in figure 3.24 there will also be a torsional force on the structure about the centre of rotation in addition to the direct forces caused by the translatory movement. The calculation procedure for this case is:

1. Determine the location of the centre of rotation by taking moments of the wall stiffnesses k about convenient axes. Such that

$$\bar{x} = \frac{\sum(k_x x)}{\sum k_x} \quad \text{and} \quad \bar{y} = \frac{\sum(k_y y)}{\sum k_y}$$

where k_x and k_y are the stiffnesses of the walls orientated in the x and y directions respectively.

2. Calculate the torsional moment M_t on the group of shear walls as

$$M_t = F \times e$$

where e is the eccentricity of the horizontal force F about the centre of rotation.

3. Calculate the force P_i in each wall as the sum of the direct component P_d and the torsional rotation component P_r

$$P_i = P_d + P_r$$

$$= F \times \frac{k_x}{\sum k_x} \pm M_t \times \frac{k_i r_i}{\sum(k_i r_i^2)}$$

where r_i is the perpendicular distance between the axis of each wall and the centre of rotation.

EXAMPLE 3.7

Unsymmetrical layout of shear walls

Determine the distribution of the 100 kN horizontal force F into the shear walls A, B, C, D and E as shown in figure 3.24. The relative stiffness of each shear wall is shown in the figure in terms of multiples of k.

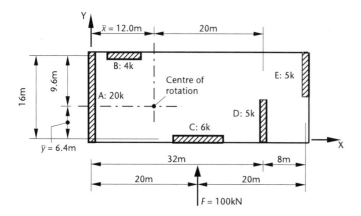

Figure 3.24
Unsymmetrical arrangement
of shear walls

Centre of rotation

$$\sum k_x = 20 + 5 + 5 = 30$$

Taking moments for k_x about YY at wall A

$$\bar{x} = \frac{\sum(k_x x)}{\sum k} = \frac{20 \times 0 + 5 \times 32 + 5 \times 40}{30}$$

$$= 12.0 \text{ metres}$$

$$\sum k_y = 6 + 4 = 10$$

Taking moments for k_y about XX at wall C

$$\bar{y} = \frac{\sum(k_y y)}{\sum k_y} = \frac{6 \times 0 + 4 \times 16}{10}$$

$$= 6.4 \text{ metres}$$

The torsional moment M_t is

$$M_t = F \times (20 - \bar{x}) = 100 \times (20 - 12)$$
$$= 800 \text{ kN m}$$

The remainder of these calculations are conveniently set out in tabular form:

Wall	k_x	k_y	r	kr	kr^2	P_d	P_r	P_i
A	20	0	12	240	2880	66.6	− 20.4	46.2
B	0	4	9.6	38.4	369	0	− 3.3	− 3.3
C	0	6	6.4	38.4	246	0	3.3	3.3
D	5	0	20	100	2000	16.7	8.5	25.2
E	5	0	28	140	3920	16.7	11.9	28.6
$\sum$	30	10			9415	100	0	100

As an example for wall A:

$$P_A = P_t + P_r = F \times \frac{k_A}{\sum k} - M_t \times \frac{k_A r_A}{\sum(k_i r_i^2)}$$

$$= 100 \times \frac{20}{30} - 800 \times \frac{20 \times 12}{9415} = 66.6 - 20.4 = 46.2 \text{ kN}$$

Figure 3.25
Shear wall with openings

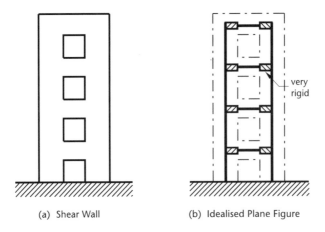

(a) Shear Wall   (b) Idealised Plane Figure

3.5.3 Shear walls with openings

Shear walls with openings can be idealised into equivalent plane frames as shown in figure 3.25. In the plane frame the second moment of area I_c of the columns is equivalent to that of the wall on either side of the openings. The second moment of area I_b of the beams is equivalent to that part of the wall between the openings.

The lengths of beam that extend beyond the openings as shown shaded in figure 3.25 are given a very large stiffnesses so that their second moment of area would be say $100I_b$.

The equivalent plane frame would be analysed by computer with a plane frame program.

3.5.4 Shear walls combined with structural frames

For simplicity in the design of low or medium-height structures shear walls or a lift shaft are usually considered to resist all of the horizontal load. With higher rise structures for reasons of stiffness and economy it often becomes necessary to include the combined action of the shear walls and the structural frames in the design.

A method of analysing a structure with shear walls and structural frames as one equivalent linked-plane frame is illustrated by the example in figure 3.26.

In the actual structure shown in plan there are four frames of type A and two frames of type B which include shear walls. In the linked frame shown in elevation the four type A frames are lumped together into one frame whose member stiffnesses are multiplied by four. Similarly the two type B frames are lumped together into one frame whose member stiffnesses are doubled. These two equivalent frames are then linked together by beams pinned at each end.

The two shear walls are represented by one column having the sectional properties of the sum of the two shear walls. For purposes of analysis this column is connected to the rest of its frame by beams with a very high bending stiffness, say 1000 times that of the other beams so as to represent the width and rigidity of the shear wall.

The link beams transfer the loads axially between the two types of frames A and B so representing the rigid diaphragm action of the concrete floor slabs. These link beams, pinned at their ends, would be given a cross-sectional area of say 1000 times that of the other beams in the frame.

As all the beams in the structural frames are pressing against the rigid shear wall in the computer model the effects of axial shortening in these beams will be exaggerated,

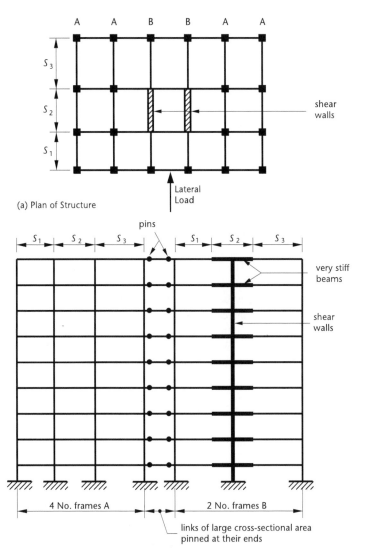

Figure 3.26
Idealised link frame for a
structure with shear walls and
structural frames

A   A   B   B   A   A

S_3

S_2 shear
walls

S_1

Lateral
Load

(a) Plan of Structure

pins

S_1 S_2 S_3 S_1 S_2 S_3

very stiff
beams

shear
walls

4 No. frames A   2 No. frames B

links of large cross-sectional area
pinned at their ends

(b) Elevation of Link-Frame Model

whereas this would normally be of a secondary magnitude. To overcome this the cross-sectional areas of all the beams in the model may be increased say to 1000 m² and this will virtually remove the effects of axial shortening in the beams.

In the computer output the member forces for type A frames would need to be divided by a factor of four and those for type B frames by a factor of two.

3.6 Redistribution of moments

Some method of elastic analysis is generally used to calculate the forces in a concrete structure, despite the fact that the structure does not behave elastically near its ultimate load. The assumption of elastic behaviour is reasonably true for low stress levels; but as a section approaches its ultimate moment of resistance, plastic deformation will occur.

Figure 3.27
Typical moment–curvature
diagram

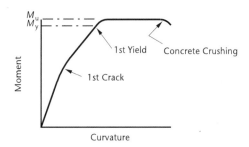

This is recognised in EC2, by allowing redistribution of the elastic moments subject to certain limitations.

Reinforced concrete behaves in a manner midway between that of steel and concrete. The stress–strain curves for the two materials (figures 1.5 and 1.2) show the elastoplastic behaviour of steel and the plastic behaviour of concrete. The latter will fail at a relatively small compressive strain. The exact behaviour of a reinforced concrete section depends on the relative quantities and the individual properties of the two materials. However, such a section may be considered virtually elastic until the steel yields; and then plastic until the concrete fails in compression. Thus the plastic behaviour is limited by the concrete failure; or more specifically, the concrete failure limits the rotation that may take place at a section in bending. A typical moment–curvature diagram for a reinforced concrete member is shown in figure 3.27

Thus, in an indeterminate structure, once a beam section develops its ultimate moment of resistance, M_u, it then behaves as a plastic hinge resisting a constant moment of that value. Further loading must be taken by other parts of the structure, with the changes in moment elsewhere being just the same as if a real hinge existed. Provided rotation of a hinge does not cause crushing of the concrete, further hinges will be formed until a mechanism is produced. This requirement is considered in more detail in chapter 4.

EXAMPLE 3.8

Moment redistribution – single span fixed-end beam

The beam shown in figure 3.28 is subjected to an increasing uniformly distributed load:

$$\text{Elastic support moment} = \frac{wL^2}{12}$$

$$\text{Elastic span moment} = \frac{wL^2}{24}$$

In the case where the ultimate bending strengths are equal at the span and at the supports, and where adequate rotation is possible, then the additional load w_a, which the member can sustain by plastic behaviour, can be found.

At collapse

$$M_u = \frac{wL^2}{12}$$

$$= \frac{wL^2}{24} + \text{ additional mid-span moment } m_B$$

where $m_B = (w_a L^2)/8$ as for a simply supported beam with hinges at A and C.

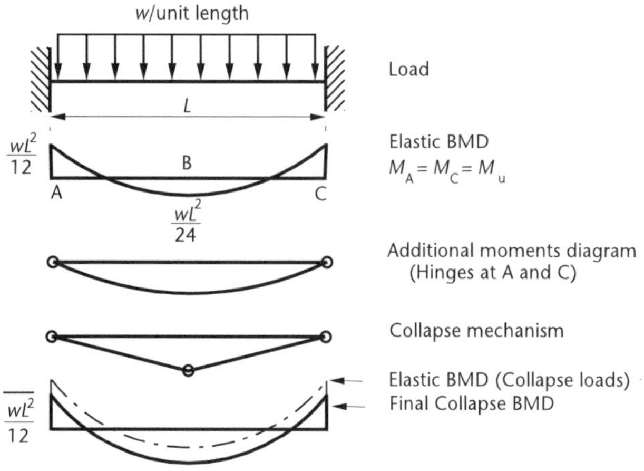

Figure 3.28
Moment redistribution,
one-span beam

Load

Elastic BMD
$M_A = M_C = M_u$

Additional moments diagram
(Hinges at A and C)

Collapse mechanism

Elastic BMD (Collapse loads)
Final Collapse BMD

Thus $\quad \dfrac{wL^2}{12} = \dfrac{wL^2}{24} + \dfrac{w_a L^2}{8}$

Hence $\quad w_a = \dfrac{w}{3}$

where w is the load to cause the first plastic hinge; thus the beam may carry a load of $1.33w$ with redistribution.

From the design point of view, the elastic bending-moment diagram can be obtained for the required ultimate loading in the ordinary way. Some of these moments may then be reduced; but this will necessitate increasing others to maintain the static equilibrium of the structure. Usually it is the maximum support moments which are reduced, so economising in reinforcing steel and also reducing congestion at the columns. The requirements for applying moment redistribution are:

1. Equilibrium between internal and external forces must be maintained, hence it is necessary to recalculate the span bending moments and the shear forces for the load case involved.

2. The continuous beams or slabs are predominately subject to flexure.

3. The ratio of adjacent spans be in the range of 0.5 to 2.

4. The column design moments must not be reduced.

There are other restrictions on the amount of moment redistribution in order to ensure ductility of the beams or slabs. This entails limitations on the grade of reinforcing steel and of the areas of tensile reinforcement and hence the depth of the neutral axis as described in Chapter Four –'Analysis of the Section'.

EXAMPLE 3.9

Moment redistribution

In example 3.3, figure 3.13, it is required to reduce the maximum support moment of $M_{BA} = 147\,\text{kN}\,\text{m}$ as much as possible, but without increasing the span moment above the present maximum value of $118\,\text{kN}\,\text{m}$.

Figure 3.29
Moments and shears after
redistribution

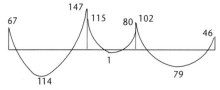

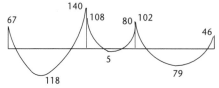

(a) Original Moments (kN m)

(b) Redistributed Moments (kN m)

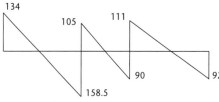

(c) Shears (kN)

Figure 3.29a duplicates the original bending-moment diagram (part 3 of figure 3.13) of example 3.3 while figure 3.29b shows the redistributed moments, with the span moment set at 118kN m. The moment at support B can be calculated, using a rearrangement of equations 3.4 and 3.1.

Thus

$$V_{AB} = \sqrt{[(M_{max} - M_{AB})2w]}$$

and

$$M_{BA} = \left(V_{AB} - \frac{wL}{2} \right) L + M_{AB}$$

For span AB, $w = 48.75\,\text{kN m}$, therefore

$$V_{AB} = \sqrt{[(118+67) \times 2 \times 48.75]} = 134\,\text{kN}$$

$$M_{BA} = \left(134 - \frac{48.75 \times 6.0}{2} \right) 6.0 - 67 = 140\,\text{kN m}$$

and

$$V_{BA} = 292.5 - 134$$
$$= 158.5\,\text{kN}$$

Reduction in $M_{BA} = 147 - 140$
$$= 7\,\text{kN m}$$
$$= \frac{7 \times 100}{147} = 4.8\,\text{per cent}$$

In order to ensure that the moments in the columns at joint B are not changed by the redistribution, moment M_{BC} must also be reduced by 7 kN m. Therefore

$$M_{BC} = 115 - 7 = 108 \text{ kN m hogging}$$

For the revised moments in BC:

$$V_{BC} = \frac{(108 - 80)}{4} + \frac{195}{2} = 105 \text{ kN}$$
$$V_{CB} = 195 - 105 = 90 \text{ kN}$$

For span BC:

$$M_{max} = \frac{105^2}{2 \times 48.75} - 108 = 5 \text{ kN m sagging}$$

Figure 3.29c shows the revised shearing-force diagram to accord with the redistributed moments. This example illustrates how, with redistribution

1. the moments at a section of beam can be reduced without exceeding the maximum design moments at other sections;

2. the values of the column moments are not affected; and

3. the equilibrium between external loads and internal forces is maintained.

CHAPTER 4

Analysis of the section

CHAPTER INTRODUCTION

A satisfactory and economic design of a concrete structure rarely depends on a complex theoretical analysis. It is achieved more by deciding on a practical overall layout of the structure, careful attention to detail and sound constructional practice. Nevertheless the total design of a structure does depend on the analysis and design of the individual member sections.

Wherever possible the analysis should be kept simple, yet it should be based on the observed and tested behaviour of reinforced concrete members. The manipulation and juggling with equations should never be allowed to obscure the fundamental principles that unite the analysis. The three most important principles are

1. The stresses and strains are related by the material properties, including the stress–strain curves for concrete and steel.
2. The distribution of strains must be compatible with the distorted shape of the cross-section.
3. The resultant forces developed by the section must balance the applied loads for static equilibrium.

These principles are true irrespective of how the stresses and strains are distributed, or how the member is loaded, or whatever the shape of the cross-section.

This chapter describes and analyses the action of a member section under load. It derives the basic equations used in design and also those equations required for

→

→

the preparation of design charts. Emphasis has been placed mostly on the analysis associated with the ultimate limit state but the behaviour of the section within the elastic range and the serviceability limit state has also been considered.

Section 4.7 deals with the redistribution of the moments from an elastic analysis of the structure, and the effect it has on the equations derived and the design procedure. It should be noted that EC2 does not give any explicit equations for the analysis or design of sections. The equations given in this chapter are developed from the principles of EC2 in a form comparable with the equations formerly given in BS 8110.

4.1   Stress–strain relations

Short-term stress–strain curves for concrete and steel are presented in EC2. These curves are in an idealised form which can be used in the analysis of member sections.

4.1.1   Concrete

The behaviour of structural concrete (figure 4.1) is represented by a parabolic stress–strain relationship, up to a strain ε_{c2}, from which point the strain increases while the stress remains constant. The ultimate design stress is given by

$$\frac{\alpha f_{ck}}{\gamma_c} = \frac{0.85 f_{ck}}{1.5}$$
$$= 0.567 f_{ck}$$

where the factor of 0.85 allows for the difference between the bending strength and the cylinder crushing strength of the concrete, and $\gamma_c = 1.5$ is the usual partial safety factor for the strength of concrete. The ultimate strain of $\varepsilon_{cu2} = 0.0035$ is typical for classes of concrete $\leq$ C50/60. Concrete classes $\leq$ C50/60 will, unless otherwise stated, be

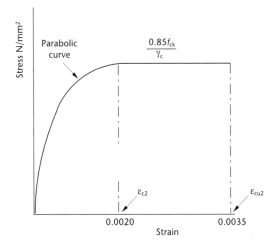

Figure 4.1
Parabolic-rectangular stress–strain diagram for concrete in compression

considered throughout this book as these are the classes most commonly used in reinforced concrete construction. Also for concrete classes higher than C50/60 the defining properties such as the ultimate strain ε_{cu2} vary for each of the higher classes. Design equations for the higher classes of concrete can in general be obtained using similar procedures to those shown in the text with the relative properties and coefficients obtained from the Eurocodes.

4.1.2 Reinforcing steel

The representative short-term design stress–strain curve for reinforcement is given in figure 4.2. The behaviour of the steel is identical in tension and compression, being linear in the elastic range up to the design yield stress of f_{yk}/γ_s where f_{yk} is the characteristic yield stress and γ_s is the partial factor of safety.

Figure 4.2
Short-term design stress–strain curve for reinforcement

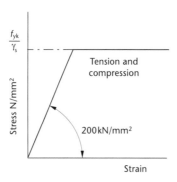

Within the elastic range, the relationship between the stress and strain is

$$\text{Stress} = \text{elastic modulus} \times \text{strain}$$
$$= E_s \times \varepsilon_s$$

(4.1)

so that the design yield strain is

$$\varepsilon_y = \left(\frac{f_{yk}}{\gamma_s}\right)/E_s$$

at the ultimate limit for $f_{yk} = 500\,\text{N/mm}^2$

$$\varepsilon_y = 500/(1.15 \times 200 \times 10^3)$$
$$= 0.00217$$

It should be noted that EC2 permits the use of an alternative design stress–strain curve to that shown in figure 4.2 with an inclined top branch and the maximum strain limited to a value which is dependent on the class of reinforcement. However the more commonly used curve shown in figure 4.2 will be used in this chapter and throughout the text.

4.2 | Distribution of strains and stresses across a section in bending

The theory of bending for reinforced concrete assumes that the concrete will crack in the regions of tensile strains and that, after cracking, all the tension is carried by the

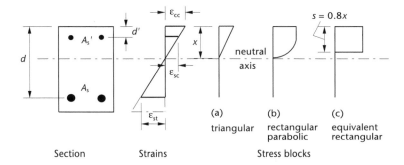

Figure 4.3
Section with strain diagram
and stress blocks

reinforcement. It is also assumed that plane sections of a structural member remain plane after straining, so that across the section there must be a linear distribution of strains.

Figure 4.3 shows the cross-section of a member subjected to bending, and the resultant strain diagram, together with three different types of stress distribution in the concrete:

1. The triangular stress distribution applies when the stresses are very nearly proportional to the strains, which generally occurs at the loading levels encountered under working conditions and is, therefore, used at the serviceability limit state.

2. The rectangular–parabolic stress block represents the distribution at failure when the compressive strains are within the plastic range, and it is associated with the design for the ultimate limit state.

3. The equivalent rectangular stress block is a simplified alternative to the rectangular–parabolic distribution.

As there is compatibility of strains between the reinforcement and the adjacent concrete, the steel strains ε_{st} in tension and ε_{sc} in compression can be determined from the strain diagram. The relationships between the depth of neutral axis (x) and the maximum concrete strain (ε_{cu2}) and the steel strains are given by

$$\varepsilon_{st} = \varepsilon_{cu2}\left(\frac{d-x}{x}\right) \tag{4.2}$$

and

$$\varepsilon_{sc} = \varepsilon_{cu2}\left(\frac{x-d'}{x}\right) \tag{4.3}$$

where d is the effective depth of the beam and d' is the depth of the compression reinforcement.

Having determined the strains, we can evaluate the stresses in the reinforcement from the stress–strain curve of figure 4.2, together with the equations developed in section 4.1.2.

For analysis of a section with known steel strains, the depth of the neutral axis can be determined by rearranging equation 4.2 as

$$x = \frac{d}{1 + \dfrac{\varepsilon_{st}}{\varepsilon_{cu2}}} \tag{4.4}$$

At the ultimate limit state the maximum compressive strain in the concrete is taken as

$\varepsilon_{cu2} = 0.0035$ for concrete class $\leq$ C50/60

For higher classes of concrete reference should be made to EC2 Table 3.1 – Strength and deformation characteristics for concrete.

For steel with $f_{yk} = 500\text{N/mm}^2$ the yield strain from section 4.1.2 is $\varepsilon_y = 0.00217$. Inserting these values for ε_{cu2} and ε_y into equation 4.4:

$$x = \frac{d}{1 + \dfrac{0.00217}{0.0035}} = 0.617d$$

Hence, to ensure yielding of the tension steel at the ultimate limit state:

$x \leq 0.617d$

At the ultimate limit state it is important that member sections in flexure should be ductile and that failure should occur with the gradual yielding of the tension steel and not by a sudden catastrophic compression failure of the concrete. Also, yielding of the reinforcement enables the formation of plastic hinges so that redistribution of maximum moments can occur, resulting in a safer and more economical structure. To ensure rotation of the plastic hinges with sufficient yielding of the tension steel and also to allow for other factors such as the strain hardening of the steel, EC2 limits the depth of neutral axis to

$x \leq 0.45d$

for concrete class $\leq$ C50/60.

This is the limiting maximum value for x given by EC2 with no redistribution applied to the moments calculated by an elastic analysis of the structure, as described in Chapter 3. When moment redistribution is applied these maximum values of x are reduced as described in Section 4.7.

The UK Annex to EC2 can give different limiting values for x. The EC2 value of $x = 0.45d$ is within the Annex's required limits and it ensures that a gradual tension failure of the steel occurs at the ultimate limit state, and not sudden brittle failure of the concrete in compression.

4.3　Bending and the equivalent rectangular stress block

For most reinforced concrete structures it is usual to commence the design for the conditions at the ultimate limit state, followed by checks to ensure that the structure is adequate for the serviceability limit state without excessive deflection or cracking of the concrete. For this reason the analysis in this chapter will first consider the simplified rectangular stress block which can be used for the design at the ultimate limit state.

The rectangular stress block as shown in figure 4.4 may be used in preference to the more rigorous rectangular–parabolic stress block. This simplified stress distribution will facilitate the analysis and provide more manageable design equations, in particular when dealing with non-rectangular cross-sections or when undertaking hand calculations.

It can be seen from figure 4.4 that the stress block does not extend to the neutral axis of the section but has a depth $s = 0.8x$. This will result in the centroid of the stress block being $s/2 = 0.40x$ from the top edge of the section, which is very nearly the same location as for the more precise rectangular–parabolic stress block. Also the areas of the

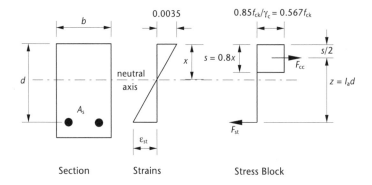

Figure 4.4
Singly reinforced section with
rectangular stress block

Section   Strains   Stress Block

two types of stress block are approximately equal (see section 4.9). Thus the moment of resistance of the section will be similar using calculations based on either of the two stress blocks.

The design equations derived in sections 4.4 to 4.6 are for zero redistribution of moments. When moment redistribution is applied, reference should be made to section 4.7 which describes how to modify the design equations.

4.4 Singly reinforced rectangular section in bending at the ultimate limit state

4.4.1 Design equations for bending

Bending of the section will induce a resultant tensile force F_{st} in the reinforcing steel, and a resultant compressive force in the concrete F_{cc} which acts through the centroid of the effective area of concrete in compression, as shown in figure 4.4.

For equilibrium, the ultimate design moment, M, must be balanced by the moment of resistance of the section so that

$$M = F_{cc}z = F_{st}z \tag{4.5}$$

where z the lever arm between the resultant forces F_{cc} and F_{st}

$$F_{cc} = \text{stress} \times \text{area of action}$$
$$= 0.567f_{ck} \times bs$$

and

$$z = d - s/2 \tag{4.6}$$

so that substituting in equation 4.5

$$M = 0.567f_{ck}bs \times z$$

and replacing s from equation 4.6 gives

$$M = 1.134f_{ck}b(d - z)z \tag{4.7}$$

Rearranging and substituting $K = M/bd^2f_{ck}$:

$$(z/d)^2 - (z/d) + K/1.134 = 0$$

Solving this quadratic equation:

$$z = d\left[0.5 + \sqrt{(0.25 - K/1.134)}\right] \tag{4.8}*$$

$K = M/bd^2 f_{ck}$	0.05	0.06	0.07	0.08	0.09	0.10	0.11	0.12	0.13	0.14	0.15	0.16	0.167
$l_a = z/d$	0.954	0.945	0.934	0.924	0.913	0.902	0.891	0.880	0.868	0.856	0.843	0.830	0.820

Figure 4.5
Lever-arm curve

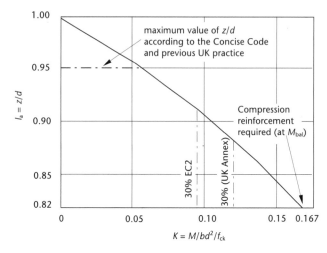

The percentage values on the K axis mark the limits for singly reinforced sections with moment redistribution applied (see Section 4.7 and Table 4.2)

in equation 4.5

$$F_{st} = (f_y/\gamma_s)A_s \quad \text{with } \gamma_s = 1.15$$
$$= 0.87f_{yk}A_s$$

Hence

$$A_s = \frac{M}{0.87f_{yk}z} \tag{4.9}*$$

Equations 4.8 and 4.9 can be used to design the area of tension reinforcement in a singly reinforced concrete section to resist an ultimate moment, M.

Equation 4.8 for the lever arm z can also be used to set up a table and draw a lever-arm curve as shown in figure 4.5. This curve may be used to determine the lever arm, z instead of solving equation 4.8.

The lower limit of $z = 0.82d$ in figure 4.5 occurs when the depth of the neutral axis equals $0.45d$. This is the maximum value allowed by EC2 for a singly reinforced section with concrete class less than or equal to C50/60 in order to provide a ductile section that will have a gradual tension type failure as already described in section 4.2.

4.4.2 The balanced section

The concrete section with the depth of neutral axis at the specified maximum depth of $0.45d$ is often referred to as the balanced section because at the ultimate limit state the concrete and tension steel reach their ultimate strains at the same time. This occurs at the maximum moment of resistance for a singly reinforced section, that is a section with no compression steel. So for this section with

$$x_{bal} = 0.45d$$

the depth of the stress block is

$$s = 0.8x_{\text{bal}} = 0.8 \times 0.45d = 0.36d$$

The force in the concrete stress block is

$$F_{\text{ccbal}} = 0.567f_{\text{ck}} \times bs = 0.204f_{\text{ck}}bd$$

For equilibrium the force in the concrete F_{ccbal} must be balanced by the force F_{stbal} in the steel. So that

$$F_{\text{stbal}} = 0.87f_{\text{yk}}A_{\text{sbal}} = F_{\text{ccbal}} = 0.204f_{\text{ckbd}}$$

Therefore

$$A_{\text{sbal}} = 0.234f_{\text{ck}}bd/f_{\text{yk}}$$

So that

$$\frac{100A_{\text{sbal}}}{bd} = 23.4\frac{f_{\text{ck}}}{f_{\text{yk}}} \text{ per cent}$$

which is the steel percentage for a balanced section which should not be exceeded for a ductile singly reinforced section.

Thus, for example, with $f_{\text{ck}} = 25\,\text{N/mm}^2$ and $f_{\text{yk}} = 500\,\text{N/mm}^2$

$$\frac{100A_{\text{sbal}}}{bd} = 23.4 \times \frac{25}{500} = 23.4 \times \frac{25}{500} = 1.17 \text{ per cent}$$

The ultimate moment of resistance of the balanced section is $M_{\text{bal}} = F_{\text{ccbal}}z_{\text{bal}}$ where

$$z_{\text{bal}} = d - s/2 = 0.82d$$

Substituting for F_{ccbal} and z:

$$M_{\text{bal}} = 0.167f_{\text{ck}}bd^2 \tag{4.10}$$

and

$$\frac{M_{\text{d}}}{f_{\text{ck}}bd^2} = 0.167 = K_{\text{bal}}$$

When the design moment M_{d} is such that $\dfrac{M_{\text{d}}}{f_{\text{ck}}bd^2} > K_{\text{bal}} = 0.167$ then the section cannot be singly reinforced and compression reinforcing steel is required in the compression zone of the section. This is the limiting value of $K = 0.167$ marked on the horizontal axis of the lever arm curve shown in figure 4.5.

EXAMPLE 4.1

Design of a singly reinforced rectangular section

The ultimate design moment to be resisted by the section in figure 4.6 is $185\,\text{kN m}$. Determine the area of tension reinforcement (A_{s}) required given the characteristic material strengths are $f_{\text{yk}} = 500\,\text{N/mm}^2$ and $f_{\text{ck}} = 25\,\text{N/mm}^2$.

$$K = \frac{M}{bd^2f_{\text{ck}}}$$

$$= \frac{185 \times 10^6}{260 \times 440^2 \times 25} = 0.147 < 0.167$$

therefore compression steel is not required.

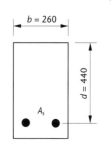

Figure 4.6
Design example – singly reinforced section

Lever arm:

$$z = d\left\{0.5 + \sqrt{\left(0.25 - \frac{K}{1.134}\right)}\right\}$$

$$= 440\left\{0.5 + \sqrt{\left(0.25 - \frac{0.147}{1.134}\right)}\right\}$$

$$= 373\,\text{mm}$$

(Or alternatively, the value of $z = l_a d$ be obtained from the lever-arm diagram, figure 4.5.)

$$A_s = \frac{M}{0.87f_{yk}z}$$

$$= \frac{185 \times 10^6}{0.87 \times 500 \times 373}$$

$$= 1140\,\text{mm}^2$$

Analysis equations for a singly reinforced section

The following equations may be used to calculate the moment of resistance of a given section with a known area of steel reinforcement.

For equilibrium of the compressive force in the concrete and the tensile force in the steel in figure 4.4:

$$F_{cc} = F_{st}$$

or

$$0.567f_{ck}b \times s = 0.87f_{yk}A_s$$

Therefore depth of stress block is

$$s = \frac{0.87f_{yk}A_s}{0.567f_{ck}b} \tag{4.11}$$

and

$$x = s/0.80$$

Therefore the moment of resistance of the section is

$$M = F_{st} \times z$$

$$= 0.87f_{yk}A_s(d - s/2)$$

$$= 0.87f_{yk}A_s\left(d - \frac{0.87f_{yk}A_s}{1.134f_{ck}b}\right) \tag{4.12}$$

These equations assume the tension reinforcement has yielded, which will be the case if $x < 0.617d$. If this is not the case, the problem would require solving by trying successive values of x until

$$F_{cc} = F_{st}$$

with the steel strains and hence stresses being determined from equations 4.2 and 4.1, to be used in equation 4.12 instead of $0.87f_{yk}$.

EXAMPLE 4.2

Analysis of a singly reinforced rectangular section in bending

Determine the ultimate moment of resistance of the cross-section shown in figure 4.7 given that the characteristic strengths are $f_{yk} = 500$ N/mm^2 for the reinforcement and $f_{ck} = 25$ N/mm^2 for the concrete.

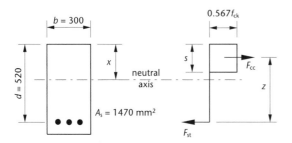

Figure 4.7
Analysis example – singly reinforced section

For equilibrium of the compressive and tensile forces on the section

$$F_{cc} = F_{st}$$

therefore

$$0.567 f_{ck} bs = 0.87 f_{yk} A_s$$
$$0.567 \times 25 \times 300 \times s = 0.87 \times 500 \times 1470$$

therefore

$$s = 150 \text{ mm}$$

and

$$x = s/0.8 = 150/0.8$$
$$= 188 \text{ mm}$$

This value of x is less than the value of $0.617d$ derived from section 4.2, and therefore the steel has yielded and $f_{st} = 0.87 f_{yk}$ as assumed.

Moment of resistance of the section is

$$M = F_{st} \times z$$
$$= 0.87 f_{yk} A_s (d - s/2)$$
$$= 0.87 \times 500 \times 1470 (520 - 150/2) \times 10^{-6} = 284 \text{ kN m}$$

4.5 | Rectangular section in bending with compression reinforcement at the ultimate limit state

(a) Derivation of basic equations

It should be noted that the equations in this section have been derived for the case of zero moment redistribution. When this is not the case, reference should be made to section 4.7 which deals with the effect of moment redistribution.

Figure 4.8
Section with compression
reinforcement

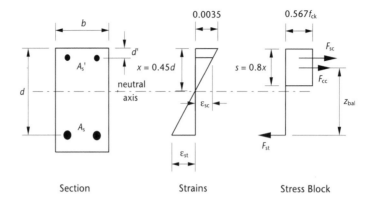

Section   Strains   Stress Block

From the section dealing with the analysis of a singly reinforced section and for concrete class not greater than C50/60 when

$$M > 0.167f_{ck}bd^2$$

the design ultimate moment exceeds the moment of resistance of the concrete(M_{bal}) and therefore compression reinforcement is required. For this condition the depth of neutral axis, $x < 0.45d$, the maximum value allowed by the code in order to ensure a tension failure with a ductile section. Therefore

$$z_{bal} = d - s_{bal}/2 = d - 0.8x_{bal}/2$$
$$= d - 0.8 \times 0.45d/2$$
$$= 0.82d$$

For equilibrium of the section in figure 4.8

$$F_{st} = F_{cc} + F_{sc}$$

so that with the reinforcement at yield

$$0.87f_{yk}A_s = 0.567f_{ck}bs + 0.87f_{yk}A'_s$$

or with

$$s = 0.8 \times 0.45d = 0.36d$$
$$0.87f_{yk}A_s = 0.204f_{ck}bd + 0.87f_{yk}A'_s \qquad (4.13)$$

and taking moments about the centroid of the tension steel,

$$M = F_{cc} \times z_{bal} + F_{sc}(d - d')$$
$$= 0.204f_{ck}bd \times 0.82d + 0.87f_{yk}A'_s(d - d')$$
$$= 0.167f_{ck}bd^2 + 0.87f_{yk}A'_s(d - d') \qquad (4.14)$$

From equation 4.14

$$A'_s = \frac{M - 0.167f_{ck}bd^2}{0.87f_{yk}(d - d')} \qquad (4.15)*$$

Multiplying both sides of equation 4.13 by $z = 0.82d$ and rearranging gives

$$A_s = \frac{0.167f_{ck}bd^2}{0.87f_{yk} \times z_{bal}} + A'_s \qquad (4.16)*$$

with $z_{bal} = 0.82d$.

Hence the areas of compression steel, A'_s, and tension steel, A_s, can be calculated from equations 4.15 and 4.16.

Substituting $K_{bal} = 0.167$ and $K = M/bd^2 f_{ck}$ into these equations would convert them into:

$$A'_s = \frac{(K - K_{bal}) f_{ck} bd^2}{0.87 f_{yk}(d - d')} \tag{4.17}*$$

$$A_s = \frac{K_{bal} f_{ck} bd^2}{0.87 f_{yk} z_{bal}} + A'_s \tag{4.18}*$$

In this analysis it has been assumed that the compression steel has yielded so that the steel stress $f_{sc} = 0.87 f_{yk}$. From the proportions of the strain distribution diagram:

$$\frac{\varepsilon_{sc}}{x - d'} = \frac{0.0035}{x} \tag{4.19}$$

so that

$$\frac{x - d'}{x} = \frac{\varepsilon_{sc}}{0.0035}$$

or

$$\frac{d'}{x} = 1 - \frac{\varepsilon_{sc}}{0.0035}$$

At yield with $f_{yk} = 500\,\text{N/mm}^2$, the steel strain $\varepsilon_{sc} = \varepsilon_y = 0.00217$. Therefore for yielding of the compression steel

$$\frac{d'}{x} < 1 - \frac{0.00217}{0.0035} < 0.38 \tag{4.20}*$$

or with $x = 0.45d$

$$\frac{d'}{d} < 0.171 \tag{4.21}$$

The ratio of d'/d for the yielding of other grades of steel can be determined by using their yield strain in equation 4.19, but for values of f_{yk} less than $500\,\text{N/mm}^2$, the application of equation 4.21 will provide an adequate safe check.

If $d'/d > 0.171$, then it is necessary to calculate the strain ε_{sc} from equation 4.19 and then determine f_{sc} from

$$f_{sc} = E_s \times \varepsilon_{sc}$$
$$= 200000 \varepsilon_{sc}$$

This value of stress for the compressive steel must then be used in the denominator of equation 4.15 in place of $0.87 f_{yk}$ in order to calculate the area A'_s of compression steel. The area of tension steel is calculated from a modified equation 4.16 such that

$$A_s = \frac{0.167 f_{ck} bd^2}{0.87 f_{yk} z_{bal}} + A'_s \times \frac{f_{sc}}{0.87 f_{yk}}$$

The above equations apply for the case where the concrete class is less than or equal to C50/60. For concrete classes greater than C50/60 similar equations, with different constants, can be derived based on the EC2 requirement for these classes. The constants for concretes up to class C50/60 are tabulated in table 4.1.

Table 4.1 Limiting constant values

	Concrete class $\leq$ C50/60
Limiting x_{bal}/d	0.45
Maximum z_{bal}	0.82d
K_{bal} = limiting K	0.167
Limiting d'/d	0.171
Maximum percentage steel area 100A_{bal}/bd	23.4f_{ck}/f_{yk}

(b) Design charts

The equations for the design charts are obtained by taking moments about the neutral axis. Thus

$$M = 0.567f_{ck}0.8x(x - 0.8x/2) + f_{sc}A'_s(x - d') + f_{st}A_s(d - x)$$

This equation and 4.13 may be written in the form

$$f_{st}\frac{A_s}{bd} = 0.454f_{ck}\frac{x}{d} + f_{sc}\frac{A'_s}{bd}$$

$$\frac{M}{bd^2} = 0.454f_{ck}\frac{x^2}{d^2}(1 - 0.40) + f_{sc}\frac{A'_s}{bd}\left(\frac{x}{d} - \frac{d'}{d}\right) + f_{st}\frac{A_s}{bd}\left(1 - \frac{x}{d}\right)$$

For specified ratios of A'_s/bd, x/d and d'/d, the two non-dimensional equations can be solved to give values for A_s/bd and M/bd^2 so that a set of design charts such as the one shown in figure 4.9 may be plotted. Before the equations can be solved, the steel stresses f_{st} and f_{sc} must be calculated for each value of x/d. This is achieved by first determining the relevant strains from the strain diagram (or by applying equations 4.2 and 4.3) and then by evaluating the stresses from the stress–strain curve of figure 4.2. Values of x/d below 0.45 will apply when moments are redistributed. It should be noted that EC2 does not give design charts for bending. Hence although it is possible to derive charts as indicated, it may be simpler to use the equations derived earlier in this chapter or simple computer programs.

Figure 4.9
Typical design chart for doubly reinforced beams

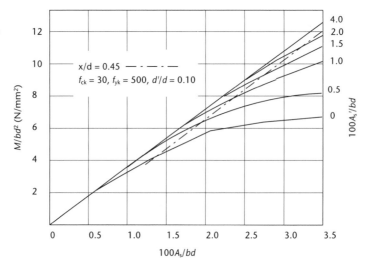

EXAMPLE 4.3

Design of a rectangular section with compression reinforcement (no moment redistribution)

The section shown in figure 4.10 is to resist an ultimate design moment of 285kNm. The characteristic material strengths are $f_{yk} = 500\,\text{N/mm}^2$ and $f_{ck} = 25\,\text{N/mm}^2$. Determine the areas of reinforcement required.

$$K = \frac{M}{bd^2 f_{ck}}$$

$$= \frac{285 \times 10^6}{260 \times 440^2 \times 25} = 0.226$$

$$> 0.167$$

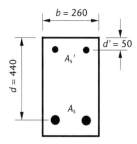

Figure 4.10
Design example with compression reinforcement, no moment redistribution

therefore compression steel is required

$$d'/d = 50/440 = 0.11 < 0.171$$

as in equation 4.21 and the compression steel will have yielded.

Compression steel:

$$A'_s = \frac{(K - K_{bal})f_{ck}bd^2}{0.87f_{yk}(d - d')}$$

$$= \frac{(0.226 - 0.167)25 \times 260 \times 440^2}{0.87 \times 500(440 - 50)}$$

$$= 438\,\text{mm}^2$$

Tension steel:

$$A_s = \frac{K_{bal}f_{ck}bd^2}{0.87f_{yk}z_{bal}} + A'_s$$

$$= \frac{0.167 \times 25 \times 260 \times 440^2}{0.87 \times 500(0.82 \times 440)} + 438$$

$$= 1339 + 438 = 1777\,\text{mm}^2$$

EXAMPLE 4.4

Analysis of a doubly reinforced rectangular section

Determine the ultimate moment of resistance of the cross-section shown in figure 4.11 given that the characteristic strengths are $f_{yk} = 500\,\text{N/mm}^2$ for the reinforcement and $f_{ck} = 25\,\text{N/mm}^2$ for the concrete.

For equilibrium of the tensile and compressive forces on the section:

$$F_{st} = F_{cc} + F_{sc}$$

Assuming initially that the steel stresses f_{st} and f_{sc} are the design yield values, then

$$0.87f_{yk}A_s = 0.567f_{ck}bs + 0.87f_{yk}A'_s$$

Figure 4.11
Analysis example, doubly
reinforced section

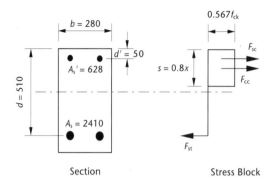

Section Stress Block

therefore

$$s = \frac{0.87 f_{yk}(A_s - A'_s)}{0.567 f_{ck} b}$$

$$= \frac{0.87 \times 500(2410 - 628)}{0.567 \times 25 \times 280}$$

$$= 195 \, \text{mm}$$

$$x = s/0.8 = 244 \, \text{mm}$$

$$x/d = 244/510 = 0.48 < 0.617 \text{ (see section 4.2)}$$

so the tension steel will have yielded. Also

$$d'/x = 50/225 = 0.22 < 0.38 \text{ (see equation 4.20)}$$

so the compression steel will also have yielded, as assumed.

Taking moments about the tension steel

$$M = F_{cc}(d - s/2) + F_{sc}(d - d')$$

$$= 0.567 f_{ck} bs(d - s/2) + 0.87 f_{yk} A'_s(d - d')$$

$$= [0.567 \times 25 \times 280 \times 195(510 - 195/2) + 0.87 \times 500 \times 620(510 - 50)] \times 10^{-6}$$

$$= 319 + 124 = 443 \, \text{kN m}$$

If the depth of neutral axis was such that the compressive or tensile steel had not yielded, it would have been necessary to try successive values of x until

$$F_{st} = F_{cc} + F_{sc}$$

balances with the steel strains and stresses being calculated from equations 4.2, 4.3 and 4.1. The steel stresses at balance would then be used to calculate the moment of resistance.

4.6 | Flanged section in bending at the ultimate limit state

T-sections and L-sections which have their flanges in compression can both be designed or analysed in a similar manner, and the equations which are derived can be applied to either type of cross-section. As the flanges generally provide a large compressive area, it is usually unnecessary to consider the case where compression steel is required; if it should be required, the design would be based on the principles derived in section 4.6.3

For the singly reinforced section it is necessary to consider two conditions:

1. the stress block lies within the compression flange, and
2. the stress block extends below the flange.

4.6.1 Flanged section – the depth of the stress block lies within the flange, $s < h_f$ (figure 4.12)

For this depth of stress block, the beam can be considered as an equivalent rectangular section of breadth b_f equal to the flange width. This is because the non-rectangular section below the neutral axis is in tension and is, therefore, considered to be cracked and inactive. Thus $K = M/b_f d^2 f_{ck}$ can be calculated and the lever arm determined from the lever arm curve of figure 4.5 or equation 4.8. The relation between the lever arm, z, and depth, x, of the neutral axis is given by

$$z = d - s/2$$

or

$$s = 2(d - z)$$

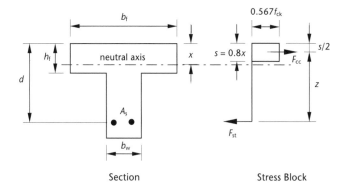

Section   Stress Block

Figure 4.12
T-section, stress block within the flange, $s < h_f$

If s is less than the flange thickness (h_f), the stress block does lie within the flange as assumed and the area of reinforcement is given by

$$A_s = \frac{M}{0.87 f_{yk} z}$$

The design of a T-section beam is described further in section 7.2.3 with a worked example.

EXAMPLE 4.5

Analysis of a flanged section

Determine the ultimate moment of resistance of the T-section shown in figure 4.13. The characteristic material strengths are $f_{yk} = 500\,\text{N/mm}^2$ and $f_{ck} = 25\,\text{N/mm}^2$.

Assume initially that the stress block depth lies within the flange and the reinforcement is strained to the yield, so that $f_{st} = 0.87 f_{yk}$.

Figure 4.13
Analysis example of a
T-section, $s < h_f$

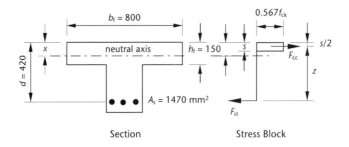

For equilibrium of the section

$$F_{cc} = F_{st}$$

therefore

$$0.567 f_{ck} b_f s = 0.87 f_{yk} A_s$$

and solving for the depth of stress block

$$s = \frac{0.87 \times 500 \times 1470}{0.567 \times 25 \times 800}$$

$$= 56\,\text{mm} < h_f = 150\,\text{mm}$$

$$x = s/0.8$$

$$= 70\,\text{mm}$$

Hence the stress block does lie within the flange and with this depth of neutral axis the steel will have yielded as assumed.

Lever arm:

$$z = d - s/2$$

$$= 420 - 56/2$$

$$= 392\,\text{mm}$$

Taking moments about the centroid of the reinforcement the moment of resistance is

$$M = F_{cc} \times z$$

$$= 0.567 f_{ck} b_f s z$$

$$= 0.567 \times 25 \times 800 \times 56 \times 392 \times 10^{-6}$$

$$= 249\,\text{kN m}$$

If in the analysis it had been found that $s > h_f$, then the procedure would have been similar to that in example 4.7.

4.6.2 Flanged section – the depth of the stress block extends below the flange, $s > h_f$

For the design of a flanged section, the procedure described in section 4.6.1 will check if the depth of the stress block extends below the flange. An alternative procedure is to calculate the moment of resistance, M_f, of the section with $s = h_f$, the depth of the flange (see equation 4.22 of example 4.6 following). Hence if the design moment, M_d, is such that

$$M_d > M_f$$

then the stress block must extend below the flange, and

$$s > h_f$$

In this case the design can be carried out by either:

(a) using an exact method to determine the depth of the neutral axis, as in example 4.6
or

(b) designing for the conservative condition of $x = 0.45d$, which is the maximum value of x for a singly reinforced section and concrete class $\leq$ C50/60.

EXAMPLE 4.6

Design of a flanged section with the depth of the stress block below the flange

The T-section beam shown in figure 4.14 is required to resist an ultimate design moment of 180 kN m. The characteristic material strengths are $f_{yk} = 500\,\text{N/mm}^2$ and $f_{ck} = 25\,\text{N/mm}^2$. Calculate the area of reinforcement required.

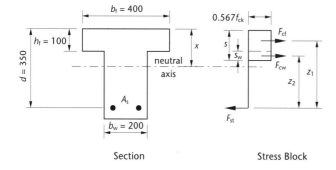

Figure 4.14
Design example of a T-section, $s > h_f$

In figure 4.14

F_{cf} is the force developed in the flange

F_{cw} is the force developed in the area of web in compression

Moment of resistance, M_f, of the flange is

$$M_f = F_{cf} \times z_1$$

or

$$M_f = 0.567 f_{ck} b_f h_f (d - h_f/2) \qquad (4.22)*$$
$$= 0.567 \times 25 \times 400 \times 100 (350 - 100/2) \times 10^{-6}$$
$$= 170\,\text{kN m} < 180\ \text{kN m, the design moment}$$

Therefore, the stress block must extend below the flange.
It is now necessary to determine the depth, s_w, of the web in compression, where

$$s_w = s - h_f.$$

For equilibrium:

Applied moment

$$180 = F_{cf} \times z_1 + F_{cw} \times z_2$$

$$= 170 + 0.567 f_{ck} b_w s_w \times z_2$$

$$= 170 + 0.567 \times 25 \times 200 s_w (250 - s_w/2) \times 10^{-6}$$

$$= 170 + 2835 s_w (250 - s_w/2) \times 10^{-6}$$

This equation can be rearranged into

$$s_w{}^2 - 500 s_w + 7.05 \times 10^3 = 0$$

Solving this quadratic equation

$$s_w = 15 \, \text{mm}$$

so that the depth of neutral axis

$$x = (h_f + s_w)/0.8 = (100 + 15)/0.8$$

$$= 144 \, \text{mm} = 0.41d$$

As $x < 0.45d$ compression reinforcement is not required.
For the equilibrium of the section

$$F_{st} = F_{cf} + F_{cw}$$

or

$$0.87 f_{yk} A_s = 0.567 f_{ck} b_f h_f + 0.567 f_{ck} b_w s_w$$

$$0.87 \times 500 \times A_s = 0.567 \times 25 (400 \times 100 + 200 \times 15) = 610 \times 10^3$$

Therefore

$$A_s = \frac{610 \times 10^3}{0.87 \times 500}$$

$$= 1402 \, \text{mm}^2$$

EXAMPLE 4.7

Analysis of a flanged section

Determine the ultimate moment of resistance of the T-beam section shown in figure 4.15 given $f_{yk} = 500 \, \text{N/mm}^2$ and $f_{ck} = 25 \, \text{N/mm}^2$.

The compressive force in the flange is

$$F_{cf} = 0.567 f_{ck} b_f h_f$$

$$= 0.567 \times 25 \times 450 \times 150 \times 10^{-3} = 957 \, \text{kN}$$

Then tensile force in the reinforcing steel, assuming it has yielded, is

$$F_{st} = 0.87 f_{yk} A_s$$

$$= 0.87 \times 500 \times 2592 \times 10^{-3} = 1128 \, \text{kN}$$

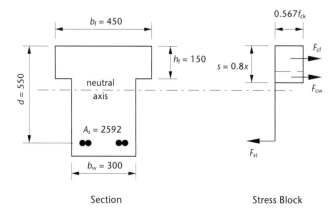

Figure 4.15
Analysis example of a
T-section, $s > h_f$

Therefore $F_{st} > F_{cf}$ so that $s > h_f$ and the force in the web is

$$F_{cw} = 0.567 f_{ck} b_w (s - h_f)$$
$$= 0.567 \times 25 \times 300(s - 150) \times 10^{-3}$$
$$= 4.25(s - 150)$$

For equilibrium

$$F_{cw} = F_{st} - F_{cf}$$

or

$$4.25(s - 150) = 1128 - 957$$

Hence

$$s = 190\,\text{mm}$$
$$x = s/0.8 = 238\,\text{mm} = 0.43d$$

With this depth of neutral axis the reinforcement has yielded, as assumed, and

$$F_{cw} = 4.25(190 - 150) = 170\,\text{kN}$$

(If $F_{cf} > F_{st}$, the the stress block would not extend beyond the flange and the section would be analysed as in example 4.2 for a rectangular section of dimensions $b_f \times d$.)
Taking moments about the centroid of the reinforcement

$$M = F_{cf}(d - h_f/2) + F_{cw}(d - s/2 - h_f/2)$$
$$= [957(550 - 150/2) + 170(550 - 190/2 - 150/2)] \times 10^{-3}$$
$$= 519\,\text{kN m}$$

EXAMPLE 4.8

Design of a flanged section with depth of neutral axis $x = 0.45d$

A safe but conservative design for a flanged section with $s > h_f$ can be achieved by setting the depth of neutral axis to $x = 0.45d$, the maximum depth allowed in the code. Design equations can be derived for this condition as follows.

Depth of stress block, $s = 0.8x = 0.8 \times 0.45d = 0.36d$

Figure 4.16
Flanged section with depth of
neutral axis $x = 0.45d$

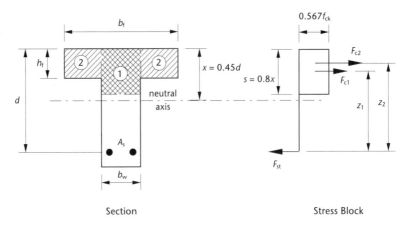

Section   Stress Block

Divide the flanged section within the depth of the stress block into areas 1 and 2 as shown in figure 4.16, so that

Area $1 = b_w \times s = 0.36 b_w d$

Area $2 = (b_f - b_w) \times h_f$

and the compression forces developed by these areas are

$$F_{c1} = 0.567 f_{ck} \times 0.36 b_w d = 0.2 f_{ck} b_w d$$

$$F_{c2} = 0.567 f_{ck} h_f (b_f - b_w)$$

Taking moments about F_{c2} at the centroid of the flange

$$M = F_{st}(d - h_f/2) - F_{c1}(s/2 - h_f/2)$$

$$= 0.87 f_{yk} A_s (d - h_f/2) - 0.2 f_{ck} b_w d (0.36d - h_f)/2$$

Therefore

$$A_s = \frac{M + 0.1 f_{ck} b_w d (0.36d - h_f)}{0.87 f_{yk}(d - 0.5 h_f)} \qquad (4.23)*$$

This equation should not be used when $h_f > 0.36d$.

Applying this equation to example 4.6:

$$A_s = \frac{180 \times 10^6 + 0.1 \times 25 \times 200 \times 350(0.36 \times 350 - 100)}{0.87 \times 500(350 - 100/2)}$$

$$= 1414 \, \text{mm}^2 \text{ (compared with 1407 mm}^2 \text{ in example 4.6)}$$

Before using equation 4.23 for calculating A_s, it is necessary to confirm that compression reinforcement is not required. This is achieved by using equation 4.24 to check that the moment of resistance of the concrete, M_{bal}, is greater than the design moment, M.

4.6.3 Flanged section with compression reinforcement

With $x = 0.45d$ in figure 4.16 and taking moments about A_s, the maximum resistance moment of the concrete is

$$M_{bal} = F_{c1} \times z_1 + F_{c2} \times z_2$$

$$= 0.167 f_{ck} b_w d^2 + 0.567 f_{ck}(b_f - b_w)(d - h_f/2) \qquad (4.24)$$

(Note that the value of 0.167 was derived in equation 4.10 for the rectangular section.)

Dividing through by $f_{ck}b_f d^2$

$$\frac{M_{bal}}{f_{ck}b_f d^2} = 0.156\frac{b_w}{b_f} + 0.567\frac{h_f}{d}\left(1 - \frac{b_w}{b_f}\right)\left(1 - \frac{h_f}{2d}\right) \tag{4.25}*$$

If the applied design moment, $M > M_{bal}$, compression reinforcement is required. In this case the area of compression steel can be calculated from

$$A'_s = \frac{M - M_{bal}}{0.87f_{yk}(d - d')} \tag{4.26}$$

and considering the equilibrium of forces on the section

$$F_{st} = F_{c1} + F_{c2} + F_{sc}$$

so that the area of tension steel is

$$A_s = \frac{0.2f_{ck}b_w d + 0.567f_{ck}h_f(b_f - b_w)}{0.87f_{yk}} + A'_s \tag{4.27}$$

Again, $d'/x < 0.38$, otherwise the design compressive steel stress is less than $0.87f_{yk}$.

4.7 Moment redistribution and the design equations

The plastic behaviour of reinforced concrete at the ultimate limit state affects the distribution of moments in a structure. To allow for this, the moments derived from an elastic analysis may be redistributed based on the assumption that plastic hinges have formed at the sections with the largest moments (see section 3.6). The formation of plastic hinges requires relatively large rotations with yielding of the tension reinforcement. To ensure large strains in the tension steel, the code of practice restricts the depth of the neutral axis according to the magnitude of the moment redistribution carried out.

The equations for this, given by EC2 for concrete class less than or equal to C50/60 is

$$\delta \geq k_1 + k_2 \frac{x_{bal}}{d}$$

or

$$\frac{x_{bal}}{d} \leq (\delta - k_1)/k_2 \tag{4.28}$$

where

$$\delta = \frac{\text{moment at section after redistribution}}{\text{moment at section before redistribution}} < 1.0$$

k_1 and k_2 are constants from the EC2 code and the UK Annex and x_{bal} is the maximum value of the depth of the neutral axis which will take the limiting value of the equality of equation (4.28) but should be less than $0.45d$ for concrete class $\leq$ C50/60.

The depth of the stress block is

$$s_{bal} = 0.8x_{bal}$$

and the level arm is

$$z_{bal} = d - s_{bal}/2 \tag{4.29}$$

The moment of resistance of the concrete in compression is

$$M_{bal} = F_{cc} \times z_{bal} = 0.567 f_{ck} b s_{bal} \times z_{bal}$$

and

$$K_{bal} = M_{bal}/bd^2 f_{ck} = 0.567 s_{bal} \times z_{bal}/d^2$$

This equation for K_{bal} and the previous equations from 4.28 to 4.29 can be arranged to give

$$K_{bal} = 0.454(\delta - k_1)/k_2 - 0.182[(\delta - k_1)/k_2]^2 \qquad (4.30)$$

or alternatively

$$K_{bal} = 0.454 \left(\frac{x_{bal}}{d}\right)\left(\frac{z_{bal}}{d}\right)$$

From EC2 clause 5.5 the constants k_1 and k_2 are given as: $k_1 = 0.44$ and $k_2 = 1.25$, but from the UK Annex to EC2 $k_1 = 0.4$ and $k_2 = 1.0$.

The relevant values of x_{bal}, z_{bal} and K_{bal} for varying percentages of moment redistribution and concrete class $\leq$ C50/60 are shown in table 4.2.

When the ultimate design moment is such that

$$M > K_{bal} bd^2 f_{ck}$$

or $K > K_{bal}$

then compression steel is required such that

$$A_s' = \frac{(K - K_{bal}) f_{ck} bd^2}{0.87 f_{yk}(d - d')} \qquad (4.31)*$$

and

$$A_s = \frac{K_{bal} f_{ck} bd^2}{0.87 f_{yk} z_{bal}} + A_s' \qquad (4.32)*$$

where $K = \dfrac{M_{bal}}{bd^2 f_{ck}} \qquad (4.33)*$

These equations are identical in form to those derived previously for the design of a section with compression reinforcement and no moment redistribution. If the value of d'/d for the section exceeds that shown in table 4.2, the compression steel will not have yielded and the compressive stress will be less than $0.87 f_{yk}$. In such cases, the compressive stress f_{sc} will be $E_s \varepsilon_{sc}$ where the strain ε_{sc} is obtained from the proportions of the strain diagram. This value of f_{sc} should replace $0.87 f_{yk}$ in equation 4.31, and equation 4.32 becomes

$$A_s = \frac{K_{bal} f_{ck} bd^2}{0.87 f_{yk} z_{bal}} + A_s' \times \frac{f_{sc}}{0.87 f_{yk}}$$

It should be noted that for a singly reinforced section ($K < K_{bal}$), the lever arm is calculated from equation 4.8.

For a section requiring compression steel, the lever arm can be calculated from equation 4.29 or by using the equation

$$z = d\left[0.5 + \sqrt{(0.25 - K_{bal}/1.134)}\right] \qquad (4.34)$$

which is similar to equation 4.8 but with K_{bal} replacing K.

Table 4.2 Moment redistribution design factors

Redistribution (%)	δ	x_{bal}/d	z_{bal}/d	K_{bal}	d'/d
According to EC2, $k_1 = 0.44$ and $k_2 = 1.25$					
0	1.0	0.448	0.821	0.167	0.171
10	0.9	0.368	0.853	0.142	0.140
15	0.85	0.328	0.869	0.129	0.125
20[a]	0.8	0.288	0.885	0.116	0.109
25	0.75	0.248	0.900	0.101	0.094
30[b]	0.70	0.206	0.917	0.087	0.079
According to EC2, UK Annex, $k_1 = 0.4$ and $k_2 = 1.0$					
0	1.0	0.45	0.82	0.167	0.171
10	0.9	0.45	0.82	0.167	0.171
15	0.85	0.45	0.82	0.167	0.171
20[a]	0.8	0.40	0.84	0.152	0.152
25	0.75	0.35	0.86	0.137	0.133
30[b]	0.70	0.30	0.88	0.120	0.114

[a] Maximum permitted redistribution for class A normal ductility steel
[b] Maximum permitted redistribution for class B and C higher ductility steel, see section 1.6.2

EXAMPLE 4.9

Design of a section with moment redistribution applied and $\delta = 0.8$

The section shown in figure 4.17 is subject to an ultimate design moment of 230 kN m after a 20% reduction due to moment redistribution. The characteristic material strengths are $f_{yk} = 500\,\text{N/mm}^2$ and $f_{ck} = 25\,\text{N/mm}^2$. Determine the areas of reinforcement required using the constants k_1 and k_2 from (a) the EC2 and (b) the UK annex to EC2.

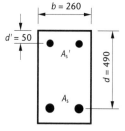

Figure 4.17
Design example with moment redistribution, $\delta = 0.8$

(a) Using EC2

(i) From first principles

Limiting neutral axis depth, $x_{bal} = (\delta - k_1)d/k_2$

From EC2 clause 5.5 $k_1 = 0.44$ and $k_2 = 1.25$,

therefore $\qquad x_{bal} = (0.8 - 0.44)490/1.25 = 141\,\text{mm}$

Stress block depth $\quad s_{bal} = 0.8x_{bal} = 113\,\text{mm}$

Lever arm $\qquad z_{bal} = d - s_{bal}/2$

$\qquad\qquad\qquad\quad = 490 - 113/2 = 434\,\text{mm}$

Moment of resistance of the concrete

$$M_{bal} = F_{cc} \times z_{bal} = 0.567 f_{ck} b s_{bal} \times z_{bal}$$

$$= 0.567 \times 25 \times 260 \times 113 \times 434 \times 10^{-6}$$

$$= 181\,\text{kN m}$$

$$< 230\,\text{kN m, the applied moment}$$

therefore compression steel is required.

$\quad d'/x_{bal} = 50/141 = 0.35 < 0.38 \qquad$ (see equation 4.20 in section 4.5)

therefore the compression steel has yielded.

Compression steel:

$$A'_s = \frac{M - M_{bal}}{0.87 f_{yk}(d - d')}$$

$$= \frac{(230 - 181) \times 10^6}{0.87 \times 500(490 - 50)}$$

$$= 256 \, \text{mm}^2$$

Tension steel:

$$A_s = \frac{M_{bal}}{0.87 f_{yk} z_{bal}} + A'_s$$

$$= \frac{181 \times 10^6}{0.87 \times 500 \times 434} + 256$$

$$= 959 + 256 = 1215 \, \text{mm}^2$$

(ii) Alternative solution applying equations developed in Section 4.7

From equations 4.30 to 4.34:

$$K_{bal} = 0.454(\delta - k_1)/k_2 - 0.182[(\delta - k_1)/k_2]^2$$

$$= 0.454(0.8 - 0.44)/1.25 - 0.182[(0.8 - 0.44)/1.25]^2$$

$$= 0.131 - 0.015 = 0.116$$

which agrees with the value given in table 4.2.

$$K = \frac{M}{bd^2 f_{ck}}$$

$$= \frac{228 \times 10^6}{260 \times 490^2 \times 25}$$

$$= 0.146 > K_{bal} = 0.116$$

Therefore compression steel is required.

Compression steel:

$$A'_s = \frac{(K - K_{bal})f_{ck} bd^2}{0.87 f_{yk}(d - d')}$$

$$= \frac{(0.146 - 0.116)25 \times 260 \times 490^2}{0.87 \times 500(490 - 50)}$$

$$= 244 \, \text{mm}^2$$

Tension steel:

$$z_{bal} = d\left[0.5 + \sqrt{(0.25 - K_{bal}/1.134)}\right]$$

$$= d\left[0.5 + \sqrt{(0.25 - 0.116/1.134)}\right] = 0.89d$$

$$A_s = \frac{K_{bal}f_{ck} bd^2}{0.87 f_{yk} z_{bal}} + A'_s$$

$$= \frac{0.116 \times 25 \times 260 \times 490^2}{0.87 \times 500 \times 0.89 \times 490} + 244$$

$$= 954 + 244 = 1198 \, \text{mm}^2$$

(b) Using the UK Annex of EC2 and applying the equations developed in section 4.7

From the UK Annex of EC2 clause 5.5 $k_1 = 0.4$ and $k_2 = 1.0$

From equations 4.30 to 4.34:

$$K_{\text{bal}} = 0.454(\delta - k_1)/k_2 - 0.182[(\delta - k_1)/k_2]^2$$
$$= 0.454(0.8 - 0.4)/1.0 - 0.182[(0.8 - 0.4)/1.0]^2$$
$$= 0.182 - 0.029 = 0.153$$

which agrees with the value given in table 4.2.

$$K = \frac{M}{bd^2 f_{\text{ck}}}$$
$$= \frac{230 \times 10^6}{260 \times 490^2 \times 25}$$
$$= 0.147 < K_{\text{bal}} = 0.153$$

Therefore compression steel is not required.

Tension steel:

Using equation 4.8 in section 4.4

$$z_{\text{bal}} = d\left[0.5 + \sqrt{(0.25 - K_{\text{bal}}/1.134)}\right]$$
$$= 490\left[0.5 + \sqrt{(0.25 - 0.146/1.134)}\right] = 490 \times 0.847 = 415 \, \text{mm}$$

$$A_{\text{s}} = \frac{M}{0.87 f_{\text{yk}} z}$$
$$= \frac{230 \times 10^6}{0.87 \times 500 \times 415} = 1274 \, \text{mm}^2$$

4.8 Bending plus axial load at the ultimate limit state

The applied axial force may be tensile or compressive. In the analysis that follows, a compressive force is considered. For a tensile load the same basic principles of equilibrium, compatibility of strains, and stress–strain relationships would apply, but it would be necessary to change the sign of the applied load (N) when we consider the equilibrium of forces on the cross-section. (The area of concrete in compression has not been reduced to allow for the concrete displaced by the compression steel. This could be taken into account by reducing the stress f_{sc} in the compression steel by an amount equal to $0.567 f_{\text{ck}}$.)

Figure 4.18 represents the cross-section of a member with typical strain and stress distributions for varying positions of the neutral axis. The cross-section is subject to a moment M and an axial compressive force N, and in the figure the direction of the moment is such as to cause compression on the upper part of the section and tension on the lower part. For cases where there is tension in the section (figure 4.18a) the limiting concrete strain is taken as 0.0035 – the value used in the design and analysis of sections for bending. However for cases where there is no tension in the section (figure 4.18b) the limiting strain is taken as a value of 0.002 at the level of 3/7 of the depth of the section.

Figure 4.18
Bending plus axial load with
varying position of the
neutral axis

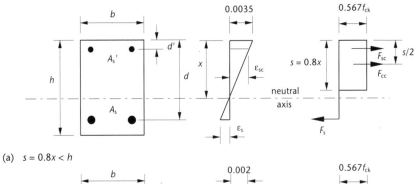

(a) $s = 0.8x < h$

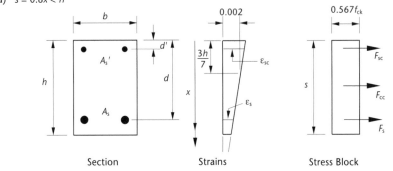

Section   Strains   Stress Block

(b) $s = h : 0.8x > h$

Let

F_{cc} be the compressive force developed in the concrete and acting through the
centroid of the stress block

F_{sc} be the compressive force in the reinforcement area A'_s and acting through its
centroid

F_s be the tensile or compressive force in the reinforcement area A_s and acting
through its centroid.

(i) Basic equations and design charts

The applied force (N) must be balanced by the forces developed within the cross-
section, therefore

$$N = F_{cc} + F_{sc} + F_s$$

In this equation, F_s will be negative whenever the position of the neutral axis is such
that the reinforcement A_s is in tension, as in figure 4.18a. Substituting into this equation
the terms for the stresses and areas

$$N = 0.567 f_{ck} bs + f_{sc} A'_s + f_s A_s \tag{4.35}*$$

where f_{sc} is the compressive stress in reinforcement A'_s and f_s is the tensile or
compressive stress in reinforcement A_s.

The design moment M must be balanced by the moment of resistance of the forces
developed within the cross-section. Hence, taking moments about the mid-depth of the
section

$$M = F_{cc}\left(\frac{h}{2} - \frac{s}{2}\right) + F_{sc}\left(\frac{h}{2} - d'\right) + F_s\left(\frac{h}{2} - d\right)$$

or

$$M = 0.567 f_{ck} bs \left(\frac{h}{2} - \frac{s}{2} \right) + f_{sc} A_s' \left(\frac{h}{2} - d' \right) - f_s A_s \left(d - \frac{h}{2} \right) \qquad (4.36)*$$

When the depth of neutral axis is such that $0.8x \geq h$, as in part (b) of figure 4.18, then the whole concrete section is subject to a uniform compressive stress of $0.567 f_{ck}$. In this case, the concrete provides no contribution to the moment of resistance and the first term on the right side of the equation 4.36 disappears.

For a symmetrical arrangement of reinforcement ($A_s' = A_s = A_{sc}/2$ and $d' = h - d$), equations 4.35 and 4.36 can be rewritten in the following form

$$\frac{N}{bhf_{ck}} = \frac{0.567 s}{h} + \frac{f_{sc} A_s}{f_{ck} bh} + \frac{f_s A_s}{f_{ck} bh} \qquad (4.37)$$

$$\frac{M}{bh^2 f_{ck}} = \frac{0.567 s}{h} \left(0.5 - \frac{s}{2h} \right) + \frac{f_{sc} A_s}{f_{ck} bh} \left(\frac{d}{h} - 0.5 \right) - \frac{f_s A_s}{f_{ck} bh} \left(\frac{d}{h} - 0.5 \right) \qquad (4.38)$$

In these equations the steel strains, and hence the stresses f_{sc} and f_s, vary with the depth of the neutral axis (x). Thus N/bhf_{ck} and $M/bh^2 f_{ck}$ can be calculated for specified ratios of A_s/bh and x/h so that column design charts for a symmetrical arrangement of reinforcement such as the one shown in figure 4.19 can be plotted.

The direct solution of equations 4.37 and 4.38 for the design of column reinforcement would be very tedious and, therefore, a set of design charts for the usual case of symmetrical sections is available in publications such as *The Designers Guide* (ref. 20). Examples showing the design of column steel are given in chapter 9.

(ii) Modes of failure

The relative magnitude of the moment (M) and the axial load (N) governs whether the section will fail in tension or in compression. With large effective eccentricity ($e = M/N$) a tensile failure is likely, but with a small eccentricity a compressive failure is more likely. The magnitude of the eccentricity affects the position of the neutral axis and hence the strains and stresses in the reinforcement.

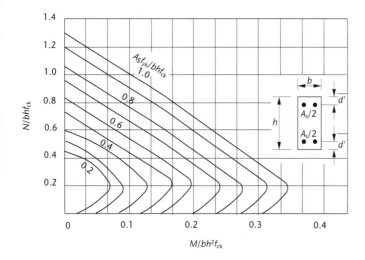

Figure 4.19
Typical column design chart

Let

ε_{sc} be the compressive strain in reinforcement A'_s

ε_s be the tensile or compressive strain in reinforcement A_s

ε_y be the tensile yield strain of steel as shown in the stress–strain curve of figure 4.2.

From the linear strain distribution of figure 4.18(a)

$$\varepsilon_{sc} = 0.0035 \left(\frac{x - d'}{x} \right)$$

and   (4.39)*

$$\varepsilon_s = 0.0035 \left(\frac{d - x}{x} \right)$$

For values of x greater than h, when the neutral axis extends below the section, as shown in figure 4.18b, the steel strains are given by the alternative expressions:

$$\varepsilon_{sc} = 0.002 \frac{7(x - d')}{(7x - 3h)}$$

and

$$\varepsilon_s = 0.002 \frac{7(x - d)}{(7x - 3h)}$$

The steel stresses and strains are then related according to the stress–strain curve of figure 4.2.

Consider the following modes of failure of the section as shown on the interaction diagram of figure 4.20.

(a) Tension failure, $\varepsilon_s > \varepsilon_y$

This type of failure is associated with large eccentricities (e) and small depths of neutral axis (x). Failure begins with yielding of the tensile reinforcement, followed by crushing of the concrete as the tensile strains rapidly increase.

(b) Balanced failure, $\varepsilon_s = \varepsilon_y$, point b on figure 4.20

When failure occurs with yielding of the tension steel and crushing of the concrete at the same instant it is described as a 'balanced' failure. With $\varepsilon_s = \varepsilon_y$ and from equation 4.39

$$x = x_{bal} = \frac{d}{1 + \dfrac{\varepsilon_y}{0.0035}}$$

For example, substituting the values of $\varepsilon_y = 0.00217$ for grade 500 steel

$$x_{bal} = 0.617d$$

Equations 4.35 and 4.36 become

$$N_{bal} = F_{cc} + F_{sc} - F_s$$
$$= 0.567 f_{ck} b \times 0.8 x_{bal} + f_{sc} A'_s - 0.87 f_{yk} A_s \qquad (4.40)$$

and

$$M_{bal} = F_{cc} \left(\frac{h}{2} - \frac{0.8 x_{bal}}{2} \right) + F_{sc} \left(\frac{h}{2} - d' \right) + F_s \left(d - \frac{h}{2} \right)$$

where

$$f_{sc} \leq 0.87 f_{yk}$$

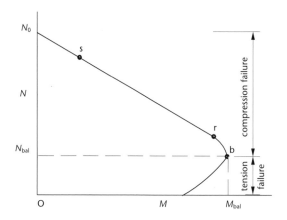

Figure 4.20
Bending plus axial load chart
with modes of failure

At point b on the interaction diagram of figure 4.20, $N = N_{bal}$, $M = M_{bal}$ and $f_s = -0.87 f_{yk}$. When the design load $N > N_{bal}$ the section will fail in compression, while if $N < N_{bal}$ there will be an initial tensile failure, with yielding of reinforcement A_s.

(c) Compression failure

In this case $x > x_{bal}$ and $N > N_{bal}$. The change in slope at point r in figure 4.20 occurs when

$$\varepsilon_{sc} = \varepsilon_y$$

and from equation 4.39

$$x_r = 0.0035 d'/(0.0035 - \varepsilon_y)$$
$$= 2.63 d' \text{ for grade 500 steel}$$

Point r will occur in the tension failure zone of the interaction diagram if $x_r < x_{bal}$. When $x < d$

$$f_s \leq 0.87 f_{yk} \text{ and tensile}$$

When $x = d$

$$f_s = 0$$

When $x > d$

$$f_s \leq 0.87 f_{yk} \text{ and compressive}$$

When x becomes very large and the section approaches a state of uniform axial compression

$$\varepsilon_s = 0.00217 = \varepsilon_y \text{ for grade 500 steel}$$

At this stage, both layers of steel will have yielded and there will be zero moment of resistance with a symmetrical section, so that

$$N_0 = 0.567 f_{ck} bh + 0.87 f_{yk} (A'_s + A_s)$$

At the stage where the neutral axis coincides with the bottom of the section the strain diagram changes from that shown in figure 4.18a to the alternative strain diagram shown in figure 4.18b. To calculate N and M at this stage, corresponding to point s in figure 4.20, equations 4.35 and 4.36 should be used, taking the neutral axis depth equal to the overall section depth, h.

Such *M–N* interaction diagrams can be constructed for any shape of cross-section which has an axis of symmetry by applying the basic equilibrium and strain compatibility equations with the stress–strain relations, as demonstrated in the following examples. These diagrams can be very useful for design purposes.

EXAMPLE 4.10

M–N interaction diagram for a non-symmetrical section

Construct the interaction diagram for the section shown in figure 4.21 with $f_{ck} = 25\,\text{N/mm}^2$ and $f_{yk} = 500\,\text{N/mm}^2$. The bending causes maximum compression on the face adjacent to the steel area A_s'.

For a symmetrical cross-section, taking moments about the centre-line of the concrete section will give $M = 0$ with $N = N_0$ and both areas of steel at the yield stress. This is no longer true for unsymmetrical steel areas as $F_{sc} \neq F_s$ at yield therefore, theoretically, moments should be calculated about an axis referred to as the 'plastic centroid'. The ultimate axial load N_0 acting through the plastic centroid causes a uniform strain across the section with compression yielding of all the reinforcement, and thus there is zero moment of resistance. With uniform strain the neutral-axis depth, x, is at infinity.

Figure 4.21
Non-symmetrical section *M–N* interaction example

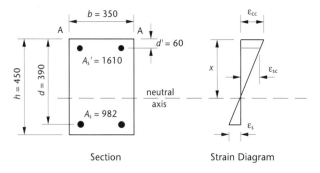

Section   Strain Diagram

The location of the plastic centroid is determined by taking moments of all the stress resultants about an arbitrary axis such as AA in figure 4.21 so that

$$
\begin{aligned}
\bar{x}_p &= \frac{\sum(F_{cc}h/2 + F_{sc}d' + F_s d)}{\sum(F_{cc} + F_{sc} + F_s)} \\
&= \frac{0.567 f_{ck} A_{cc} \times 450/2 + 0.87 f_{yk} A_s' \times 60 + 0.87 f_{yk} A_s \times 390}{0.567 f_{ck} A_{cc} + 0.87 f_{yk} A_s' + 0.87 f_{yk} A_s} \\
&= \frac{0.567 \times 25 \times 350 \times 450^2/2 + 0.87 \times 500(1610 \times 60 + 982 \times 390)}{0.567 \times 25 \times 350 \times 450 + 0.87 \times 500(1610 + 982)} \\
&= 212\,\text{mm from AA}
\end{aligned}
$$

The fundamental equations for calculating points on the interaction diagram with varying depths of neutral axis are:

(i) Compatibility of strains (used in table 4.3, columns 2 and 3):

$$
\varepsilon_{sc} = 0.0035\left(\frac{x - d'}{x}\right)
$$

$$
\varepsilon_s = 0.0035\left(\frac{d - x}{x}\right)
$$

(4.41)

Table 4.3 *M–N interaction values for example 4.10*

(1) x (mm)	(2) ε_{sc}	(3) ε_s	(4) f_{sc} (N/mm²)	(5) f_s (N/mm²)	(6) N (kN)	(7) M (kNm)
$d' = 60$	0	>0.00217	0	$-0.87 f_{yk}$	-189	121
$2.63d' = 158$	0.00217	>0.00217	$0.87 f_{yk}$	$-0.87 f_{yk}$	899	275
$x_{bal} = 0.617d = 241$	>0.00217	0.00217	$0.87 f_{yk}$	$-0.87 f_{yk}$	1229	292
$d = 390$	>0.00217	0	$0.87 f_{yk}$	0	2248	192
$h = 450$	>0.00217	0.00047	$0.87 f_{yk}$	93.3	2580	146
∞	0.00217	0.00217	$0.87 f_{yk}$	$0.87 f_{yk}$	3361	0

or when the neutral axis depth extends below the bottom of the section ($x > h$):

$$\varepsilon_{sc} = 0.002 \frac{7(x - d')}{(7x - 3h)} \quad \text{and} \quad \varepsilon_s = 0.002 \frac{7(x - d)}{(7x - 3h)}$$

(ii) Stress–strain relations for the steel (table 4.3, columns 4 and 5):

$$\begin{aligned} \varepsilon \geq \varepsilon_y = 0.00217 \qquad & f = 0.87 f_{yk} \\ \varepsilon < \varepsilon_y \qquad\qquad & f = E \times \varepsilon \end{aligned}$$

(4.42)

(iii) Equilibrium (table 4.3, columns 6 and 7):

$$N = F_{cc} + F_{sc} + F_s$$

$$0.8x < h \qquad N = 0.567 f_{ck}b \times 0.8x + f_{sc}A'_s + f_s A_s$$

$$0.8x \geq h \qquad N = 0.567 f_{ck}bh + f_{sc}A'_s + f_s A_s$$

Taking moments about the plastic centroid

$$0.8x < h \qquad M = F_{cc}(\bar{x}_p - 0.8x/2) + F_{sc}(\bar{x}_p - d') - F_s(d - \bar{x}_p)$$

$$0.8x \geq h \qquad M = F_{cc}(\bar{x}_p - h/2) + F_{sc}(\bar{x}_p - d') - F_s(d - \bar{x}_p)$$

F_s is negative when f_s is a tensile stress.

These equations have been applied to provide the values in table 4.3 for a range of key values of x. Then the $M–N$ interaction diagram has been plotted in figure 4.22 from the values in table 4.3 as a series of straight lines. Of course, N and M could have been calculated for intermediate values of x to provide a more accurate curve.

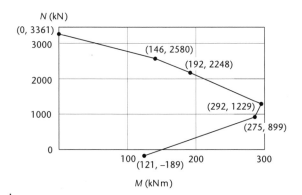

Figure 4.22
M–N interaction diagram for a non-symmetrical section

EXAMPLE 4.11

M–N interaction diagram for a non-rectangular section

Construct the interaction diagram for the equilateral triangular column section in figure 4.23 with $f_{ck} = 25\,\text{N/mm}^2$ and $f_{yk} = 500\,\text{N/mm}^2$. The bending is about an axis parallel to the side AA and causes maximum compression on the corner adjacent to the steel area A_s'.

Figure 4.23
Non-rectangular M–N interaction example

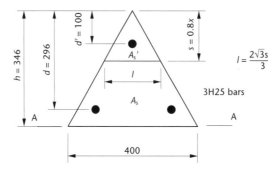

For this triangular section, the plastic centroid is at the same location as the geometric centroid, since the moment of F_{sc} equals the moment of F_s about this axis when all the bars have yielded in compression.

The fundamental equations for strain compatibility and the steel's stress–strain relations are as presented in example 4.10 and are used again in this example. The equilibrium equations for the triangular section become

$$N = F_{cc} + F_{sc} + F_s$$

or

$$0.8x < h \quad N = 0.567 f_{ck} sl/2 + f_{sc} A_s' + f_s A_s$$

$$0.8x \geq h \quad N = 0.567 f_{ck} h \times 400/2 + f_{sc} A_s' + f_s A_s$$

$$0.8x < h \quad M = F_{cc} 2(h - 0.8x)/3 + F_{sc}(2h/3 - d') - F_s(d - 2h/3)$$

$$0.8x \geq h \quad M = F_{sc}(2h/3 - d') - F_s(d - 2h/3)$$

F_s is negative when f_s is a tensile stress, and from the geometry of figure 4.23 the width of the section at depth $s = 0.8x$ is $l = \frac{2}{3} s \sqrt{3}$.

Table 4.4 M–N interaction values for example 4.11

x (mm)	ε_{sc}	ε_{st}	f_{sc} (N/mm²)	f_s (N/mm²)	N (kN)	M (kN m)
$d' = 100$	0	>0.00217	0	$-0.87 f_{yk}$	-375	37
$x_{bal} = 0.617d$ = 183	0.00158	0.00217	317	$-0.87 f_{yk}$	-96	72
$2.63d' = 263$	0.00217	0.00044	$0.87 f_{yk}$	-88	490	66
$d = 296$	>0.00217	0	$0.87 f_{yk}$	0	672	61
$h = 346$	>0.00217	0.00051	$0.87 f_{yk}$	101	940	50
∞	>0.00217	>0.00217	$0.87 f_{yk}$	$0.87 f_{yk}$	1622	0

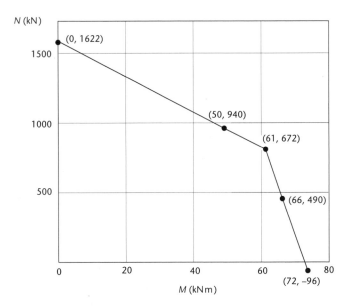

Figure 4.24
M–N interaction diagram for a
non-rectangular section

Table 4.4 has been calculated using the fundamental equations with the values of x shown. The interaction diagram is shown constructed in figure 4.24.

With a non-rectangular section, it could be advisable to construct a more accurate interaction diagram using other intermediate values of x. This would certainly be the case with, say, a flanged section where there is sudden change in breadth.

4.9 ▎ Rectangular–parabolic stress block

A rectangular–parabolic stress block may be used to provide a more rigorous analysis of the reinforced concrete section. The stress block is similar in shape to the stress–strain curve for concrete in figure 4.1, having a maximum stress of $0.567 f_{ck}$ at the ultimate strain of 0.0035.

In figure 4.25

ε_0 = the concrete strain at the end of the parabolic section

w = the distance from the neutral axis to strain ε_{c2}

x = depth of the neutral axis

k_1 = the mean concrete stress

$k_2 x$ = depth to the centroid of the stress block.

(a) To determine the mean concrete stress, k_1

From the strain diagram

$$\frac{x}{0.0035} = \frac{w}{\varepsilon_0}$$

therefore

$$w = \frac{x \varepsilon_{c2}}{0.0035}$$

Figure 4.25
Section in bending with a
rectangular–parabolic stress
block

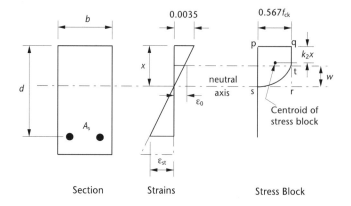

Section   Strains   Stress Block

Substituting for $\varepsilon_{c2} = 0.002$ (figure 4.1)

$$w = 0.571x \tag{4.43}$$

For the stress block

$$k_1 = \frac{\text{area of stress block}}{x}$$
$$= \frac{\text{area pqrs} - \text{area rst}}{x}$$

Thus, using the area properties of a parabola as shown in figure 4.26, we have

$$k_1 = \frac{0.567f_{ck}x - 0.567f_{ck}.w/3}{x}$$

Substituting for w from equation 4.43 gives

$$k_1 = 0.459f_{ck} \tag{4.44}*$$

(b) To determine the depth of the centroid k_2x

k_2 is determined for a rectangular section by taking area moments of the stress block about the neutral axis – see figures 4.25 and 4.26. Thus

$$(x - k_2x) = \frac{\text{area pqrs} \times x/2 - \text{area rst} \times w/4}{\text{area of stress block}}$$
$$= \frac{(0.567f_{ck}x)x/2 - (0.567f_{ck}w/3)w/4}{k_1x}$$
$$= \frac{0.567f_{ck}(x^2/2 - w^2/12)}{k_1x}$$

Figure 4.26
Properties of a parabola

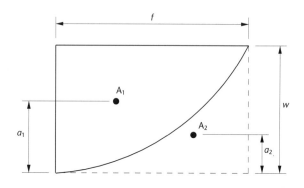

Areas:

$$A_1 = \frac{2wf}{3}$$
$$A_2 = \frac{wf}{3}$$

Position of centroids:

$$a_1 = \frac{5w}{8}$$
$$a_2 = \frac{w}{4}$$

Substituting for w from equation 4.43

$$(x - k_2 x) = \frac{0.567 f_{ck} x^2}{k_1 x} \left[0.5 - \frac{0.571^2}{12} \right]$$

hence

$$k_2 = 1 - \frac{0.268 f_{ck}}{k_1} = 1 - \frac{0.268 f_{ck}}{0.459 f_{ck}} = 0.416 \qquad (4.45)*$$

Once we know the properties of the stress block, the magnitude and position of the resultant compressive force in the concrete can be determined, and hence the moment of resistance of the section calculated using procedures similar to those for the rectangular stress block.

Comparison of the rectangular–parabolic and the rectangular stress blocks provides

(i)   Stress resultant, F_{cc}

  rectangular–parabolic: $k_1 bx \approx 0.459 f_{ck} bx$

  rectangular: $0.567 f_{ck} \times 0.8bx \approx 0.454 f_{ck} bx$

(ii) Lever arm, z

  rectangular parabolic: $d - k_2 x \approx d - 0.416x$

  rectangular: $d - \frac{1}{2} \times 0.8x = d - 0.40x$

So both stress blocks have almost the same moment of resistance, $F_{cc} \times z$, showing it is adequate to use the simpler rectangular stress block for design calculations.

4.10   Triangular stress block

The triangular stress block applies to elastic conditions during the serviceability limit state. In practice it is not generally used in design calculations except for liquid-retaining structures, or for the calculations of crack widths and deflections as described in chapter 6. With the triangular stress block, the cross-section can be considered as

(i)   cracked in the tension zone, or

(ii) uncracked with the concrete resisting a small amount of tension.

4.10.1   Cracked section

A cracked section is shown in figure 4.27 with a stress resultant F_{st} acting through the centroid of the steel and F_{cc} acting through the centroid of the triangular stress block. For equilibrium of the section

$$F_{cc} = F_{st}$$

or $0.5 bx f_{cc} = A_s f_{st}$ $\qquad (4.46)*$

and the moment of resistance

$$M = F_{cc} \times z = F_{st} \times z$$

or $M = 0.5 bx f_{cc}(d - x/3) = A_s f_{st}(d - x/3)$ $\qquad (4.47)*$

Figure 4.27
Triangular stress block –
cracked section

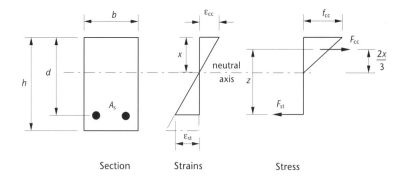

Section   Strains   Stress

(i) Analysis of a specified section

The depth of the neutral axis, x, can be determined by converting the section into an 'equivalent' area of concrete as shown in figure 4.28, where $\alpha_e = E_s/E_c$, the modular ratio.

Figure 4.28
Equivalent transformed
section with the concrete
cracked

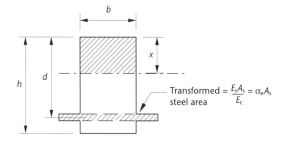

Transformed steel area $= \dfrac{E_s A_s}{E_c} = \alpha_e A_s$

Taking the area moments about the upper edge:

$$x = \frac{\sum(Ax)}{\sum A}$$

Therefore

$$x = \frac{bx \times x/2 + \alpha_e A_s d}{bx + \alpha_e A_s}$$

or

$$\frac{1}{2}bx^2 + \alpha_e A_s x - \alpha_e A_s d = 0$$

Solving this quadratic equation gives

$$x = \frac{-\alpha_e A_s \pm \sqrt{\left[(\alpha_e A_s)^2 + 2b\alpha_e A_s d\right]}}{b} \qquad (4.48)*$$

Equation 4.48 may be solved using a chart such as the one shown in figure 4.29.

Equations 4.46 to 4.48 can be used to analyse a specified reinforced concrete section.

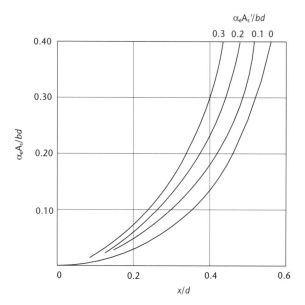

Figure 4.29
Neutral-axis depths for cracked rectangular sections – elastic behaviour

(ii) Design of steel area, A$_s$, with stresses f$_{st}$ and f$_{cc}$ specified

The depth of the neutral axis can also be expressed in terms of the strains and stresses of the concrete and steel.

From the linear strain distribution of figure 4.27:

$$\frac{x}{d} = \frac{\varepsilon_{cc}}{\varepsilon_{cc} + \varepsilon_{st}} = \frac{f_{cc}/E_c}{f_{cc}/E_c + f_{st}/E_s}$$

Therefore

$$\frac{x}{d} = \frac{1}{1 + \dfrac{f_{st}}{\alpha_e f_{cc}}} \tag{4.49}*$$

Equations 4.47 and 4.49 may be used to design the area of tension steel required, at a specified stress, in order to resist a given moment.

EXAMPLE 4.12

Analysis of a cracked section using a triangular stress block

For the section shown in figure 4.30, determine the concrete and steel stresses caused by a moment of 120 kN m, assuming a cracked section. Take $E_s/E_c = \alpha_e = 15$.

$$\alpha_e \frac{A_s}{bd} = \frac{15 \times 1470}{300 \times 460} = 0.16$$

Using the chart of figure 4.29 or equation 4.48 gives $x = 197$ mm.

From equation 4.47

$$M = \frac{1}{2}bx f_{cc}\left(d - \frac{x}{3}\right)$$

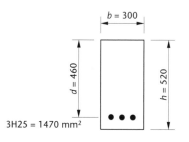

Figure 4.30
Analysis example with triangular stress block

b = 300

d = 460

h = 520

3H25 = 1470 mm²

therefore

$$120 \times 10^6 = \frac{1}{2} \times 3000 \times 197 \times f_{cc} \left(460 - \frac{197}{3}\right)$$

therefore

$$f_{cc} = 10.3 \text{ N/mm}^2$$

From equation 4.46

$$f_{st}A_s = \frac{1}{2}bx\,f_{cc}$$

therefore

$$f_{st} = 300 \times 197 \times \frac{10.3}{2} \times \frac{1}{1470} = 207 \text{ N/mm}^2$$

4.10.2 Triangular stress block – uncracked section

The concrete may be considered to resist a small amount of tension. In this case a tensile stress resultant F_{ct} acts through the centroid of the triangular stress block in the tension zone as shown in figure 4.31.

For equilibrium of the section

$$F_{cc} = F_{ct} + F_{st} \qquad (4.50)$$

where $F_{cc} = 0.5bx\,f_{cc}$

$F_{ct} = 0.5b(h - x)f_{ct}$

and $F_{st} = A_s \times f_{st}$

Taking moments about F_{cc}, the moment of resistance of the section is given by

$$M = F_{st} \times (d - x/3) + F_{ct} \times \left(\frac{2}{3}x + \frac{2}{3}(h - x)\right) \qquad (4.51)^*$$

The depth of the neutral axis, x, can be determined by taking area moments about the upper edge AA of the equivalent concrete section shown in figure 4.32, such that

$$x = \frac{\sum(Ax)}{\sum A}$$

$\alpha_e = \dfrac{E_s}{E_c}$ is termed the modular ratio

Figure 4.31
Triangular stress block – uncracked section

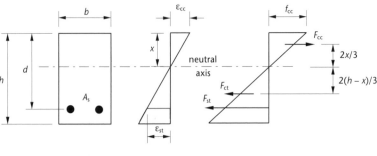

Section   Strains   Stress

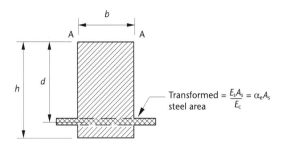

Figure 4.32
Equivalent transformed
section with the concrete
uncracked

Therefore

$$x = \frac{bh \times h/2 + \alpha_e A_s \times d}{bh + \alpha_e A_s}$$

$$= \frac{h + 2\alpha_e rd}{2 + 2\alpha_e r} \qquad (4.52)*$$

where $r = A_s/bh$

From the linear proportions of the strain diagram in figure 4.31:

$$\varepsilon_{cc} = \frac{x}{h - x} \times \varepsilon_{ct}$$

$$\varepsilon_{st} = \frac{d - x}{h - x} \times \varepsilon_{ct} \qquad (4.53)*$$

Therefore as stress $= E \times$ strain:

$$f_{ct} = E_c \varepsilon_{ct}$$

$$f_{cc} = \frac{x}{h - x} \times f_{ct} \qquad (4.54)*$$

$$f_{st} = \frac{d - x}{h - x} \times \alpha_e f_{ct}$$

Hence if the maximum tensile strain or stress is specified, it is possible to calculate the corresponding concrete compressive and steel tensile stresses from equations 4.54.

The equations derived can be used to analyse a given cross-section in order to determine the moment of resistance of the uncracked section.

EXAMPLE 4.13

Analysis of an uncracked section

For the section shown in figure 4.30, calculate the serviceability moment of resistance with no cracking of the concrete, given $f_{ct} = 3\,\text{N/mm}^2$, $E_c = 30\,\text{kN/mm}^2$ and $E_s = 200\,\text{kN/mm}^2$.

$$r = \frac{A_s}{bh}$$

$$= \frac{1470}{300 \times 520} = 0.0094$$

$$\alpha_e = \frac{E_s}{E_c}$$

$$= \frac{200}{30} = 6.67$$

$$x = \frac{h + 2\alpha_e rd}{2 + 2\alpha_e r}$$

$$= \frac{520 + 2 \times 6.67 \times 0.0094 \times 460}{2 + 2 \times 6.67 \times 0.0094} = 272 \text{ mm}$$

$$f_{st} = \left(\frac{d - x}{h - x}\right)\alpha_e f_{ct}$$

$$= \frac{(460 - 272)6.67 \times 3}{(420 - 272)} = 15.2 \text{ N/mm}^2$$

$$M = A_s f_{st}\left(d - \frac{x}{3}\right) + \frac{1}{2}b(h - x)f_{ct}\left(\frac{2}{3}x + \frac{2}{3}(h - x)\right)$$

$$= 1470 \times 15.2\left(460 - \frac{272}{3}\right)10^{-6} + \frac{1}{2} \times 300(520 - 272) \times 3$$

$$\times \left(\frac{2}{3} \times 272 + \frac{2}{3}(520 - 272)\right)10^{-6}$$

$$= 8.3 + 38.7 = 47 \text{ kN m}$$

CHAPTER 5

Shear, bond and torsion

CHAPTER INTRODUCTION

This chapter deals with the theory and derivation of the design equations for shear, bond and torsion. Some of the more practical factors governing the choice and arrangement of the reinforcement are dealt with in the chapters on member design, particularly chapter 7, which contains examples of the design and detailing of shear and torsion reinforcement in beams. Punching shear caused by concentrated loads on slabs is covered in section 8.1.1 of the chapter on slab design.

5.1   Shear

Figure 5.1 represents the distribution of principal stresses across the span of a homogeneous concrete beam. The direction of the principal compressive stresses takes the form of an arch, while the tensile stresses have the curve of a catenary or suspended chain. Towards mid-span, where the shear is low and the bending stresses are dominant, the direction of the stresses tends to be parallel to the beam axis. Near the supports, where the shearing forces are greater, the principal stresses become inclined and the greater the shear force the greater the angle of inclination. The tensile stresses due to shear are liable to cause diagonal cracking of the concrete near to the support so that shear reinforcement must be provided. This reinforcement is either in the form of (1) stirrups, or (2) inclined bars (used in conjunction with stirrups) as shown in figures 5.4 and 5.5. The steel stirrups are also often referred to as links.

Figure 5.1
Principal stresses in a beam

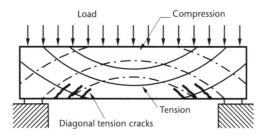

The concrete itself can resist shear by a combination of the un-cracked concrete in the compression zone, the dowelling action of the bending reinforcement and aggregate interlock across tension cracks but, because concrete is weak in tension, the shear reinforcement is designed to resist all the tensile stresses caused by the shear forces. Even where the shear forces are small near the centre of span of a beam a minimum amount of shear reinforcement in the form of links must be provided in order to form a cage supporting the longitudinal reinforcement and to resist any tensile stresses due to factors such as thermal movement and shrinkage of the concrete.

The actual behaviour of reinforced concrete in shear is complex, and difficult to analyse theoretically, but by applying the results from many experimental investigations, reasonable simplified procedures for analyses and design have been developed.

In EC2 a method of shear design is presented which will be unfamiliar to those designers who have been used to design methods based on previous British Standard design codes. This method is known as *The Variable Strut Inclination Method*. The use of this method allows the designer to seek out economies in the amount of shear reinforcement provided, but recognising that any economy achieved may be at the expense of having to provide additional curtailment and anchorage lengths to the tension steel over and above that normally required for resistance to bending as described in section 7.9.

5.1.1   Concrete sections that do not require design shear reinforcement

The concrete sections that do not require shear reinforcement are mainly lightly loaded floor slabs and pad foundations. Beams are generally more heavily loaded and have a smaller cross-section so that they nearly always require shear reinforcement. Even lightly loaded beams are required to have a minimum amount of shear links. The only exceptions to this are very minor beams such as short span, lightly loaded lintels over windows and doors.

Where shear forces are small the concrete section on its own may have sufficient shear capacity ($V_{Rd,c}$) to resist the ultimate shear force (V_{Ed}) resulting from the worst combination of actions on the structure, although in most cases a nominal or minimum amount of shear reinforcement will usually be provided.

In those sections where $V_{Ed} \leq V_{Rd,c}$ then no calculated shear reinforcement is required.

The shear capacity of the concrete, $V_{Rd,c}$, in such situations is given by an empirical expression:

$$V_{Rd,c} = \left[0.12k(100\rho_1 f_{ck})^{1/3} \right] b_w d \tag{5.1}$$

with a minimum value of:

$$V_{Rd,c} = \left[0.035k^{3/2} f_{ck}^{1/2} \right] b_w d \tag{5.2}$$

where:

$V_{Rd,c}$ = the design shear resistance of the section without shear reinforcement

$$k = \left(1 + \sqrt{\frac{200}{d}} \right) \leq 2.0 \text{ with } d \text{ expressed in mm}$$

$$\rho_1 = \frac{A_{s1}}{b_w d} \leq 0.02$$

A_{s1} = the area of tensile reinforcement that extends beyond the section being considered by at least a full anchorage length plus one effective depth (d)

b_w = the smallest width of the section in the tensile area (mm)

Some typical values of the corresponding shear stress capacities ($v_{Rd,c} = V_{Rd,c}/b_w d$) are given in chapter 8 (table 8.2).

5.1.2 The variable strut inclination method for sections that do require shear reinforcement

In order to derive the design equations the action of a reinforced concrete beam in shear is represented by an analogous truss as shown in figure 5.2. The concrete acts as the top

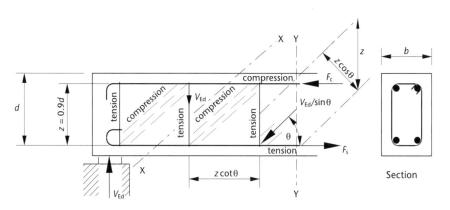

Figure 5.2
Assumed truss model for the variable strut inclination method

compression member and as the diagonal compression members inclined at an angle θ to the horizontal. The bottom chord is the horizontal tension steel and the vertical links are the transverse tension members. It should be noted that in this model of shear behaviour all shear will be resisted by the provision of links *with no direct contribution from the shear capacity of the concrete itself.*

The angle θ increases with the magnitude of the maximum shear force on the beam and hence the compressive forces in the diagonal concrete members. It is set by EC2 to have a value between 22 and 45 degrees. For most cases of predominately uniformly distributed loading the angle θ will be 22 degrees but for heavy and concentrated loads it can be higher in order to resist crushing of the concrete diagonal members.

The analysis of the truss to derive the design equations will be carried out in the following order:

1. Consideration of the compressive strength of the diagonal concrete strut and its angle θ;

2. Calculation of the required shear reinforcement A_{sw}/s for the vertical ties;

3. Calculation of the additional tension steel A_{sl} required in the bottom chord member.

The following notation is used in the equations for the shear design

$$A_{sw} = \text{the cross-sectional area of the two legs of the link}$$
$$s = \text{the spacing of the links}$$
$$z = \text{the lever arm between the upper and lower chord members of the analogous truss}$$
$$f_{ywd} = \text{the design yield strength of the link reinforcement}$$
$$f_{yk} = \text{the characteristic strength of the link reinforcement}$$
$$V_{Ed} = \text{the shear force due to the actions at the ultimate limit state}$$
$$V_{Ef} = \text{the ultimate shear force at the face of the support}$$
$$V_{wd} = \text{the shear force in the link}$$
$$V_{Rd,s} = \text{the shear resistance of the links}$$
$$V_{Rd,max} = \text{the maximum design value of the shear which can be resisted by the concrete strut}$$

(1) The diagonal compressive strut and the angle θ

The shear force applied to the section must be limited so that excessive compressive stresses do not occur in the diagonal compressive struts, leading to compressive failure of the concrete. Thus the maximum design shear force $V_{Rd,max}$ is limited by the ultimate crushing strength of the diagonal concrete member in the analogous truss and its vertical component.

With reference to figure 5.2, the effective cross sectional area of concrete acting as the diagonal strut is taken as $b_w \times z \cos \theta$ and the design concrete stress $f_{cd} = f_{ck}/1.5$.

The ultimate strength of the strut = ultimate design stress × cross-sectional area
$$= (f_{ck}/1.5) \times (b_w \times z \cos \theta)$$
and its vertical component $\quad = [(f_{ck}/1.5) \times (b_w \times z \cos \theta)] \times \sin \theta$
so that $\qquad\qquad V_{Rd,max} = f_{ck} b_w z \cos \theta \sin \theta / 1.5$

which by conversion of the trigometrical functions can also be expressed as

$$V_{Rd, max} = \frac{f_{ck} b_w z}{1.5(\cot \theta + \tan \theta)}$$

In EC2 this equation is modified by the inclusion of a *strength reduction factor* (v_1) for concrete cracked in shear.

Thus

$$V_{Rd, max} = \frac{f_{ck} b_w z v_1}{1.5(\cot \theta + \tan \theta)})$$ (5.3)

where the strength reduction factor takes the value of $v_1 = 0.6(1 - f_{ck}/250)$ and, putting $z = 0.9d$, equation 5.3 becomes

$$V_{Rd, max} = \frac{0.9d \times b_w \times 0.6(1 - f_{ck}/250)f_{ck}}{1.5(\cot \theta + \tan \theta)}$$

$$= \frac{0.36 b_w d(1 - f_{ck}/250)f_{ck}}{(\cot \theta + \tan \theta)}$$ (5.4)*

and to ensure that there is no crushing of the diagonal compressive strut:

$$V_{Rd, max} \geq V_{Ed}$$ (5.5)

This must be checked for the maximum value of shear on the beam, which is usually taken as the shear force, V_{Ef}, at the face of the beam's supports so that

$$V_{Rd, max} \geq V_{Ef}$$

As previously noted EC2 limits θ to a value between 22 and 45 degrees.

(i) With $\theta = 22$ degrees (this is the usual case for uniformly distributed loads)

From equation 5.4:

$$V_{Rd, max(22)} = 0.124 b_w d(1 - f_{ck}/250)f_{ck}$$ (5.6)*

If $V_{Rd, max(22)} < V_{Ef}$ then a larger value of the angle θ must be used so that the diagonal concrete strut has a larger vertical component to balance V_{Ed}.

(ii) With $\theta = 45$ degrees (the maximum value of θ as allowed by EC2)

From equation 5.4:

$$V_{Rd, max(45)} = 0.18 b_w d(1 - f_{ck}/250)f_{ck}$$ (5.7)*

which is the upper limit on the compressive strength of the concrete diagonal member in the analogous truss . When $V_{Ef} > V_{Rd, max(45)}$, from equation 5.7 the diagonal strut will be over stressed and the beam's dimensions must be increased or a higher class of concrete be used.

(iii) With θ between 22 degrees and 45 degrees

The required value for θ can be obtained by equating V_{Ed} to $V_{Rd, max}$ and solving for θ in equation 5.4 as follows:

$$V_{Ed} = V_{Rd, max} = \frac{0.36 b_w d(1 - f_{ck}/250)f_{ck}}{(\cot \theta + \tan \theta)}$$

and

$$1/(\cot\theta + \tan\theta) = \sin\theta \times \cos\theta$$
$$= 0.5\sin 2\theta \quad \text{(see proof in the Appendix)}$$

therefore by substitution

$$\theta = 0.5\sin^{-1}\left\{\frac{V_{Ed}}{0.18b_w d(1 - f_{ck}/250)f_{ck}}\right\} \leq 45° \tag{5.8a}*$$

which alternatively can be expressed as:

$$\theta = 0.5\sin^{-1}\left\{\frac{V_{Ef}}{V_{Rd,\,max(45)}}\right\} \leq 45° \tag{5.8b}$$

where V_{Ef} is the shear force at the face of the support and the calculated value of the angle θ can then be used to determine $\cot\theta$ and calculate the shear reinforcement A_{sw}/s from equation 5.9 below (when $22° < \theta < 45°$).

(2) The vertical shear reinforcement

As previously noted, all shear will be resisted by the provision of links *with no direct contribution from the shear capacity of the concrete itself*. Using the method of sections it can be seen that, at section X-X in figure 5.2, the force in the vertical link member (V_{wd}) must equal the shear force (V_{Ed}), that is

$$V_{wd} = V_{Ed} = f_{ywd}A_{sw}$$
$$= \frac{f_{yk}A_{sw}}{1.15}$$
$$= 0.87f_{yk}A_{sw}$$

If the links are spaced at a distance s apart, then the force in each link is reduced proportionately and is given by

$$V_{wd}\frac{s}{z\cot\theta} = 0.87f_{yk}A_{sw}$$

or

$$V_{wd} = V_{Ed}$$
$$= 0.87\frac{A_{sw}}{s}zf_{yk}\cot\theta$$
$$= 0.87\frac{A_{sw}}{s}0.9df_{yk}\cot\theta$$

thus rearranging

$$\frac{A_{sw}}{s} = \frac{V_{Ed}}{0.78df_{yk}\cot\theta} \tag{5.9}*$$

EC2 specifies a minimum value for A_{sw}/s such that

$$\frac{A_{sw,\,min}}{s} = \frac{0.08f_{ck}^{0.5}b_w}{f_{yk}} \tag{5.10}*$$

Equation 5.9 can be used to determine the amount and spacing of the shear links and will depend on the value of θ used in the design. For most cases of beams with predominately uniformly distributed loads the angle θ will be 22 degrees with $\cot\theta = 2.5$. Otherwise the value for θ can be calculated from equation 5.8.

EC2 also specifies that, for beams with predominately uniformly distributed loads, the design shear force V_{Ed} need not be checked at a distance less than d from the face of the support but the shear reinforcement calculated must be continued to the support.

Equation 5.9 can be rearranged to give the shear resistance $V_{Rd,s}$ of a given arrangement of links A_{sw}/s.

Thus:

$$V_{Rd,s} = \frac{A_{sw}}{s} \times 0.78 d f_{yk} \cot\theta \tag{5.11}*$$

(3) Additional longitudinal force

When using this method of shear design it is necessary to allow for the additional longitudinal force in the tension steel caused by the shear V_{Ed}. This longitudinal tensile force ΔF_{td} is caused by the horizontal component required to balance the compressive force in the inclined concrete strut.

Resolving forces horizontally in the section YY shown in figure 5.2, the longitudinal component of the force in the compressive strut is given by

$$\text{Longitudinal force} = (V_{Ed}/\sin\theta) \times \cos\theta$$
$$= V_{Ed} \cot\theta$$

It is assumed that half of this force is carried by the reinforcement in the tension zone of the beam then the additional tensile force to be provided in the tensile zone is given by

$$\Delta F_{td} = 0.5 V_{Ed} \cot\theta \tag{5.12}*$$

To provide for this longitudinal force, at any cross-section it is necessary to provide longitudinal reinforcement *additional* to that required at that section to resist bending. In practice, increasing the curtailment lengths of the bottom-face tension reinforcement can usually provide the required force. This reinforcement provides bending resistance in sections of high sagging bending moment and then, when no longer required to resist bending, can provide the additional tensile force to resist shear in those sections away from mid-span and towards the supports where sagging bending moments reduce but shear forces increase. This is discussed further and illustrated in section 7.9 (Anchorage and curtailment of reinforcement)

The total force given by $M_{Ed}/z + \Delta F_{td}$ should not be taken as greater than $M_{Ed,max}/z$, where $M_{Ed,max}$ is the maximum hogging or maximum sagging moment along the beam.

Equations 5.3 to 5.12 can be used together to design a section for shear with a value of θ chosen by the designer within limits of 22 and 45 degrees as specified in EC2. From these equations it is obvious that the steel ratio (A_{sw}/s) is a function of the inverse of $\cot\theta$ (equation 5.9), the maximum shear force governed by diagonal compression failure is a function of the inverse of $(\cot\theta + \tan\theta)$ (equation 5.4) and the additional longitudinal tensile force, ΔF_{td} varies with $\cot\theta$ (equation 5.12).

Figure 5.3 shows the variation of these functions from which it can be seen that as θ is reduced less shear reinforcement is required, but this is compensated for by an increase in the necessary longitudinal reinforcement. At values of θ greater or less than 45° the shear capacity of the section, based on compressive failure in the diagonal struts, is also reduced. The designer should calculate the value of θ but, as previously stated, for practical reasons EC2 places a lower and upper limit of 1.0 and 2.5 respectively on the value of $\cot\theta$. This corresponds to limiting θ to 45° and 22° respectively.

Figure 5.3
Variation of $V_{Rd, max}$, ΔF_{td}
and A_{sw}/s

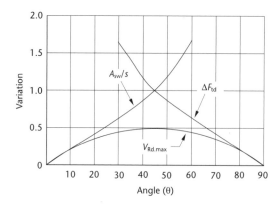

Summary of the design procedure with vertical links

1. Calculate the ultimate design shear forces V_{Ed} along the beam's span.

2. Check the crushing strength $V_{Rd, max}$ of the concrete diagonal strut at the section of maximum shear, usually V_{Ef} at the face of the beam's support.

   For most cases the angle of inclination of the strut is $\theta = 22°$, with $\cot\theta = 2.5$ and $\tan\theta = 0.4$ so that from equation 5.4

   $$V_{Rd, max} = \frac{0.36b_w d(1 - f_{ck}/250)f_{ck}}{(\cot\theta + \tan\theta)}$$

   and if $V_{Rd, max} \geq V_{Ef}$ with $\theta = 22°$ and $\cot\theta = 2.5$ then go directly to step (3). However, if $V_{Rd, max} < V_{Ef}$ then $\theta > 22°$ and therefore θ must be calculated from equation 5.8 as:

   $$\theta = 0.5 \sin^{-1}\left\{\frac{V_{Ef}}{0.18b_w d(1 - f_{ck}/250)f_{ck}}\right\} \leq 45°$$

   If this calculation gives a value of θ greater than $45°$ then the beam should be re-sized or a higher class of concrete could be used.

3. The shear links required can be calculated from equation 5.9

   $$\frac{A_{sw}}{s} = \frac{V_{Ed}}{0.78df_{yk}\cot\theta}$$

   where A_{sw} is the cross-sectional area of the legs of the links ($2 \times \pi\phi^2/4$ for single stirrups).

   For a predominately uniformly distributed load the shear V_{Ed} should be calculated at a distance d from the face of the support and the shear reinforcement should continue to the face of the support.

   The shear resistance for the links actually specified is

   $$V_{min} = \frac{A_{sw}}{s} \times 0.78df_{yk}\cot\theta$$

   and this value will be used together with the shear force envelope to determine the curtailment position of each set of designed links

4. Calculate the minimum links required by EC2 from

   $$\frac{A_{sw, min}}{s} = \frac{0.08f_{ck}^{0.5}b_w}{f_{yk}}$$

5. Calculate the additional longitudinal tensile force caused by the shear

$$\Delta F_{td} = 0.5 V_{Ed} \cot \theta$$

This additional tensile force can usually be allowed for by increasing the curtailment length of the tension bars as described in section 7.9.

Examples illustrating the design of shear reinforcement for a beam are given in Chapter 7.

EXAMPLE 5.1

Shear resistance of a beam

The beam in figure 5.4 spans 8.0 metres on 300 mm wide supports. It is required to support a uniformly distributed ultimate load, w_u of 200 kN/m. The characteristic material strengths are $f_{ck} = 30 \text{ N/mm}^2$ for the concrete and $f_{yk} = 500 \text{ N/mm}^2$ for the steel. Check if the shear reinforcement in the form of the vertical links shown can support, in shear, the given ultimate load.

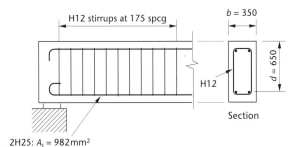

Figure 5.4
Beam with stirrups

Total ultimate load on beam	$= 200 \times 8.0 = 1600 \text{ kN}$
Support reaction	$= 1600/2 = 800 \text{ kN}$
Shear, V_{Ef} at face of support	$= 800 - 200 \times 0.3/2 = 770 \text{ kN}$
Shear, V_{Ed} distance d from face of support	$= 770 - 200 \times 0.65 = 640 \text{ kN}$

1. Check the crushing strength $V_{Rd, max}$ of the concrete diagonal strut at the face of the beams support.

From equation 5.6 with $\theta = 22°$

$$
\begin{aligned}
V_{Rd, max(22)} &= 0.124 b_w d (1 - f_{ck}/250) f_{ck} \\
&= 0.124 \times 350 \times 650 (1 - 30/250) 30 \\
&= 745 \text{ kN} \quad (< V_{Ef} = 770 \text{ kN})
\end{aligned}
$$

From equation 5.7 with $\theta = 45°$

$$
\begin{aligned}
V_{Rd, max(45)} &= 0.18 b_w d (1 - f_{ck}/250) f_{ck} \\
&= 0.18 \times 350 \times 650 (1 - 30/250) 30 \\
&= 1081 \text{ kN} \quad (> V_{Ef} = 770 \text{ kN})
\end{aligned}
$$

Therefore: $22° < \theta < 45°$.

2. Determine angle θ

From equation 5.8(a)

$$\theta = 0.5\sin^{-1}\left\{\frac{V_{Ef}}{0.18b_wd(1 - f_{ck}/250)f_{ck}}\right\} \le 45°$$

or alternatively from equation 5.8(b)

$$\theta = 0.5\sin^{-1}\left\{\frac{V_{Ef}}{V_{Rd,\,max(45)}}\right\} = 0.5\sin^{-1}\left\{\frac{770}{1081}\right\} = 22.7°$$

From which $\cot\theta = 2.39$ and $\tan\theta = 0.42$.

3. Determine shear resistance of the links

The cross-sectional area A_{sw} of a 12mm bar $= 113\,\text{mm}^2$. Thus for the two legs of the link and a spacing of 175 mm

$$\frac{A_{sw}}{s} = \frac{2 \times 113}{175} = 1.29$$

(or alternatively the value could have been obtained from table A4 in the Appendix)

From equation 5.11 the shear resistance, $V_{Rd,\,s}$ of the links is given by

$$V_{Rd,\,s} = \frac{A_{sw}}{s} \times 0.78df_{yk}\cot\theta$$
$$= 1.29 \times 0.78 \times 650 \times 500 \times 2.39 \times 10^{-3} = 781\,\text{kN}$$

Therefore shear resistance of links $= 781\,\text{kN}$.

Design shear, V_{Ed} distance d from the face of the support $= 640\,\text{kN}\ (< 781\ \text{kN})$. Therefore, the beam can support, in shear, the ultimate load of 200 kN/m.

4. Additional longitudinal tensile force in the tension steel

It is necessary to check that the bottom tension steel has a sufficient length of curtailment and anchorage to resist the additional horizontal tension ΔF_{td} caused by the design shear. These additional tension forces are calculated from equation 5.12. Therefore

$$\Delta F_{td} = 0.5V_{Ed}\cot\theta$$
$$= 0.5 \times 640 \times 2.39 = 765\,\text{kN}$$

This force is added to the M_{Ed}/z diagram, as described in section 7.9, to ensure there is sufficient curtailment of the tension reinforcement and its anchorage bond length at the supports, as described in section 5.2.

5.1.3 Bent-up bars

To resist shearing forces, longitudinal tension bars may be bent up near to the supports as shown in figure 5.5. The bent-up bars and the concrete in compression are considered to act as an analogous lattice girder and the shear resistance of the bars is determined by taking a section X–X through the girder.

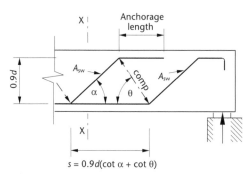

X

Anchorage length

A_{sw}

A_{sw}

comp.

α

θ

0.9d

X

$s = 0.9d(\cot \alpha + \cot \theta)$

(a) Single System

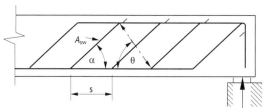

A_{sw}

α

θ

s

(b) Multiple System

Figure 5.5
Bent up bars

From the geometry of part (a) of figure 5.5, the spacing of the bent-up bars is:

$$s = 0.9d(\cot \alpha + \cot \theta)$$

and at the section X–X the shear resistance of a single bent-up bar (V_{wd}) must equal the shear force (V_{Ed}).

$$V_{wd} = V_{Ed} = f_{ywd}A_{sw} \sin \alpha = \frac{f_{yk}}{1.15}A_{sw} \sin \alpha = 0.87f_{yk}A_{sw} \sin \alpha$$

where A_{sw} is the cross-sectional area of the bent-up bar.

For a multiple system of bent-up bars, as in part (b) of figure 5.5, the shear resistance is increased proportionately to the spacing, s. Hence:

$$V_{Ed} = 0.87f_{yk}A_{sw} \sin \alpha \times \frac{0.9d(\cot \alpha + \cot \theta)}{s}$$

or

$$\frac{A_{sw}}{s} = \frac{V_{Ed}}{0.78df_{yk}(\cot \alpha + \cot \theta) \sin \alpha} \tag{5.13}$$

This equation is analogous to equation (5.9) for the shear resistance of shear links. In a similar way it can be shown that, based on crushing of the concrete in the compressive struts, the analogous equation to (5.4) is given by:

$$V_{Rd,\,max} \leq 0.36b_w d(1 - f_{ck}/250)f_{ck} \times \frac{(\cot \theta + \cot \alpha)}{(1 + \cot^2 \theta)} \tag{5.14}$$

and the *additional* tensile force to be provided by the provision of additional tension steel is given by a modified version of equation 5.12:

$$\Delta F_{td} = 0.5V_{Ed}(\cot \theta - \cot \alpha) \tag{5.15}$$

EC2 also requires that the maximum longitudinal spacing of bent-up bars is limited to $0.6d(1 + \cot \alpha)$ and specifies that at least 50 per cent of the required shear reinforcement should be in the form of shear links.

5.1.4 Shear between the web and flange of a flanged section

The provision of shear links to resist vertical shear in a flanged beam is identical to that previously described for a rectangular section, on the assumption that the web carries all of the vertical shear and that the web width, b_w, is used as the minimum width of the section in the relevant calculations.

Longitudinal complementary shear stresses also occur in a flanged section along the interface between the web and flange as shown in figure 5.6. This is allowed for by providing transverse reinforcement over the width of the flange on the assumption that this reinforcement acts as ties combined with compressive struts in the concrete. It is necessary to check the possibility of failure by excessive compressive stresses in the struts and to provide sufficient steel area to prevent tensile failure in the ties. The variable strut inclination method is used in a similar manner to that for the design to resist vertical shear in a beam described in section 5.1.2.

The design is divided into the following stages:

1. Calculate the longitudinal design shear stresses, v_{Ed} at the web-flange interface.

The longitudinal shear stresses are at a maximum in the regions of the maximum changes in bending stresses that, in turn, occur at the steepest parts of the bending moment diagram. These occur at the lengths up to the maximum hogging moment over the supports and at the lengths away from the zero sagging moments in the span of the beam.

The change in the longitudinal force ΔF_d in the flange outstand at a section is obtained from

$$\Delta F_d = \frac{\Delta M}{(d - h_f/2)} \times \frac{b_{fo}}{b_f}$$

where $b_f =$ the effective breadth of the flange

         $b_{fo} =$ the breadth of the outstand of the flange $= (b_f - b_w)/2$

         $b_w =$ the breadth of the web

         $h_f =$ the thickness of the flange

and $\Delta M =$ the change in moment over the distance Δx

Therefore

$$\Delta F_d = \frac{\Delta M}{(d - h_f/2)} \times \frac{(b_f - b_w)/2}{b_f}$$

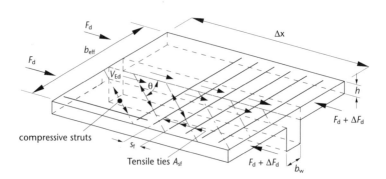

Figure 5.6
Shear between flange and web

The longitudinal shear stress, v_{Ed}, at the vertical section between the outstand of the flange and the web is caused by the change in the longitudinal force, ΔF_d, which occurs over the distance Δx, so that

$$v_{Ed} = \frac{\Delta F_d}{(h_f \times \Delta x)} \tag{5.16}$$

The maximum value allowed for Δx is half the distance between the section with zero moment and that where maximum moment occurs. Where point loads occur Δx should not exceed the distance between the loads.

If v_{Ed} is less than or equal to 40 per cent of the design tensile cracking strength of the concrete, f_{ctd}, i.e. $v_{Ed} \leq 0.4f_{ctd} = 0.4f_{ctk}/1.5 = 0.27f_{ctk}$, then no shear reinforcement is required and proceed directly to step 4.

2. Check the shear stresses in the inclined strut

As before, the angle θ for the inclination of the concrete strut is restricted to a lower and upper value and EC2 recommends that, in this case:

$$26.5° \leq \theta_f \leq 45° \qquad \text{i.e } 2.0 \geq \cot \theta_f \geq 1.0 \text{ for flanges in compression}$$
$$38.6° \leq \theta_f \leq 45° \qquad \text{i.e } 1.25 \geq \cot \theta_f \geq 1.0 \text{ for flanges in tension.}$$

To prevent crushing of the concrete in the compressive struts the longitudinal shear stress is limited to:

$$v_{Ed} \leq \frac{v_1 f_{ck}}{1.5(\cot \theta_f + \tan \theta_f)} \tag{5.17}$$

where the strength reduction factor $v_1 = 0.6(1 - f_{ck}/250)$.

The lower value of the angle θ is first tried and if the shear stresses are too high the angle θ is calculated from the following equation:

$$\theta_f = 0.5 \sin^{-1}\left\{\frac{v_{Ed}}{0.2(1 - f_{ck}/250)f_{ck}}\right\} \leq 45°$$

3. Calculate the transverse shear reinforcement required

The required transverse reinforcement per unit length, A_{sf}/s_f, may be calculated from the equation:

$$\frac{A_{sf}}{s_f} \geq \frac{v_{Ed}h_f}{0.87f_{yk} \cot \theta_f} \tag{5.18}$$

which is derived by considering the tensile force in each tie.

4. The requirements of transverse steel.

EC2 requires that the area of transverse steel should be the greater of (a) that given by equation 5.18 or (b) half that given by equation 5.18 plus the area of steel required by transverse bending of the flange.

The minimum amount of transverse steel required in the flange is $A_{s,min} = 0.26bd_f \times f_{ctm}/f_{yk} (> 0.0013bd_f) \text{ mm}^2/\text{m}$, where $b = 1000 \text{ mm}$ (see table 6.8).

Example 7.5 (p. 184) illustrates the approach to calculating transverse shear reinforcement in flanged beams.

5.2 | Anchorage bond

The reinforcing bar subject to direct tension shown in figure 5.7 must be firmly anchored if it is not to be pulled out of the concrete. Bars subject to forces induced by flexure must be similarly anchored to develop their design stresses. The anchorage depends on the bond between the bar and the concrete, the area of contact and whether or not the bar is located in a region where good bond conditions can be expected. Let:

$l_{b,rqd}$ = basic required anchorage length to prevent pull out

ϕ = bar size or nominal diameter

f_{bd} = ultimate anchorage bond stress

f_s = the direct tensile or compressive stress in the bar.

Figure 5.7
Anchorage bond

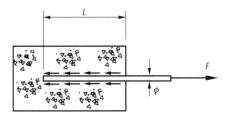

Considering the forces on the bar:

Tensile pull-out force = cross-sectional area of bar × direct stress

$$= \frac{\pi\phi^2}{4}f_s$$

anchorage force = contact area × anchorage bond stress

$$= (l_{b,rqd}\pi\phi) \times f_{bd}$$

therefore

$$(l_{b,rqd}\pi\phi) \times f_{bd} = \frac{\pi\phi^2}{4}f_s$$

hence

$$l_{b,rqd} = \frac{f_s\phi}{4f_{bd}}$$

and when $f_s = f_{yd}$, the design yield strength of the reinforcement ($= f_{yk}/1.15$) the anchorage length is given by

$$l_{b,rqd} = (\phi/4)([f_{yk}/1.15]/f_{bd})$$
$$l_{b,rqd} = (f_{yk}/4.6f_{bd})\phi \qquad\qquad (5.19)*$$

Basic anchorage length

Equation 5.19 may be used to determine the *basic anchorage length* of bars which are either in tension or compression. For the calculation of anchorage lengths, design values of ultimate anchorage bond stresses are specified according to whether the bond conditions are good or otherwise.

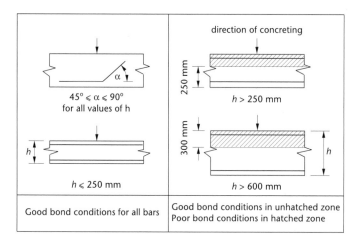

Figure 5.8
Definition of good and poor bond conditions

Good bond conditions are considered to be when (a) bars are inclined at an angle of between 45° and 90° to the horizontal or (b) zero to 45° provided that in this second case additional requirements are met. These additional conditions are that bars are

1. either placed in members whose depth in the direction of concreting does not exceed 250 mm or

2. embedded in members with a depth greater than 250 mm and are either in the lower 250 mm of the member or at least 300 mm from the top surface when the depth exceeds 600 mm.

These conditions are illustrated in figure 5.8. When bond conditions are poor then the specified ultimate bond stresses should be reduced by a factor of 0.7.

The design value of the ultimate bond stress is also dependent on the bar size. For all bar sizes (ϕ) greater than 32 mm the bond stress should additionally be multiplied by a factor $(132 - \phi)/100$.

Table 5.1 gives the design values of ultimate bond stresses for 'good' conditions. These depend on the class of concrete and are obtained from the equation $f_{bd} = 1.50 f_{ctk}$ where f_{ctk} is the characteristic tensile strength of the concrete.

Design anchorage length

The *basic* anchorage length discussed above must be further modified to give the *minimum* design anchorage length taking into account factors not directly covered by table 5.1.

Table 5.1 Design values of bond stresses f_{bd} (N/mm^2)

f_{ck} N/mm^2	12	16	20	25	30	35	40	45	50	55	60
Bars ≤ 32 mm diameter and good bond conditions	1.6	2.0	2.3	2.7	3.0	3.4	3.7	4.0	4.3	4.5	4.7
Bars ≤ 32 mm diameter and poor bond conditions	1.1	1.4	1.6	1.9	2.1	2.4	2.6	2.8	3.0	3.1	3.3

Table 5.2 Coefficients α

Value of α	α allows for the effect of:	Type of anchorage	Reinforcement in	
			Tension	Compression
α_1	The shape of the bars	Straight	1.0	1.0
		Other than straight	0.7 if $c_d > 3.0\phi$ or 1.0 if not	1.0
α_2	Concrete cover to the reinforcement	Straight	$1 - 0.15(c_d - \phi)/\phi$ but ≥ 0.7 and ≤ 1.0	1.0
		Other than straight	$1 - 0.15(c_d - 3\phi)/\phi$ but ≥ 0.7 and ≤ 1.0	1.0
α_3	Confinement of transverse reinforcement not welded to the main reinforcement	All types of reinforcement	$1 - K\lambda$ but ≥ 0.7 and ≤ 1.0	1.0
α_4	Confinement of transverse reinforcement welded to the main reinforcement	All types, position and sizes of reinforcement	0.7	0.7
α_5	Confinement by transverse pressure	All types of reinforcement	$1 - 0.04p$ but ≥ 0.7 and ≤ 1.0	–

Note: the product $\alpha_2 \times \alpha_3 \times \alpha_5$ should be greater than or equal to 0.7

The required minimum anchorage length (l_{bd}) is given by

$$l_{bd} = \alpha_1, \alpha_2, \alpha_3, \alpha_4, \alpha_5 l_{b,rqd} A_{s,req}/A_{s,prov} \tag{5.20)*}$$

where $A_{s,req}$, $A_{s,prov}$ = area of reinforcement required and provided at that section

α (1 to 5) = set of coefficients as given in Table 5.2

In Table 5.2:

c_d = concrete cover coefficient as shown in figure 5.9

K = values as shown in figure 5.10

$\lambda = \left(\sum A_{st} - \sum A_{st,min}\right)/A_s$

$\sum A_{st}$ = the cross-sectional area of the transverse reinforcement along the design anchorage length

$\sum A_{st,min}$ = the cross-sectional area of the minimum transverse reinforcement ($= 0.25A_s$ for beams and zero for slabs)

A_s = the area of a single anchored bar with maximum bar diameter

This minimum design length must not be less than:

for tension bars: $0.3l_{b,rqd}$

for compression bars: $0.6l_{b,rqd}$

In both cases the minimum value must also exceed both 10 bar diameters and 100 mm.
Anchorages may also be provided by hooks or bends in the reinforcement. Hooks and bends are considered adequate forms of anchorage to the main reinforcement if they

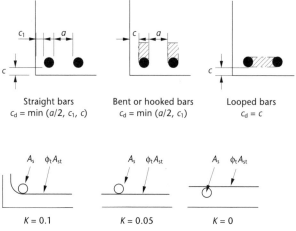

Figure 5.9
Values of c_d for beams and slabs (see table 5.2)

Straight bars
$c_d = \min (a/2, c_1, c)$

Bent or hooked bars
$c_d = \min (a/2, c_1)$

Looped bars
$c_d = c$

Figure 5.10
Values of K for beams and slabs (see table 5.2)

$K = 0.1$ $K = 0.05$ $K = 0$

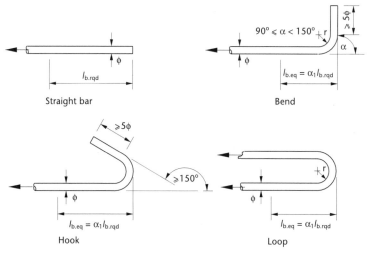

Figure 5.11
Equivalent anchorage lengths for bends and hooks

Straight bar   Bend

Hook   Loop

Minimum internal radius of a hook, bend or loop = 2ϕ or 3.5ϕ for $\phi \geqslant 16$ mm

satisfy the minimum dimensions shown in figure 5.11. Bends and hooks are not recommended for use as compression anchorages. In the case of the hooks and bends shown in figure 5.11 the anchorage length (shown as $l_{b, eq}$) which is equivalent to that required by the straight bar can be simply calculated from the expression: $l_{b, eq} = \alpha_1 l_{b, rqd}$ where α_1 is taken as 0.7 or 1.0 depending on the cover conditions (see table 5.2).

The internal diameter of any bent bar (referred to as the mandrel size) is limited to avoid damage to the bar when bending. For bars less than or equal to 16 mm diameter the internal diameter of any bend should be a minimum of 4 times the bar diameter. For larger bar sizes the limit is 7 times the bar diameter.

To give a general idea of the full anchorage lengths required for $f_{ck} = 30$ N/mm^2 and $f_{yk} = 500$ N/mm^2, with bar diameters, $\phi \leq 32$ mm, $l_{b, req}$ can vary between 25 bar diameters (25ϕ) and 52 bar diameters (52ϕ), depending on good and poor bond conditions, and the value of the coefficients α from table 5.2.

EXAMPLE 5.2

Calculations of anchorage length

Determine the anchorage length required for the top reinforcement of 25mm bars in the beam at its junction with the external column as shown in figure 5.12. The reinforcing bars are in tension resisting a hogging moment. The characteristic material strengths are $f_{ck} = 30\,\text{N/mm}^2$ and $f_{yk} = 500\,\text{N/mm}^2$.

Figure 5.12
Anchorage for a beam framing into an end column

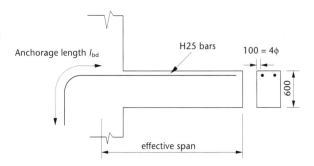

Assuming there is a construction joint in the column just above the beam and, as the bars are in the top of the beam, from figure 5.8 the bond conditions are poor and from table 5.1 the ultimate anchorage bond stress is $2.1\,\text{N/mm}^2$.

As the bars are bent into the column and the concrete cover coefficient, c_d (figure 5.9) is equivalent to 4ϕ, which is greater than 3ϕ, from table 5.2 coefficient α_1 is 0.7. Also from table 5.2, coefficient $\alpha_2 = 1 - 0.15(c_d - 3\phi)/\phi = 1 - 0.15(4\phi - 3\phi)/\phi = 0.85$.

Hence the required anchorage length is

$$l_{bd} = \alpha_1 \alpha_2 \left(\frac{f_{yk}}{4.6 f_{bd}}\right)\phi$$

$$= 0.7 \times 0.85 \left(\frac{500}{4.6 \times 2.1}\right)\phi = 31\phi$$

$$= 31 \times 25 = 775\,\text{mm}.$$

See also table A.6 in the Appendix for tabulated values of anchorage lengths.

5.3 Laps in reinforcement

Lapping of reinforcement is often necessary to transfer the forces from one bar to another. Laps between bars should be staggered and should not occur in regions of high stress. The length of the lap should be based on the minimum anchorage length modified to take into account factors such as cover, etc. The lap length l_o required is given by

$$l_o = l_{b,\text{rqd}} \times \alpha_1 \times \alpha_2 \times \alpha_3 \times \alpha_5 \times \alpha_6 \tag{5.21}*$$

where α_1, α_2, α_3, and α_5, are obtained from table 5.2. $\alpha_6 = (\rho_1/25)^{0.5}$ (with an upper and lower limit of 1.5 and 1.0 respectively) and ρ_1 is the percentage of reinforcement lapped within $0.65 l_o$ from the centre of the lap length being considered. Values of α_6 can be conveniently taken from table 5.3.

Table 5.3 Values of the coefficient α_6

Percentage of lapped bars relative to the total cross-sectional area of bars at the section being considered	<25%	33%	50%	>50%
α_6	1	1.15	1.4	1.5

Intermediate values may be interpolated from the table

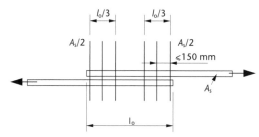

(a) tension lap

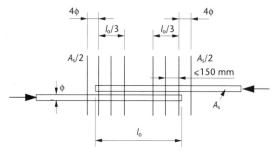

(b) compression lap

Figure 5.13
Transverse reinforcement
for lapped bars

Notwithstanding the above requirements, the absolute minimum lap length is given as

$$l_{o,\,min} = 0.3\alpha_6 l_{b,\,reqd}$$
$$\geq 15 \text{ diameters} \geq 200\,mm \tag{5.22}$$

Transverse reinforcement must be provided around laps unless the lapped bars are less than 20 mm diameter or there is less than 25 per cent lapped bars. In these cases minimum transverse reinforcement provided for other purposes such as shear links will be adequate. Otherwise transverse reinforcement must be provided, as shown in figure 5.13, having a total area of not less than the area of one spliced bar.

The arrangement of lapped bars must also conform to figure 5.14. The clear space between lapped bars should not be greater than 4ϕ or 50mm otherwise an additional lap length equal to the clear space must be provided. In the case of adjacent laps the clear distance between adjacent bars should not be greater than 2ϕ or 20 mm. The longitudinal distance between two adjacent laps should be greater than $0.3l_o$. If all these conditions are complied with then 100% of all tension bars in one layer at any section may be lapped; otherwise, where bars are in several layers, this figure should be reduced to 50%. In the case of compression steel, up to 100% of the reinforcement at a section may be lapped.

Figure 5.14
Lapping of adjacent bars

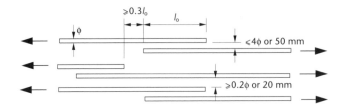

Figure 5.14
Lapping of adjacent bars

5.4 | Analysis of section subject to torsional moments

5.4.1 Development of torsional equations

Torsional moments produce shear stresses that result in principal tensile stresses inclined at approximately 45° to the longitudinal axis of the member. Diagonal cracking occurs when these tensile stresses exceed the tensile strength of the concrete. The cracks will form a spiral around the member as in figure 5.15.

Reinforcement in the form of closed links and longitudinal bars will carry forces from increasing torsional moment after cracking, by a truss action with reinforcement acting as tension members and concrete as compressive struts between links. Failure will eventually occur by reinforcement yielding, coupled with crushing of the concrete along line A–A as cracks on the other face open up. It is assumed that once the torsional shear stress on a section exceeds the value to cause cracking, tension reinforcement in the form of closed links must be provided to resist the full torsional moment.

The equations for torsional design are developed from a structural model where it is assumed that the concrete beam in torsion behaves in a similar fashion to a thin walled box section. The box is reinforced with longitudinal bars in each corner with closed loop stirrups as transverse tension ties and the concrete providing diagonal compression struts. It is assumed that the concrete cannot provide any tensile resistance.

EC2 gives the principles and some limited design equations for a generalised shape of a hollow box section. In this section of the text we will develop the equations that can be used for the design and analysis of a typical solid or hollow rectangle box section.

Consider figure 5.16a. The applied torque (T_{Ed}) at the far end of the section produces a *shear flow* (q) around the perimeter of the box section at the near end of the diagram. The shear flow is the product of the shear stress (τ) and the thickness of the hollow section. Hence from classical elastic theory the applied torque can be related to the shear flow by the expression

$$T = 2A_k q$$

where A_k is the area enclosed within the centre line of the hollow box section, hence

$$q = T/(2A_k) \tag{5.23}$$

Figure 5.15
Torsional reinforcement

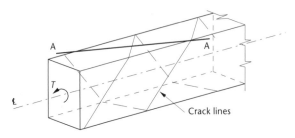

Crack lines

Figure 5.16
Structural model for torsion

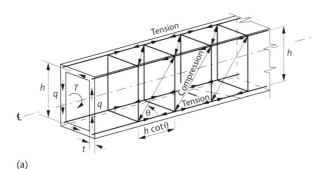

(a)

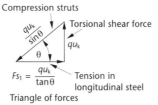

(b) Forces acting on whole body (one face shown representative of all four faces)

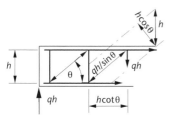

(c) Forces acting on one face of the section

As q is the shear force per unit length of the circumference of the box section, the *force* produced by the shear flow is the product of q and the circumference (u_k) of the area A_k. Hence, if it assumed that this force is resisted by the truss action of the concrete compressive struts acting at an angle, θ, together with tension in the longitudinal steel, from figure 5.16b the force (F_{s1}) in the longitudinal tension steel is given by

$$F_{s1} = \frac{qu_k \cos \theta}{\sin \theta} = \frac{qu_k}{\tan \theta} = \frac{Tu_k}{2A_k \tan \theta} \qquad (5.24)$$

The required area of longitudinal tension steel to resist torsion (A_{s1}), acting at its design strength ($f_{yk}/1.15$), is therefore given by

$$\frac{A_{s1}f_{y1k}}{1.15} = \frac{Tu_k}{2A_k \tan \theta} = \frac{Tu_k \cot \theta}{2A_k} \qquad (5.25)$$

In the above equation the torque, T, is the maximum that can be resisted by the longitudinal reinforcement and is therefore equivalent to the design ultimate torsional moment, T_{Ed}. Hence

$$\frac{A_{s1}f_{y1k}}{1.15} = \frac{T_{Ed}u_k \cot \theta}{2A_k} \qquad (5.26)^*$$

The required cross-sectional area of torsional links can be determined by considering one face of the box section as shown in figure 5.16c. If it is assumed that the area of one leg of a link (A_{sw}) is acting at its design yield strength ($f_{yk}/1.15$) the force in one link is given by

$$A_{sw}f_{yk}/1.15 = q \times h$$

However if the links are spaced at a distance s apart the force in each link is reduced proportionately and is given by

$$\frac{A_{sw}f_{yk}}{1.15} = q \times h \times \frac{s}{h\cot\theta} = \frac{qs}{\cot\theta}$$
$$= \frac{T_{Ed}s}{2A_k \cot\theta} \qquad (5.27)*$$

Equations 5.26 and 5.27 can be used to design a section to resist torsion and an example of their use is given in chapter 7. The calculated amount of reinforcement must be provided in addition to the full bending and shear reinforcement requirements for the ultimate load combinations corresponding to the torsional load case considered. Where longitudinal bending reinforcement is required the additional torsional steel area may either be provided by increasing the size of the bars, or by additional bars. Torsional links must consist of fully anchored closed links spaced longitudinally no more than $u_k/8$ apart. The longitudinal steel must consist of at least one bar in each corner of the section with other bars distributed around the inner periphery of the links at no more than 350 mm centres. Where the reinforcement is known equations 5.26 and 5.27 can be rearranged for analysis purposes to give T_{Ed} and θ as follows:

$$T_{Ed} = 2A_k \left(\frac{A_{sw}}{s} 0.87 f_{yk} \frac{A_{sl}}{u_k} 0.87 f_{ylk} \right)^{0.5} \qquad (5.28)*$$

and

$$\tan^2\theta = \left(\frac{A_{sw}}{s} f_{yk} \right) \Big/ \left(\frac{A_{sl}}{u_k} f_{ylk} \right) \qquad (5.29)*$$

The use of all the above equations assumes that the section is replaced by an equivalent hollow box section. To determine the thickness of the section an equivalent thickness (t_{ef}) is used, defined as equal to the total area of the cross-section divided by the outer circumference of the section. In the case of an actual hollow section the cross-section area would include any inner hollow areas and the calculated thickness should not be taken as greater than the actual wall thickness. In no case should the thickness be taken as less than twice the cover to the longitudinal bars.

When analysing or designing a section it is also necessary to check that excessive compressive stresses do not occur in the diagonal compressive struts, leading possibly to compressive failure of the concrete. With reference to figure 5.16c and taking the limiting torsional moment for strut compressive failure as $T_{rd,max}$:

Force in strut $= (q \times h)/\sin\theta$

Area of strut $= t_{ef} \times (h\cos\theta)$

Stress in strut $=$ Force/Area $= \dfrac{q}{t_{ef}\sin\theta\cos\theta} \le f_{ck}/1.5$

where f_{ck} is the characteristic compressive stress in the concrete. As $q = T_{Rd,max}/(2A_k)$ then the above equation can be expressed as

$$\frac{T_{Rd,max}/(2A_k)}{t_{ef}\sin\theta\cos\theta} \le f_{ck}/1.5$$

or

$$T_{Rd,max} \leq 1.33 f_{ck} t_{ef} A_k \sin\theta \cos\theta)$$

which can also be expressed as

$$T_{Rd,max} \leq 1.33 f_{ck} t_{ef} A_k / (\cot\theta + \tan\theta) \qquad (5.30)$$

In EC2 this equation is modified by the inclusion of *a strength reduction factor* (v_1) to give

$$T_{Rd,max} \leq 1.33 v_1 f_{ck} t_{ef} A_k / (\cot\theta + \tan\theta) \qquad (5.31)*$$

where the strength reduction factor takes the value of $0.6(1 - f_{ck}/250)$.

In using the above equations to design for torsion, the designer is free to choose a value of θ which will permit a reduction in link requirements balanced by a corresponding increase in longitudinal steel, as for the *Variable Strut Inclination Method* for shear design. However there are practical limitations on the values of θ that can be used and EC2 recommends that $1.0 \leq \cot\theta \leq 2.5$ representing limiting values of θ of $45°$ and $22°$ respectively.

The approach to design for torsion is therefore:

(a) Based on the calculated ultimate torsional moment (T_{Ed}), check the maximum torsional moment that can be carried by the section $(T_{Rd,max})$ which is governed by compression in the concrete struts, as given by equation 5.31:

$$T_{Ed} < T_{Rd,max} = 1.33 v_1 f_{ck} t_{ef} A_k / (\cot\theta + \tan\theta)$$

(b) Calculate the torsional reinforcement required from equation 5.27:

$$A_{sw}/s = T_{Ed} / (2A_k 0.87 f_{yk} \cot\theta)$$

where A_{sw} is the area of *one leg* of a link.

(c) Calculate the additional longitudinal reinforcement (A_{sl}) from equation 5.26:

$$A_{sl} = (T_{Ed} u_k / 2A_k) \cot\theta / (0.87 f_{ylk})$$

Further information on the practical details of design for torsion and a design example are given in chapter 7.

5.4.2 Torsion in complex shapes

A section consisting of a T, L or I shape should be divided into component rectangles and each component is then designed separately to carry a proportion of the torque (T_{Ed}). The torsion carried by each rectangle (T_i) can be determined elastically by calculating the torsional stiffness of each part according to its *St Venant* torsional stiffness from the expression

$$T_i = \frac{T_{Ed} k_i \left(h_{min}^3 h_{max} \right)_i}{\sum \left(K h_{min}^3 h_{max} \right)} \qquad (5.32)*$$

where h_{min} and h_{max} are the minimum and maximum dimension of each section. K is the *St Venant's torsional constant* that varies according to the ratio h_{max}/h_{min}; typical values of which are shown in table 5.4. The subdivision of a shape into its component rectangles should be done in order to maximise the stiffness expression $\sum \left(K h_{min}^3 h_{max} \right)$.

Table 5.4 St Venant's torsional constant K

h_{max}/h_{min}	K	h_{max}/h_{min}	K
1.0	0.14	3.0	0.26
1.2	0.17	4.0	0.28
1.5	0.20	5.0	0.29
2.0	0.23	10.0	0.31
2.5	0.25	>10	0.33

5.4.3 Torsion combined with bending and shear stresses

Torsion is seldom present alone, and in most practical cases will be combined with shear and bending stresses.

(a) Shear stresses

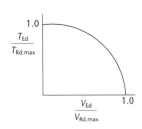

Diagonal cracking will start on the side face of a member where torsion and shear stresses are additive. Figure 5.17 shows a typical ultimate torsion and ultimate shear interaction diagram from which it can be seen that the beam's resistance to combined shear and torsion is less than that when subject to either effect alone. The effect of combined shear stress and torsional stress may therefore need to be considered and in this case both shear and torsion are calculated on the basis of the same equivalent thin walled section previously described for torsional design.

Figure 5.17
Combined shear and torsion

The recommended simplified approach to design is to ensure that the ultimate shear force (V_{Ed}) and the ultimate torsional moment (T_{Ed}) satisfy the interaction formula

$$T_{Ed}/T_{Rd,\,max} + V_{Ed}/V_{Rd,\,max} \leq 1 \qquad (5.33)^*$$

where

$T_{Rd,\,max}$ = the design torsional resistance (equation 5.31)

$V_{Rd,\,max}$ = the design shear resistance (equation 5.8)

If this interaction equation is satisfied, the design of the shear and torsional links can be carried out separately providing that the assumed angle of the compressive struts (θ) is the same for both torsional and shear design.

However for a solid rectangular section, subject to relatively small torsional and shear stresses, neither shear nor torsional reinforcement is necessary if

$$\frac{T_{Ed}}{T_{Rd,\,c}} + \frac{V_{Ed}}{V_{Rd,\,c}} \leq 1.0 \qquad (5.34)$$

where $V_{Rd,\,c}$ is the shear capacity of the concrete as given by equation 5.1. $T_{Rd,\,c}$ is the torsional cracking moment which can be calculated from equation 5.23 for a shear stress equal to the design tensile stress, f_{ctd}, of the concrete. i.e. from equation 5.23:

$$q = \frac{T}{2A_k}$$

where q = shear force per unit length = shear stress $\times$ ($t_{ef} \times 1$) or

T = shear stress $\times t_{ef} \times 2A_k$

so when the concrete reaches its design tensile cracking strength, f_{ctd}

$$T_{Rd,c} = f_{ctd} \times t_{ef} \times 2A_k$$
$$= \frac{f_{ctk}}{1.5} t_{ef} 2A_k$$
$$= 1.33 f_{ctk} t_{ef} A_k \tag{5.35}$$

It should also be noted that the calculation for A_{sw} for shear (equation 5.3) gives the required cross-sectional area of both legs of a link whereas equation 5.27 for torsion gives the required cross-sectional area of a single leg of a link. This needs to be taken into consideration when determining the total link requirement as in example 7.9. Furthermore, the additional area of longitudinal reinforcement for shear design (equation 5.12) must be provided in the tension zone of the beam, whereas the additional longitudinal reinforcement for torsion (equation 5.26) must be distributed around the inner periphery of the links.

(b) Bending stresses

When a bending moment is present, diagonal cracks will usually develop from the top of flexural cracks. The flexural cracks themselves only slightly reduce the torsional stiffness provided that the diagonal cracks do not develop. The final mode of failure will depend on the distribution and quantity of reinforcement present.

Figure 5.18 shows a typical ultimate moment and ultimate torsion interaction curve for a section. As can be seen, for moments up to approximately 80 per cent of the ultimate moment the section can resist the full ultimate torsional moment. Hence no calculations for torsion are generally necessary for the ultimate limit state of bending of reinforced concrete unless torsion has been included in the original analysis or is required for equilibrium.

When combined flexure and torsion is considered the longitudinal steel for both cases can be determined separately. In the flexural tension zone the longitudinal steel required for both cases can be added. However in the flexural compressive zone no additional torsional longitudinal steel is necessary if the longitudinal force due to torsion is less than the concrete compressive force due to flexure.

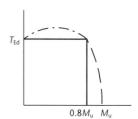

Figure 5.18
Combined bending and torsion

CHAPTER 6

Serviceability, durability and stability requirements

CHAPTER INTRODUCTION

The concept of serviceability limit states has been introduced in chapter 2, and for reinforced concrete structures these states are often satisfied by observing empirical rules which effect the detailing only. In some circumstances, however, it may be desired to estimate the behaviour of a member under working conditions, and mathematical methods of estimating deformations and cracking must be used. The design of prestressed concrete is based primarily on the avoidance or limitation of cracking and this is considered separately in chapter 11.

Where the foundations of a structure are in contact with the ground, the pressures developed will influence the amount of settlement that is likely to occur. To ensure that these movements are limited to acceptable values and are similar throughout a structure, the sizes of the foundations necessary are based on the service loads for the structure.

Consideration of durability is necessary to ensure that a structure remains serviceable throughout its lifetime. This requirement will involve aspects of design, such as concrete mix selection and

→

→

determination of cover to reinforcing bars, as well as selection of suitable materials for the exposure conditions which are expected. Good construction procedures including adequate curing are also essential if reinforced concrete is to be durable.

Simplified rules governing selection of cover, member dimensions and reinforcement detailing are given in sections 6.1 and 6.2, while more rigorous procedures for calculation of actual deflection and crack widths are described in sections 6.3 to 6.5. Durability and fire resistance are discussed in section 6.6.

The stability of a structure under accidental loading, although an ultimate limit state analysis, will usually take the form of a check to ensure that empirical rules, designed to give a minimum reasonable resistance against misuse or accident are satisfied. Like serviceability checks, this will often involve detailing of reinforcement and not affect the total quantity provided. Stability requirements are discussed in section 6.7 and considered more fully for seismic effects in section 6.8.

6.1 Detailing requirements

These requirements ensure that a structure has satisfactory durability and serviceability performance under normal circumstances. EC2 recommends simple rules concerning the concrete mix and cover to reinforcement, minimum member dimensions, and limits to reinforcement quantities, spacings and bar diameters which must be taken into account at the member sizing and reinforcement detailing stage. In some cases tabulated values are provided for typical common cases, which are based on more complex formulae given in the code of practice. Reinforcement detailing may also be affected by stability considerations as described in section 6.7, as well as rules concerning anchorage and lapping of bars which have been discussed in sections 5.2 and 5.3.

6.1.1 Minimum concrete mix and cover (exposure conditions)

These requirements are interrelated and, although not fully detailed in EC2, EN 206 *Concrete – Performance, Production, Placing and Compliance Criteria* and the complementary British Standard BS 8500 give more detailed guidance on minimum combinations of thickness of cover and mix characteristics for various classes of exposure. It should be noted that the UK national Annex to EC2 (and BS 8500) include significant modifications to EC2 itself. The mixes are expressed in terms of minimum cement content, maximum free water/cement ratio and corresponding lowest concrete strength class. Exposure classifications are given in table 6.1 which then defines the mix and cover requirements and so on which must be complied with.

Cover to reinforcement is specified and shown on drawings as a nominal value. This is obtained from

$$c_{nom} = c_{min} + \Delta c_{dev}$$

where Δc_{dev} is an allowance for construction deviations and is normally taken as 10 mm except where an approved quality control system on cover (e.g. in situ measurements) is specified – in which case it can be reduced to 5 mm.

Table 6.1 Exposure class designation

Class designation	Description	Examples of environmental conditions
XO	No risk of corrosion – Very dry	Unreinforced concrete (no freeze/thaw, abrasion or chemical attack) Reinforced concrete buildings with very low humidity
XC -1 -2 -3 -4	Carbonation-induced corrosion risk – Dry or permanently wet – Wet – rarely dry – Moderate humidity – Cyclic wet and dry	Reinforced and prestressed concrete: – inside structures (except high humidity) or permanently submerged (non-aggressive water) – completely buried in non-aggressive soil – external surfaces (including exposed to rain) – exposed to alternate wetting and drying
XD -1 -2 -3	Chloride-induced corrosion risk (not due to seawater) – Moderate humidity – Wet, rarely dry – Cyclic wet and dry	Reinforced and prestressed concrete: – exposed to airborne chlorides, bridge parts away from direct spray containing de-icing agents, occasional/slight chloride exposure – totally immersed in water containing chlorides (swimming pools, industrial waters) – exposed to de-icing salts and spray (bridges and adjacent structures, pavements, car parks)
XS -1 -2 -3	Chloride-induced corrosion risk (sea water) – Exposed to airborne salt but not in direct water contact – Permanently submerged – Tidal, splash and spray zones	Reinforced and prestressed concrete: – external in coastal areas – remaining saturated (e.g. below mid-tide level) – in upper tidal, splash and spray zones
XF -1 -2 -3 -4	Freeze/thaw attack whilst wet – Moderate water saturation – without de-icing agent – Moderate water saturation – with de-icing agent – High water saturation – without de-icing agent – High water saturation – with de-icing agent or sea water	Concrete surfaces exposed to freezing: – vertical exposed to rain – vertical (road structures) exposed to de-icing agents as spray or run-off – horizontal exposed to rain or water accumulation – horizontal exposed to de-icing agents directly or as spray or run-off. Others subject to frequent splashing
XA -1 -2 -3	Chemical attack – Slightly aggressive – Moderately aggressive – Highly aggressive	 – Defined in specialist literature

The cover is necessary to provide

1. safe transfer of bond forces;
2. adequate durability;
3. fire resistance.

The value of c_{min} should not be less than the bar diameter (or equivalent diameter of bundled bars) to ensure satisfactory bond performance. The value of c_{min} to ensure adequate durability is influenced by the exposure classification, mix characteristics and intended design life of the structure. This is a potentially complex process since there may commonly be a combination of exposure classes relating to attack of the reinforcement, whilst freeze/thaw and chemical attack apply to the concrete rather than the steel. The range of relevant mix parameters include maximum water/cement ratio, minimum cement content, cement type, aggregate size and air-entrainment.

Table 6.2 shows typical combinations of cover and mix details for commonly occurring situations. If any of the parameters of cement type, aggregate size or design life change, then adjustments will be necessary. Design should be based on the most severe exposure classification if more than one are combined. Minimum concrete mix requirements for cases where freeze/thaw apply are summarised in table 6.3, whilst exposure to chemical attack (class XA exposure) may place further limits on mix details and may also require additional protective measures. Reference should be made to the appropriate documentation in such cases (e.g. BS 8500).

Table 6.2 Cover to reinforcement (50-year design life, Portland cement concrete with 20mm maximum aggregate size) [Based on UK National Annex]

Exposure class	Nominal Cover (mm)							
XO	Not recommended for reinforced concrete							
XC1	25 ⟶							
XC2	–	35	35 ⟶					
XC3/4	–	45	40	35	35	30 ⟶		
XD1	–	–	45[1]	40[1]	40	35[1]	35	35
XD2	–	–	50[2]	45[2]	45[1]	40[2]	40	40
XD3	–	–	–	–	60[2]	55[2]	50[1]	50
XS1	–	–	–	–	50[1]	45[1]	45	40
XS2	–	–	50[2]	45[2]	45[1]	40[2]	40	40
XS3	–	–	–	–	–	60[2]	55[1]	55
Maximum free water/cement ratio	0.70	0.65	0.60	0.55	0.50	0.45	0.35	0.35
Minimum cement (kg/m^3)	240	260	280	300	320	340	360	380
Lowest concrete	C20/25	C25/30	C28/35	C32/40	C35/45	C40/50	C45/55	C50/60

Notes:

1. Cement content should be increased by 20 kg/m^3 above the values shown in the table.
2. Cement content should be increased by 40 kg/m^3 AND water–cement ratio reduced by 0.05 compared with the values shown in the table.

General Notes

These values may be reduced by 5 mm if an approved quality control system is specified.

Cover should not be less than the bar diameter + 10 mm to ensure adequate bond performance.

Table 6.3 Minimum concrete mix requirements for concrete exposed to freeze/thaw (Exposure Class XF) – 20mm aggregates

Class	Strength Class (maximum water/cement ratio)	
	No air-entrainment	3.5% air-entrainment
XF1	C25/30 (0.6)	C28/35 (0.6)
XF2	C25/30 (0.6)	C32/40 (0.55)
XF3[1]	C25/30 (0.6)	C40/50 (0.45)
XF4[1]	C28/35 (0.55)	C40/50 (0.45)

Note:
1. Freeze-thaw resisting aggregates to be specified.

6.1.2 Minimum member dimensions and cover (fire resistance)

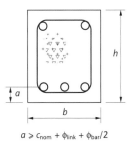

$a \geqslant c_{nom} + \phi_{link} + \phi_{bar}/2$

Figure 6.1
Definition of axis distance, a

In order that a reinforced concrete member is capable of withstanding fire for a specified period of time, it is necessary to ensure the provision of minimum dimensions and cover (here defined as nominal minimum concrete surface to main bar axis dimension as illustrated in figure 6.1) for various types of concrete member. In many practical situations with modest fire resistance periods, the cover provided for durability will govern the design.

Structural fire design in considered by part 1-2 of EC2, which gives several possible methods ranging from detailed calculations to simplified tables as presented here. Fire effects are considered further in section 6.6.2 and design is based on satisfying load bearing (R), integrity (E), and insulating (I) performance as appropriate.

The approach offers the designer permissible combinations of member dimension and axis distance as indicated in Tables 6.4 to 6.6. for a range of standard fire resistance periods (minutes). These will generally apply when normal detailing rules have been followed and when moment redistribution does not exceed 15%. For beams and slabs,

Table 6.4 Minimum dimensions and axis distance for RC beams for fire resistance

Standard fire resistance		Minimum dimensions (mm)							
		Possible combinations of a and b_{min} where a is the average axis distance and b_{min} is the width of the beam							
		Simply supported				Continuous			
		A	B	C	D	E	F	G	H
R60	$b_{min} =$	120	160	200	300	120	200		
	$a =$	40	35	30	25	25	12		
R90	$b_{min} =$	150	200	300	400	150	250		
	$a =$	55	45	40	35	35	25		
R120	$b_{min} =$	200	240	300	500	200	300	450	500
	$a =$	65	60	55	50	45	35	35	30
R240	$b_{min} =$	280	350	500	700	280	500	650	700
	$a =$	90	80	75	70	75	60	60	50

Note: The axis distance a_{sd} from the side of a beam to the corner bar should be $a + 10$ mm except where b_{min} is greater than the values in columns C and F

further detailing requirements may apply for higher fire periods, whilst effective length and axial load (relative to design capacity) may need to be specifically considered for columns.

Table 6.5 Minimum dimensions and axis distance for RC slabs for fire resistance

Standard fire resistance		Minimum dimensions (mm)						
		One-way spanning	Two-way spanning		Ribs in two-way spanning ribbed slab			
			$l_y/l_x \leq 1.5$	$1.5 < l_y/l_x \leq 2.0$				
REI 60	$h_s =$	80	80	80	$b_{min} =$	100	120	≥ 200
	$a =$	20	10	15	$a =$	25	15	10
REI 90	$h_s =$	100	100	100	$b_{min} =$	120	160	≥ 250
	$a =$	30	15	20	$a =$	35	25	15
REI 120	$h_s =$	120	120	120	$b_{min} =$	160	190	≥ 300
	$a =$	40	20	25	$a =$	45	40	30
REI 240	$h_s =$	175	175	175	$b_{min} =$	450	700	–
	$a =$	65	40	50	$a =$	70	60	–

Notes:
1. The slab thickness h_s is the sum of the slab thickness and the thickness of any non-combustible flooring
2. In two-way slabs the axis refers to the lower layer of reinforcement
3. The term 'two-way slabs' relates to slabs supported at all four edges; if this is not the case they should be treated as one-way spanning
4. For two-way ribbed slabs:
    (a) The axis distance measured to the lateral surface of the rib should be at least $(a + 10)$.
    (b) The values apply where there is predominantly uniformly distributed loading.
    (c) There should be at least one restrained edge.
    (d) The top reinforcement should be placed in the upper half of the flange.

Table 6.6 Minimum dimensions and axis distance for RC columns and walls for fire resistance

Standard fire resistance	Minimum dimensions (mm)			
	Column width b_{min}/axis distance, a, of the main bars		Wall thickness/axis distance, a, of the main bars	
	Columns exposed on more than one side	Columns exposed on one side	Wall exposed on one side	Wall exposed on two sides
R60	250/46 350/40	155/25	130/10	140/10
R90	350/53 450/40[2]	155/25	140/25	170/25
R120	350/57[2] 450/51[2]	175/35	160/35	220/35
R240	600/70	295/70	270/60	350/60

Notes:
1. Based on the ratio of the design axial load under fire conditions to the design resistance at normal temperature conditions conservatively taken as 0.7
2. Minimum of 8 bars required.

6.1.3   Maximum spacing of reinforcement

Cracking of a concrete member can result from the effect of loading or can arise because of restraint to shrinkage or thermal movement. In addition to providing a minimum area of bonded reinforcement (see section 6.1.5), cracking due to loading is minimised by ensuring that the maximum clear spacings between longitudinal reinforcing bars in beams is limited to that given in table 6.7. This will ensure that the maximum crack widths in the concrete do not exceed 0.3 mm. It can be seen that the spacing depends on the stress in the reinforcement which should be taken as the stress under the action of the *quasi-permanent* loadings. The *quasi-permanent* loading is taken as the permanent load, G_k, plus a proportion of the variable load, Q_k, depending on the type of structure. The calculation of the stress level (f_s) can be complicated and an acceptable approximation is to take f_s as

$$f_s = \frac{f_{yk}}{1.15} \times \frac{G_k + 0.3Q_k}{(1.35G_k + 1.5Q_k)} \frac{1}{\delta} \tag{6.1}$$

for office and domestic situations (see Table 2.4 for other circumstances), where f_{yk} is the characteristic strength of the reinforcement. δ will have a value of 1.0 unless moment redistribution has been carried out, in which case δ is the ratio of the distributed moment to the undistributed moment at the section at the ultimate limit.

Table 6.7 Maximum clear bar spacings (mm) for high bond bars in tension caused predominantly by loading

Steel stress (N/mm²)	Maximum bar spacing (mm)
160	300
200	250
240	200
280	150
320	100
360	50

These spacing rules do not apply to slabs with an overall thickness of 200 mm or less. In this case the spacing of longitudinal reinforcement should be no greater than three times the overall slab depth or 400 mm, whichever is the lesser, and secondary reinforcement three-and-a-half times the depth or 450 mm generally. In areas of concentrated loads or maximum moments these should be reduced to $2h < 250$ mm and $3h < 400$ mm respectively.

6.1.4   Minimum spacing of reinforcement

To permit concrete flow around reinforcement during construction, the clear distance between bars should not be less than (i) the maximum bar size, (ii) 20 mm, or (iii) the maximum aggregate size plus 5 mm, whichever is the greater figure.

6.1.5 Minimum areas of reinforcement

For most purposes, thermal and shrinkage cracking can be controlled within acceptable limits by the use of minimum reinforcement quantities. The principal requirements, to be checked at the detailing stage, are as specified in table 6.8. Requirements for shear links are given in sections 7.3 and 9.3.

In addition to the requirements of table 6.8 a minimum steel area, $A_{s,\,min}$, must be provided in all cases to control cracking. The provision of the minimum steel area ensures that the reinforcement does not yield when the concrete in the tension zone cracks with a sudden transfer of stress to the reinforcement. This could cause the uncontrolled development of a few wide cracks. Whenever this minimum area is provided, then yield should not occur and cracking will then be distributed throughout the section with a greater number of cracks but of lesser width. $A_{s,\,min}$ is given by the expression

$$A_{s,\,min} = k_c k f_{ct,\,eff} A_{ct}/f_{yk} \tag{6.2}$$

where

$A_{s,\,min}$ = minimum area of reinforcement that must be provided within the tensile zone

A_{ct} = area of concrete within tensile zone – defined as that area which is in tension just before the initiation of the first crack

$f_{ct,\,eff}$ = tensile strength of concrete at time of cracking with a suggested minimum of 3 N/mm^2 – otherwise obtained from table 6.11 using a concrete strength class appropriate to the anticipated time of cracking

k_c = stress distribution coefficient (1.0 for pure tension, 0.4 for flexure)

k = non-linear stress distribution coefficient – leading to a reduction in restraint force

= 1.0 for webs less than 300 mm deep or flanges less than 300 mm wide

= 0.65 for webs greater than 800 mm deep or flanges greater than 800 mm wide

(interpolate for intermediate values)

Table 6.8 Minimum areas of reinforcement

Tension reinforcement in beams and slabs	Concrete class ($f_{yk} = 500\,N/mm^2$)			
	C25/30	C30/35	C40/50	C50/60
$\dfrac{A_{s,\,min}}{b_t d} > 0.26\dfrac{f_{ctm}}{f_{yk}}$ (> 0.0013)	0.0013	0.0015	0.0018	0.0021

Secondary reinforcement $> 20\%$ main reinforcement

Longitudinal reinforcement in columns
$A_{s,\,min} > 0.10 N_{sd}/0.87 f_{yk} > 0.002 A_c$ where N_{sd} is the axial compression force

Vertical reinforcement in walls
$A_{s,\,min} > 0.002 A_c$

Note: b_t is the mean width of the tension zone.

6.1.6 Maximum areas of reinforcement

These are determined largely from the practical need to achieve adequate compaction of the concrete around the reinforcement. The limits specified are as follows

(a) For a slab or beam, tension or compression reinforcement

$100A_s/A_c \leq 4$ per cent other than at laps

(b) For a column

$100A_s/A_c \leq 4$ per cent other than at laps and 8 per cent at laps

(c) For a wall, vertical reinforcement

$100A_s/A_c \leq 4$ per cent

6.1.7 Maximum bar size

Section 6.1.3 described the limitations on bar spacing to ensure that crack widths due to loading are kept within acceptable limits. When considering load-induced cracking bar diameters may be restricted as indicated in table 6.9 which is based on C30/37 concrete and 25 mm cover as an alternative to limiting spacing. In calculating the steel stress, the approximation given in equation 6.1 may be used.

Table 6.9 Maximum bar diameters (0.3 mm crack width)

Steel stress (N/mm^2)	Maximum bar size (mm)
160	32
200	25
240	16
280	12
320	10
360	8
400	6
450	5

When cracking occurs as a result of restraint to shrinkage or thermal effects then the bar sizes *must* be limited as indicated in table 6.9, but the maximum spacing limits of table 6.7 do not need to be applied. The steel stress to be used in table 6.9 can be calculated from equation 6.3 where $A_{s,\,prov}$ is the steel area provided at the section under consideration and $A_{s,\,min}$ is given in equation 6.2.

$$f_s = f_{yk}A_{s,\,min}/A_{s,\,prov} \tag{6.3}$$

6.1.8 Side face and surface reinforcement in beams

In beams over 1 m deep additional reinforcement must be provided in the side faces to control cracking as indicated in figure 6.2(a). This reinforcement should be distributed evenly between the main tension steel and the neutral axis and *within* the stirrups. The

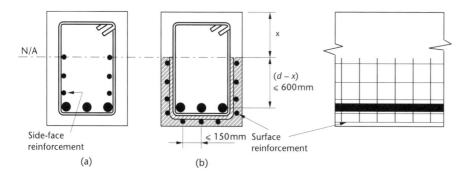

Figure 6.2
Side-face and surface
reinforcement

minimum area of this reinforcement can be calculated from equation 6.2 with k taken as 0.5. In assessing the maximum spacing and size of this reinforcement from tables 6.7 and 6.9 a stress value equal to one-half of that calculated for the main tensile reinforcement may be used and it may be assumed that the side face reinforcement is in pure tension.

In addition to the above requirement, EC2 requires that surface reinforcement is provided where it is necessary to control spalling of the concrete due to fire (axis distance > 70 mm) or where bundled bars or bars greater than 40 mm diameter are used as main reinforcement. In the UK, however, this is not adopted due to practical difficulties in providing such reinforcement. For high covers it is recommended that additional fire protection is provided and crack width calculations are recommended with large diameter bars.

The surface reinforcement, if provided, should consist of welded mesh or small diameter high bond bars located *outside* the links as indicated in figure 6.2(b). Cover to this reinforcement must comply with the requirements of section 6.1.1 and the minimum area of longitudinal surface reinforcement should be 1 per cent of the area of the concrete outside the links and in the tension zone below the neutral axis; shown as the shaded area in figure 6.2(b). The surface reinforcement bars should be spaced no further than 150 mm apart and if properly anchored can be taken into account as longitudinal bending and shear reinforcement.

6.2 | Span–effective depth ratios

The appearance and function of a reinforced concrete beam or slab may be impaired if the deflection under serviceability loading is excessive. Deflections can be calculated as indicated in section 6.3 but it is more usual to control deflections by placing a limit on the ratio of the span to the effective depth of the beam or slab. EC2 specifies equations to calculate basic span–effective depth ratios, to control deflections to a maximum of span/250. Some typical values are given in table 6.10 for rectangular sections of class C30/35 concrete and for grade 500 steel. The ratios can also be used for flanged sections except where the ratio of the width of flange to the width of web exceeds 3 when the basic values should be multiplied by 0.8. For two-way spanning slabs, the check for the basic span–effective depth ratio should be based on the shorter span whereas for flat slabs calculations should be based on the longer span.

The two columns given in table 6.10 correspond to levels of concrete stress under serviceability conditions: highly stressed when the steel ratio ρ exceeds 1.5 per cent and

Table 6.10 Basic span–effective depth ratios ($f_{yk} = 500\,\text{N/mm}^2$, C30/35 concrete)

Structural system	Factor for structural system K	Basic span–effective depth ratio	
		Concrete highly stressed ($\rho = 1.5\%$)	Concrete lightly stressed ($\rho = 0.5\%$)
1. Simply supported beam or one/two-way spanning simply supported slab	1.0	14	20
2. End span of continuous beam or one-way continuous slab or two-way slab continuous over one long side	1.3	18	26
3. Interior span of continuous beam or one-way or two-way spanning slab	1.5	20	30
4. Slab on columns without beams (flat slab) based on longer span	1.2	17	24
5. Cantilever	0.4	6	8

lightly stressed when ρ equals 0.5 per cent. ρ is given by $100A_{s,req}/bd$ where $A_{s,req}$ is the area of tension reinforcement required in the section. Interpolation between the values of ρ indicated is permissible. In the case of slabs it is reasonable to assume that they are lightly stressed.

Since the value of allowable span–effective depth ratio is affected by both reinforcement ratio and concrete strength it may be more convenient to use the chart in figure 6.3 which is for a simply supported span with no compression steel together with a modification factor K (as shown in table 6.10) according to member type. This approach is based on the same basic equations and offers greater flexibility than reliance placed on tabulated values.

Figure 6.3
Graph of basic span–effective depth ratios for different classes of concrete

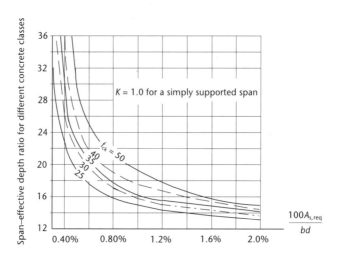

The basic ratios are modified in particular cases as follows:

(a) For spans longer than 7 m (except flat slabs) and where it is necessary to limit deflections to ensure that finishes, such as partitions, are not damaged, the basic values should be multiplied by 7/span.

(b) For flat slabs with spans in excess of 8.5 m, similarly multiply the basic ratios by 8.5/span.

(c) For characteristic steel strengths other than 500 N/mm^2, multiply the basic ratios by $500/f_{yk}$.

(d) Where more tension reinforcement is provided ($A_{s,prov}$) than that calculated ($A_{s,req}$) at the ultimate limit state, multiply the basic ratios by $A_{s,prov}/A_{s,req}$ (upper limit = 1.5).

These basic ratios assume a steel working stress of $f_s = 310$ N/mm^2 where $f_{yk} = 500$ N/mm^2

EXAMPLE 6.1

Span–effective depth ratio

A rectangular continuous beam of class C25/30 concrete spans 10 m. If the breadth is 300 mm, check the acceptability of an effective depth of 600 mm when high yield reinforcement, $f_{yk} = 500$ N/mm^2, is used. At the ultimate limit state it is determined that 1250 mm^2 of tension steel is needed and 3 No. 25 mm diameter reinforcing bars ($A_{s,prov} = 1470$ mm^2) are actually provided in an interior span.

$$\rho = 100A_{s,req}/bd$$
$$= (100 \times 1250)/(300 \times 600)$$
$$= 0.7 \text{ per cent.}$$

From table 6.10, for an interior span $K = 1.5$

Basic span–effective depth ratio (figure 6.3) = 16

Therefore for an interior span, basic span–effective depth ratio = $1.5 \times 16 = 24$

To avoid damage to finishes for span greater than 7 m:

$$\text{Modified ratio} = 24 \times \frac{7}{10} = 16.8$$

Modification for steel area provided:

$$\text{Modified ratio} = 16.8 \times \frac{1470}{1250} = 19.8$$

$$\text{Span–effective depth ratio provided} = \frac{10 \times 10^3}{600} = 16.7$$

which is less than the allowable upper limit, thus deflection requirements are likely to be satisfied.

6.3 | Calculation of deflection

The general requirement is that neither the efficiency nor the appearance of a structure is harmed by the deflections that will occur during its life. Deflections must thus be considered at various stages. The limitations necessary to satisfy the requirements will vary considerably according to the nature of the structure and its loadings, but for reinforced concrete the following may be considered as reasonable guides:

1. the final deflection of a beam, slab or cantilever should not exceed span/250
2. that part of the deflection which takes place after the application of finishes or fixing of partitions should not exceed span/500 to avoid damage to fixtures and fittings.

The code suggests that deflections should be calculated under the action of the *quasi-permanent* load combination, assuming this loading to be of long-term duration. Hence the total loading to be taken in the calculation will be the permanent load plus a proportion of the variable load which will typically be 30 per cent of the variable load for office-type construction. This is a reasonable assumption as deflection will be affected by long-term effects such as concrete creep, while not all of the variable load is likely to be long-term and hence will not contribute to the creep effects.

Lateral deflection must not be ignored, especially on tall slender structures, and limitations in these cases must be judged by the engineer. It is important to realise that there are many factors which may have significant effects on deflections, and are difficult to allow for. Thus any calculated values must be regarded as an estimate only. The most important of these factors are:

1. support restraint must be estimated on the basis of simplified assumptions, which will have varying degrees of accuracy;
2. the precise loading cannot be predicted and errors in permanent loading may have a significant effect;
3. a cracked member will behave differently from one that is uncracked – this may be a problem in lightly reinforced members where the working load may be near to the cracking limit;
4. the effects of floor screeds, finishes and partitions are very difficult to assess – frequently these are neglected despite their 'stiffening' effect.

It may be possible to allow for these factors by averaging maximum and minimum estimated effects and, provided that this is done, there are a number of calculation methods available which will give reasonable results. The method adopted by EC2 is based on the calculation of curvature of sections subjected to the appropriate moments with allowance for creep and shrinkage effects where necessary. Deflections are then calculated from these curvatures. A rigorous approach to deflection is to calculate the curvature at intervals along the span and then use numerical integration techniques to estimate the critical deflections, taking into account the fact that some sections along the span will be cracked under load and others, in regions of lesser moment, will be uncracked. Such an approach is rarely justified and the approach adopted below, based on EC2, assumes that it is acceptably accurate to calculate the curvature of the beam or slab based on both the cracked and uncracked sections and then to use an 'average' value in estimating the final deflection using standard deflection formulae or simple numerical integration based on elastic theory.

The procedure for estimating deflections involves the following stages which are illustrated in example 6.2.

6.3.1 Calculation of curvature

Curvature under the action of the quasi-permanent load should be calculated based on both the *cracked* and *uncracked* sections. An estimate of an 'average' value of curvature can then be obtained using the formula:

$$1/r = \xi(1/r)_{cr} + (1-\xi)(1/r)_{uc} \tag{6.4}$$

where

$1/r =$ average curvature

$\left.\begin{array}{c}(1/r)_{uc}\\(1/r)_{cr}\end{array}\right\} =$ values of curvature calculated for the uncracked case and cracked case respectively

$\xi =$ coefficient given by $1 - \beta(\sigma_{sr}/\sigma_s)^2$ allowing for tension stiffening

$\beta =$ load duration factor (1 for a single short-term load; 0.5 for sustained loads or cyclic loading)

$\sigma_s =$ stress in the tension steel for the cracked concrete section

$\sigma_{sr} =$ stress in the tension steel calculated on the basis of a cracked section *under the loading that will just cause cracking at the section being considered.*

Appropriate values of concrete tensile strength to be used in the calculation of σ_{sr} can be obtained from table 6.11. In calculating ξ, the ratio (σ_{sr}/σ_s) can more conveniently be replaced by (M_{cr}/M) where M_{cr} is the moment that will just cause cracking of the section and M is the design moment for the calculation of curvature and deflection.

In order to calculate the 'average' curvature, separate calculations have to be carried out for both the cracked and uncracked cases.

Uncracked section

The assumed elastic strain and stress distribution for an uncracked section is shown in figure 6.4.

For a given moment, M, and from elastic bending theory, the curvature of the section, $(1/r)_{uc}$, is given by

$$(1/r)_{uc} = \frac{M}{E_{c,eff}I_{uc}} \tag{6.5}$$

where $E_{c,eff}$ is the effective elastic modulus of the concrete allowing for creep effects and I_{uc} is the second moment of area of the uncracked concrete section.

Table 6.11 Mean tensile strengths of concrete and secant modulus of elasticity

	Strength class						
	20/25	C25/30	C30/37	C35/45	C40/50	C45/55	C50/60
f_{ctm} (N/mm^2)	2.2	2.6	2.9	3.2	3.5	3.8	4.1
E_{cm} (kN/mm^2)	30	31	33	34	35	36	37

Figure 6.4
Uncracked section – strain and
stress distribution

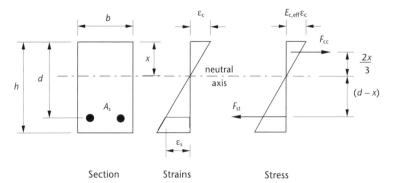

Figure 6.5
Cracked section – strain and
stress distribution

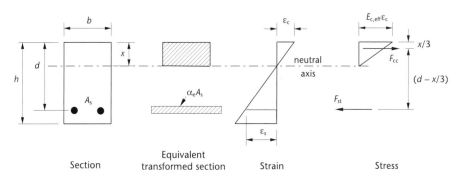

Cracked section

The assumed elastic strain and stress distribution for a cracked section is shown in figure 6.5. This is identical to that shown in figures 4.27 and 4.28, and equation 4.48 or figure 4.29 can be used to determine the neutral-axis depth. Alternatively, moments of area can be taken to establish the neutral-axis depth directly. The second moment of area of the cracked section can then be determined by taking second moments of area about the neutral axis

$$I_{cr} = bx^3/3 + a_e A_s (d - x)^2 \tag{6.6}$$

where a_e is the modular ratio equal to the ratio of the elastic modulus of the reinforcement to that of the concrete.

For a given moment, M, and from elastic bending theory, the curvature of the cracked section, $(1/r)_{cr}$, is therefore given by

$$(1/r)_{cr} = \frac{M}{E_{c,\,eff} I_{cr}} \tag{6.7}$$

6.3.2 Creep and shrinkage effects

Creep

The effect of creep will be to increase deflections with time and thus should be allowed for in the calculations by using an effective modulus, $E_{c,\,eff}$, using the equation

$$E_{c,\,eff} = E_{cm}/(1 + \phi(\infty, t_0)) \tag{6.8}$$

where ϕ is a creep coefficient equal to the ratio of creep strain to initial elastic strain.

Table 6.12 Final creep coefficient of normal weight concrete (Class C25/30)

Age at loading (days)	Notional size ($2A_C/u$) mm							
	100	200	300	500	100	200	300	500
	Dry atmosphere (inside: 50% RH)				Humid atmosphere (outside: 80% RH)			
1	5.5	5.0	4.7	4.3	3.8	3.5	3.4	3.3
3	4.6	4.0	3.8	3.6	3.1	2.9	2.8	2.8
7	3.8	3.5	3.2	2.9	2.6	2.4	2.3	2.2
28	3.0	2.8	2.6	2.3	2.1	2.0	1.9	1.9
100+	2.7	2.5	2.3	2.1	1.9	1.8	1.7	1.6

Note: A_c = cross-sectional area of concrete, u = perimeter of that area exposed to drying.

The value of ϕ, while being affected by aggregate properties, mix design and curing conditions, is also governed by age at first loading, the duration of load and the section dimensions.

Table 6.12 gives some typical long-term values of $\phi(\infty, t_0)$ as suggested by EC2 for a class C25/30 concrete made with a type N cement. These are valid if the concrete is not subjected to a compressive stress greater than $0.45f_{ck(t_0)}$ at age t_0 (age at time of loading) and will reduce as the concrete strength increases. Equations and charts are given in EC2 for a range of cement types, concrete classes, loading ages and notional member sizes. These equations include the development of creep with time and adjustments if the stress at loading exceeds that indicated above. The 'notional size' of the section is taken as twice the cross-sectional area divided by the perimeter of the area exposed to drying. An estimate of the elastic modulus of concrete, E_{cm}, can be obtained from table 6.11 or from the expression:

$$E_{cm} = 22 \left[\frac{(f_{ck} + 8)}{10} \right]^{0.3} \text{kN/mm}^2$$

Shrinkage

The effect of shrinkage of the concrete will be to increase the curvature and hence the deflection of the beam or slab. The curvature due to shrinkage can be calculated using the equation

$$1/r_{cs} = \varepsilon_{cs} a_e S / I \tag{6.9}$$

where

$1/r_{cs}$ = the shrinkage curvature

ε_{cs} = free shrinkage strain

S = first moment of area of reinforcement about the centroid of the section

I = second moment of area of section (cracked or uncracked as appropriate)

a_e = effective modular ratio ($E_s/E_{c,\,eff}$)

Shrinkage is influenced by many features of the mix and construction procedures but for most normal weight concretes values of ε_{cs} may be obtained from table 6.13. Shrinkage strains are affected by the ambient humidity and element dimensions. The total shrinkage strain can be considered as two components; the drying shrinkage

Table 6.13   Final shrinkage strains of normal weight concrete (10^{-6}) (Class C25/30)

Location of member	Relative humidity (%)	Notional size ($2A_c/u$) mm			
		100	200	300	≥ 500
Inside	50	550	470	410	390
Outside	80	330	280	250	230

Note: A_c = cross-sectional area of concrete, u = perimeter of that area exposed to drying.

strain ε_{cd} which develops slowly as water migrates through the hardened concrete and the autogenous shrinkage strain, ε_{ca} which develops during hardening at early ages. Thus:

$$\varepsilon_{cs} = \varepsilon_{cd} + \varepsilon_{ca}$$

EC2 provides formulae to evaluate these components at various ages of the concrete from which the typical long-term values in table 6.13 have been developed for a class C25/30 concrete. The total shrinkage will tend to be far less for higher strengths especially at lower relative humidities.

The 'average' shrinkage curvature can be calculated from equation 6.4 having calculated the curvature based on both the 'cracked' and 'uncracked' sections.

6.3.3   Calculation of deflection from curvature

The total curvature can be determined by adding the shrinkage curvature to the calculated curvature due to the quasi-permanent loads, having made allowance for creep effects.

The deflection of the beam or slab can be calculated from the total curvature using elastic bending theory which for small deflections is based on the expression

$$M_x = EI \frac{d^2y}{dx^2} \tag{6.10}$$

where M_x is the bending moment at a section distance x from the origin as shown in figure 6.6. For small deflections the term d^2y/dx^2 approximately equals the curvature which is the reciprocal of the radius of curvature. Double integration of equation 6.10 will yield an expression for deflection. This may be illustrated by considering the case of a pin-ended beam subjected to constant M throughout its length, so that $M_x = M$.

$$EI \frac{d^2y}{dx^2} = M \tag{6.11}$$

therefore

$$EI \frac{dy}{dx} = Mx + C$$

Figure 6.6
Pin-ended beam subject to a constant moment M

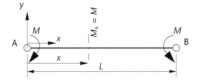

but if the slope is zero at mid-span where $x = L/2$, then

$$C = -\frac{ML}{2}$$

and

$$EI\frac{dy}{dx} = Mx - \frac{ML}{2}$$

Integrating again gives

$$EIy = \frac{Mx^2}{2} - \frac{MLx}{2} + D$$

but at support A when $x = 0$, $y = 0$. Hence

$$D = 0$$

thus

$$y = \frac{M}{EI}\left(\frac{x^2}{2} - \frac{Lx}{2}\right) \quad \text{at any section} \tag{6.12}$$

The maximum deflection in this case will occur at mid-span, where $x = L/2$, in which case

$$y_{max} = -\frac{M}{EI}\frac{L^2}{8} \tag{6.13}$$

but since at any uncracked section

$$\frac{M}{EI} = \frac{1}{r}$$

the maximum deflection may be expressed as

$$y_{max} = -\frac{1}{8}L^2\frac{1}{r}$$

In general, the bending-moment distribution along a member will not be constant, but will be a function of x. The basic form of the result will however be the same, and the deflection may be expressed as

$$\text{maximum deflection } a = kL^2\frac{1}{r_b} \tag{6.14}*$$

where

$k =$ a constant, the value of which depends on the distribution of bending moments in the member

$L =$ the effective span

$\dfrac{1}{r_b} =$ the mid-span curvature for beams, or the support curvature for cantilevers

Typical values of k are given in table 6.14 for various common shapes of bending-moment diagrams. If the loading is complex, then a value of k must be estimated for the complete load since summing deflections of simpler components will yield incorrect results.

Table 6.14 Typical deflection coefficients

Loading	B.M. diagram	k
$M \quad M$	(rectangular)	0.125
W, aL	$WaL(1-a)$	$\dfrac{4a^2 - 8a + 1}{48a}$ (if $a = 0.5$ then $k = 0.83$)
w	$wL^2/8$	0.104
aL, W	$-WaL$	End deflection $= \dfrac{a(3-a)}{6}$ (if $a = 1$ then $k = 0.33$)
aL, w	$-wa^2L^2/2$	End deflection $= \dfrac{a(4-a)}{12}$ (if $a = 1$ then $k = 0.25$)

Although the derivation has been on the basis of an uncracked section, the final expression is in a form that will deal with a cracked section simply by the substitution of the appropriate curvature.

Since the expression involves the square of the span, it is important that the true effective span as defined in chapter 7 is used, particularly in the case of cantilevers. Deflections of cantilevers may also be increased by rotation of the supporting member, and this must be taken into account when the supporting structure is fairly flexible.

EXAMPLE 6.2

Calculation of deflection

Estimate the long-term deflection of the beam shown in figure 6.7. It spans 9.5 metres and is designed to carry a uniformly distributed load giving rise to a quasi-permanent moment of 200 kNm. It is constructed with class C25/30 concrete, is made of normal aggregates and the construction props are removed at 28 days.

(a) Calculate curvature due to uncracked section

From equation 6.5:

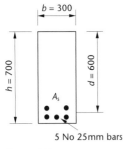

Figure 6.7
Deflection calculation example

$$(1/r)_{uc} = \frac{M}{E_{c,eff} I_{uc}}$$

where from table 6.11, $E_{cm} = 31 \, \text{kN/mm}^2$. From table 6.12, assuming loading at 28 days with indoor exposure, the creep coefficient $\phi \approx 2.8$ because

$$2A_c/u = (2 \times [700 \times 300])/2000 = 210$$

and hence from equation 6.8 the effective modulus is given by

$$E_{c,eff} = 31/(1 + 2.8) = 8.15 \, \text{kN/mm}^2$$

Hence

$$(1/r)_{uc} = \frac{200 \times 10^6}{8.15 \times 10^3 \times (300 \times 700^3/12)}$$

$$= 2.86 \times 10^{-6}/\text{mm}$$

Note that in the above calculation I_{uc} has been calculated on the basis of the gross concrete sectional area ignoring the contribution of the reinforcement. A more accurate calculation could have been performed, as in example 4.13 in chapter 4, but such accuracy is not justified and the simpler approach indicated will be sufficiently accurate.

(b) Calculate curvature due to cracked section

To calculate the curvature of the cracked section the I value of the transformed concrete section must be calculated. With reference to figure 6.5 the calculations can be set out as below.

(i) Calculate the neutral axis position

Taking area moments about the neutral axis:

$$b \times x \times x/2 = a_e A_s (d - x)$$

$$300 \times x^3/2 = \frac{200}{8.15} \times 2450(600 - x)$$

which has the solution

$$x = 329 \, \text{mm}$$

(ii) Calculate the second moment of area of the cracked section

$$I_{cr} = bx^3/3 + a_e A_s (d - x)^2$$

$$= \frac{300 \times 329^3}{3} + \frac{200}{8.15} 2450(600 - 329)^2$$

$$= 7976 \times 10^6 \, \text{mm}^4$$

(iii) Calculate the curvature of the cracked section

From equation 6.7

$$(1/r)_{cr} = \frac{M}{E_{c,\text{eff}} I_{cr}}$$

$$= \frac{200 \times 10^6}{8.15 \times 10^3 \times 7976 \times 10^6}$$

$$= 3.08 \times 10^{-6}/\text{mm}$$

(c) Calculate the 'average' of the cracked and uncracked curvature

From equation 6.4:

$$1/r = \xi(1/r)_{cr} + (1 - \xi)(1/r)_{uc}$$

where

$$\xi = 1 - \beta(\sigma_{sr}/\sigma_s)^2$$

$$= 1 - \beta(M_{cr}/M)^2$$

(i) Calculate M_{cr}

From table 6.11 the cracking strength of the concrete, f_{ctm}, is given as 2.6 N/mm^2. Hence from elastic bending theory and considering the uncracked concrete section, the moment that will just cause cracking of the section, M_{cr}, is given by

$$M_{cr} = f_{ctm} \times (b_w h^2/6)$$
$$= 2.6 \times (300 \times 700^2/6) \times 10^{-6}$$
$$= 63.7 \, \text{kN m}$$

(ii) Calculate ξ

$$\xi = 1 - \beta(M_{cr}/M)^2$$
$$= 1 - 0.5 \times (63.7/200)^2$$
$$= 0.95$$

(iii) Calculate the 'average' curvature

$$1/r = \xi(1/r)_{cr} + (1 - \xi)(1/r)_{uc}$$
$$= 0.95 \times 3.08 \times 10^{-6} + (1 - 0.95) \times 2.86 \times 10^{-6}$$
$$= 3.07 \times 10^{-6} /\text{mm}$$

(d) Calculate shrinkage curvature

(i) For the cracked section

$$1/r_{cs} = \varepsilon_{cs} a_e S/I_{cr}$$

where

$$S = A_s(d - x)$$
$$= 2450(600 - 329)$$
$$= 664 \times 10^3 \, \text{mm}^3$$

and from table 6.13, $\varepsilon_{cs} \approx 470 \times 10^{-6}$ (because $2A_c/u = 210$, as in part (a)).
Therefore

$$1/r_{cs} = \frac{470 \times 10^{-6}(200/8.15)664 \times 10^3}{7976 \times 10^6}$$
$$= 0.96 \times 10^{-6}/\text{mm}$$

(ii) For the uncracked section

$$1/r_{cs} = \varepsilon_{cs} a_e S/I_{uc}$$

where

$$S = A_s(d - x)$$
$$= 2450(600 - 700/2)$$
$$= 612.5 \times 10^3 \, \text{mm}^3$$

Therefore

$$1/r_{cs} = \frac{470 \times 10^{-6}(200/8.15)612.5 \times 10^3}{300 \times 700^3/12}$$
$$= 0.82 \times 10^{-6} /\text{mm}$$

(iii) Calculate the 'average' shrinkage curvature

$$1/r_{cs} = \xi(1/r)_{cr} + (1 - \xi)(1/r)_{uc}$$
$$= 0.95 \times 0.96 \times 10^{-6} + (1 - 0.95) \times 0.82 \times 10^{-6}$$
$$= 0.95 \times 10^{-6} \, /\text{mm}$$

(e) Calculate deflection

$$\text{Curvature due to loading} = 3.07 \times 10^{-6} \, /\text{mm}$$
$$\text{Curvature due to shrinkage} = 0.95 \times 10^{-6} \, /\text{mm}$$
$$\text{Therefore total curvature} = 4.02 \times 10^{-6} \, /\text{mm}$$

For a simply supported span subjected to a uniformly distributed load, the maximum mid-span deflection is given by

$$\text{Deflection} = 0.104L^2(1/r)$$
$$= 0.104 \times 9500^2 \times 4.02 \times 10^{-6}$$
$$= 37.8 \, \text{mm}$$

This value almost exactly matches the allowable value of span/250 (9500/250 = 38 mm) and would be considered acceptable noting the inherent uncertainty of some of the parameters used in the calculations.

6.3.4 Basis of span–effective depth ratios

The calculation of deflection has been shown to be a tedious operation. However, for general use, rules based on limiting the span–effective depth ratio of a member are adequate to ensure that the deflections are not excessive. The application of this method was described in section 6.2.

The relationship between the deflection and the span–effective depth ratio of a member can be derived from equation 6.14; thus

$$\text{deflection } a = k\frac{1}{r_b}L^2$$

and for small deflections it can be seen from figure 6.8 that for unit length, s

$$\phi = \frac{1}{r_b} = \frac{\varepsilon_{cm} + \varepsilon_{rm}}{d}$$

where

$$\varepsilon_{c,max} = \text{maximum compressive strain in the concrete}$$
$$\varepsilon_{rm} = \text{tensile strain in the reinforcement}$$
$$k = \text{a factor which depends on the pattern of loading.}$$

therefore

$$\frac{\text{span}}{\text{effective depth}} = \frac{L}{d}$$
$$= \frac{a}{L}\frac{1}{k}\frac{1}{(\varepsilon_{c,max} + \varepsilon_{rm})}$$

Figure 6.8
Curvature and strain
distribution

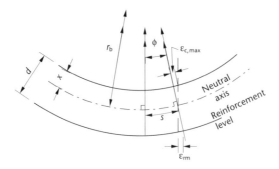

The strains in the concrete and tensile reinforcement depend on the areas of reinforcement provided and their stresses. Thus for a particular member section and a pattern of loading, it is possible to determine a span–effective depth ratio to satisfy a particular a/L or deflection/span limitation.

The span–effective depth ratios obtained in section 6.2 are based on limiting the total deflection to span/250 for a uniformly distributed loading and are presented for different stress levels depending on whether the concrete is highly or lightly stressed. This in turn depends on the percentage of tension reinforcement in the section. For spans of less than 7 m this should also ensure that the limits of span/500 after application of finishes are met but, for spans over 7 m where avoidance of damage to finishes may be important, the basic ratios of table 6.10 should be factored by 7/span.

For loading patterns that are not uniformly distributed a revised ratio is given by changing the basic ratio in proportion to the relative values of k, as shown in example 6.3. Similarly, for limiting the deflection to span/β

$$\text{revised ratio} = \text{basic ratio} \times \frac{250}{\beta}$$

In cases where the basic ratio has been modified for spans greater than 7 m, maximum deflections are unlikely to exceed span/500 after construction of partitions and finishes.

When another deflection limit is required, the ratios given should be multiplied by $500/a$ where a is the proposed maximum deflection.

EXAMPLE 6.3

Adjustment of basic span to effective depth ratio

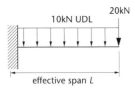

Figure 6.9
Point load on a cantilever
example

Determine the appropriate basic ratio for a cantilever beam supporting a uniform load and a concentrated point load at its tip as shown in figure 6.9. Assume that the concrete is C30/35 and is highly stressed.

Basic ratio from table 6.10 = 6.0 for u.d.l.

From table 6.14:

k for cantilever with u.d.l. over full length = 0.25

k for cantilever with point load at tip = 0.33

Thus, for the point load only, adjusted basic ratio equals

$$6.0 \times \frac{0.25}{0.33} = 4.5$$

An adjusted basic ratio to account for both loads can be obtained by factoring the moment due to the point load by the ratio of the k values as follows

$$M_{udl} = 10 \times L/2 = 5L$$

$$M_{point} = 20L$$

$$\text{Adjusted basic ratio} = \text{Basic ratio}\left(\frac{M_{udl} + M_{point} \times k_{udl}/k_{point}}{M_{udl} + M_{point}}\right)$$

$$= 6.0\left(\frac{5 + 20 \times 0.25/0.33}{5 + 20}\right)$$

$$= 4.8$$

Thus it can be seen that the effect of the point load dominates.

6.4 Flexural cracking

Members subject to bending generally exhibit a series of distributed flexural cracks, even at working loads. These cracks are unobtrusive and harmless unless the width becomes excessive, in which case appearance and durability suffer as the reinforcement is exposed to corrosion.

The actual widths of cracks in a reinforced concrete structure will vary between wide limits and cannot be precisely estimated, thus the limiting requirement to be satisfied is that the probability of the maximum width exceeding a satisfactory value is small. The maximum acceptable value suggested by EC2 is 0.3 mm for all exposure classes under the action of the *quasi-permanent* combination of loads. Other codes of practice may recommend lower values of crack widths for important members and requirements for special cases, such as water-retaining structures, may be even more stringent.

Flexural cracking is generally controlled by providing a minimum area of tension reinforcement (section 6.1.5) and limiting bar spacings (section 6.1.3) or limiting bar sizes (section 6.1.7). If calculations to estimate maximum crack widths are performed, they are based on the *quasi-permanent* combination of loads and an effective modulus of elasticity of the concrete should be used to allow for creep effects.

6.4.1 Mechanism of flexural cracking

This can be illustrated by considering the behaviour of a member subject to a uniform moment.

A length of beam as shown in figure 6.10 will initially behave elastically throughout, as the applied moment M is increased. When the limiting tensile strain for the concrete is reached, a crack will form and the adjacent tensile zone will no longer be acted on by direct tension forces. The curvature of the beam, however, causes further direct tension stresses to develop at some distance from the original crack to maintain equilibrium. This in turn causes further cracks to form, and the process continues until the distance between cracks does not permit sufficient tensile stresses to develop and cause further cracking. These initial cracks are called 'primary cracks', and the average spacing in a region of constant moment is largely independent of reinforcement detailing.

As the applied moment is increased beyond this point, the development of cracks is governed to a large extent by the reinforcement. Tensile stresses in the concrete

Figure 6.10
Bending of a length of beam

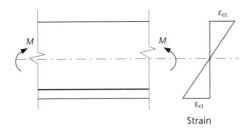

surrounding reinforcing bars are caused by bond as the strain in the reinforcement increases. These stresses increase with distance from the primary cracks and may eventually cause further cracks to form approximately mid-way between the primary cracks. This action may continue with increasing moment until the bond between concrete and steel is incapable of developing sufficient tension in the concrete to cause further cracking in the length between existing cracks. Since the development of the tensile stresses is caused directly by the presence of reinforcing bars, the spacing of cracks will be influenced by the spacing of the reinforcement.

If bars are sufficiently close for their 'zones of influence' to overlap then secondary cracks will join up across the member, while otherwise they will form only adjacent to individual bars. According to EC2 (see section 6.4.2) the average crack spacing in a flexural member depends in part on the efficiency of bond, the diameter of reinforcing bar used and the quantity and location of the reinforcement in relation to the tension face of the section.

6.4.2 Estimation of crack widths

If the behaviour of the member in figure 6.11 is examined, it can be seen that the overall extension per unit length at a depth y below the neutral axis is given by

$$\varepsilon_1 = \frac{y}{(d-x)}\varepsilon_s$$

where ε_s is the average strain in the main reinforcement over the length considered, and may be assumed to be equal to σ_s/E_s where σ_s is the steel stress at the cracked sections. ε_1 is the strain at level y which by definition is the extension over the unit length of the member. Hence, assuming any tensile strain of concrete between cracks is small, since full bond is never developed, the total width of all cracks over this unit length will equate to the extension per unit length, that is

$$\varepsilon_1 = \frac{y}{(d-x)}\frac{\sigma_s}{E_s} = \sum w$$

where $\sum w =$ the sum of all crack widths at level y.

The actual width of individual cracks will depend on the number of cracks in this unit length, the average being given by unit length/average spacing (s_{rm}). Thus

$$\text{average crack width } w_{av} = \frac{\sum w}{\text{av. number of cracks}}$$

$$= \frac{\varepsilon_1}{(1/s_{rm})}$$

$$= s_{rm}\varepsilon_1$$

Figure 6.11
Bending strains

The designer is concerned however with the maximum crack width which has an acceptably low probability of being exceeded. For design purposes the design maximum crack width, w_k, can be based on the maximum spacing, $s_{r, max}$. Hence the design crack width at any level defined by y in a member will thus be given by

$$w_k = s_{r, max} \varepsilon_1$$

The expression for the design crack width given in EC2 is of the above form and is given as

$$w_k = s_{r, max} (\varepsilon_{sm} - \varepsilon_{cm}) \tag{6.15}*$$

where

$\qquad w_k = $ the design crack width

$\qquad s_{r, max} = $ the maximum crack spacing

$\qquad \varepsilon_{sm} = $ the *mean strain in the reinforcement* allowing for the effects of tension stiffening of the concrete, shrinkage etc.

$\qquad \varepsilon_{cm} = $ the mean strain in the concrete between cracks

The mean strain, ε_{sm}, will be less than the apparent value ε_1 and $(\varepsilon_{sm} - \varepsilon_{cm})$ is given by the expression

$$\varepsilon_{sm} - \varepsilon_{cm} = \frac{\sigma_s - k_t \dfrac{f_{ct, eff}}{\rho_{p, eff}} (1 + \alpha_e \rho_{p, eff})}{E_s} \geq 0.6 \frac{\sigma_s}{E_s} \tag{6.16}*$$

where σ_s is the stress in the tension steel calculated using the cracked concrete section. k_t is a factor that accounts for the duration of loading (0.6 for short-term load, 0.4 for long-term load).

The maximum crack spacing, $s_{r, max}$, is given by the empirical formula

$$s_{r, max} = 3.4c + 0.425 k_1 k_2 \phi / \rho_{p, eff} \tag{6.17}*$$

where ϕ is the bar size in mm or an average bar size where a mixture of different sizes have been used and c is the cover to the longitudinal reinforcement. k_1 is a coefficient accounting for the bond properties of the reinforcement (0.8 for high bond, 1.6 for plain bars) and k_2 is a coefficient accounting for the nature of the strain distribution which for cracking due to flexure can be taken as 0.5. $\rho_{p, eff}$ is the effective reinforcement ratio, $A_s / A_{c, eff}$, where A_s is the area of reinforcement within an effective tension area of concrete $A_{c, eff}$, as shown in figure 6.12.

The effective tension area is that area of the concrete cross-section which will crack due to the tension developed in bending. This is the cracking which will be controlled by the presence of an appropriate type, amount and distribution of reinforcement. Generally the effective tension area should be taken as having a depth equal to 2.5 times the distance from the tension face of the concrete to the centroid of the reinforcement,

Figure 6.12
Typical examples of effective
concrete tension area

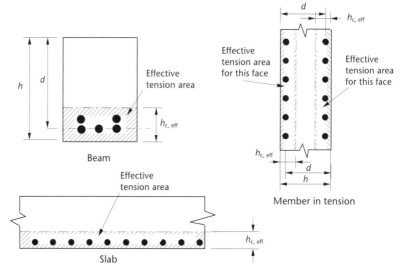

$h_{c, eff}$ = lesser of 2.5(h – d), (h – x)/3 or h/2

although for slabs the depth of this effective area should be limited to $(h - x)/3$. An overall upper depth limit of $h/2$ also applies.

Although not directly incorporated into the above formulae, it should be noted that crack widths may vary across the width of the soffit of a beam and are generally likely to be greater at positions mid-way between longitudinal reinforcing bars and at the corners of the beam. Where the maximum crack spacing exceeds $5(c + \phi/2)$ then an upper bound to crack width can be estimated by using $s_{r, max} = 1.3(h - x)$.

6.4.3 Analysis of section to determine crack widths

To use the formula of EC2 it is necessary to carry out an elastic analysis of the cracked concrete section using an effective modulus $E_{c, eff}$, as given in equation 6.8 to allow for creep effects.

The methods discussed in section 4.10.1 should be used to find the neutral axis position, x, and hence the stresses, σ_s and σ_{sr}, in the tensile reinforcement from which ε_{sm} (equation 6.16) can be obtained.

EXAMPLE 6.4

Calculation of flexural crack widths

Calculate the design flexural crack widths for the beam shown in figure 6.13 when subject to a quasi-permanent moment of 650 kN m. The concrete is class C25/30 and the reinforcement is high bond with a total cross-sectional area of 3770 mm^2.

(a) Calculate the mean strain ε_{sm}

From table 6.11, $E_{cm} = 31$ kN/mm^2. From table 6.12, assuming loading at 28 days with indoor exposure, the creep coefficient $\phi \approx 2.63$ (because $2A_c/u = 2 \times [1000 \times 400]/2800 = 285$) and hence the effective modulus is given by equation 6.8 as

$$E_{c, eff} = 31/(1 + 2.63) = 8.54 \text{ kN/mm}^2$$

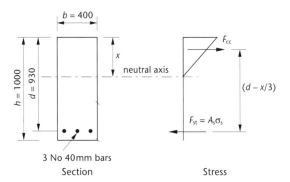

3 No 40mm bars

Section

Stress

Figure 6.13
Crack width calculation
example

(i) Calculate the neutral axis depth of the cracked section

Taking moments about the neutral axis:

$$b \times x \times x/2 = a_e A_s (d - x)$$

$$400 \times x^2/2 = \frac{200}{8.54} \times 3770(930 - x)$$

which has the solution $x = 457\,\text{mm}$.

(ii) Calculate the stress in the tension steel, σ_s

Taking moments about the level of the compressive force in the concrete:

$$\sigma_s = M/(d - x/3)A_s$$

$$= \frac{650 \times 10^6}{(930 - 457/3)3770}$$

$$= 222\,\text{N/mm}^2$$

(iii) Calculate $(\varepsilon_{sm} - \varepsilon_{cm})$

$$\varepsilon_{sm} - \varepsilon_{cm} = \frac{\sigma_s - k_t \dfrac{f_{ct,\,eff}}{\rho_{p,\,eff}}(1 + \alpha_e \rho_{p,\,eff})}{E_s} \geq 0.6\frac{\sigma_s}{E_s}$$

where:

$k_t = 0.4$ assuming long-term loading

$f_{ct,\,eff} = f_{ctm}$ (from table 6.11) $= 2.6\,\text{N/mm}^2$

$$\alpha_e = \frac{E_s}{E_{cm}} = \frac{200}{31} = 6.45$$

$$\rho_{p,\,eff} = \frac{A_s}{A_{c,\,eff}} = \frac{3770}{2.5(1000 - 930)400} = 0.0539$$

giving:

$$\varepsilon_{sm} - \varepsilon_{cm} = \frac{222 - 0.4 \times \dfrac{2.6}{0.0539}(1 + 6.45 \times 0.0539)}{200 \times 10^3} \geq 0.6\frac{222}{200 \times 10^3}$$

$$\frac{222 - 19.97}{200 \times 10^3} \geq 0.00067$$

$$= 0.001 \geq 0.00067$$

(iv) Calculate the maximum crack spacing ($s_{r,\,max}$)

$$s_{r,\,max} = 3.4c + 0.425k_1k_2\phi/\rho_{p,\,eff}$$

where:

$c = $ cover $= 1000 - 930 - 40/2 = 50\,mm$ to main bars

$k_1 = 0.8$ for ribbed bars

$k_2 = 0.5$ for flexure

$\phi = $ bar diameter $= 40\,mm$

hence

$$s_{r,\,max} = 3.4 \times 50 + \frac{0.425 \times 0.8 \times 0.5 \times 40}{0.0539}$$
$$= 296\,mm \text{ (which is less than } 5(c + \phi/2) = 350\,mm)$$

(v) Calculate crack width

$$w_k = 0.001 \times 296$$
$$= 0.30\,mm$$

which just satisfies the recommended limit.

6.4.4 Control of crack widths

It is apparent from the expressions derived above that there are four fundamental ways in which surface crack widths may be reduced:

(i) reduce the stress in the reinforcement (σ_s) which will hence reduce ε_{sm};

(ii) reduce the bar diameters (ϕ) which will reduce bar spacing and have the effect of reducing the crack spacing ($s_{r,\,max}$);

(iii) increase the effective reinforcement ratio ($\rho_{p,\,eff}$);

(iv) use high bond rather than plain bars.

The use of steel at reduced stresses is generally uneconomical and, although this approach is used in the design of water-retaining structures where cracking must often be avoided altogether, it is generally easier to limit the bar diameters, increase $\rho_{p,\,eff}$ or use high bond bars in preference to plain bars.

To increase $\rho_{p,\,eff}$ the effective tension area $A_{c,\,eff}$ should be made as small as possible. This is best achieved by placing the reinforcement close to the tension face such that the depth of tension area $\{2.5(h - d)\}$ is made as small as possible recognising, nevertheless, that durability requirements limit the minimum value of cover.

The calculation of the design crack widths indicated above only applies to regions within the effective tension zone. Since cracking can also occur in the side face of a beam it is also good practice to consider the provision of longitudinal steel in the side faces of beams. The critical position for the width of such cracks is likely to be approximately mid-way between the main tension steel and neutral axis. Recommendations regarding this and requirements for the main reinforcement are discussed in section 6.1. If these recommendations are followed, it is not necessary to calculate crack widths except in unusual circumstances. Reinforcement detailing, however, has been shown to have a large effect on flexural cracking, and must in practice be a compromise between the requirements of cracking, durability and constructional ease and costs.

6.5 | Thermal and shrinkage cracking

Thermal and shrinkage effects, and the stresses developed prior to cracking of the concrete, were discussed in chapter 1. The rules for providing minimum areas of reinforcement and limiting bar sizes to control thermal and shrinkage cracking were discussed in sections 6.1.5, 6.1.7 and 6.1.8. In this section, further consideration will be given to the control of such cracking and the calculations that can be performed, if necessary, to calculate design crack widths.

Consider the concrete section of figure 6.14 which is in a state of stress owing to thermal contraction and concrete shrinkage and the effects of external restraint. After cracking, the equilibrium of concrete adjacent to a crack is as illustrated.

Figure 6.14
Forces adjacent to a crack

Equating tension and compression forces:

$$A_s f_{st} = A_{ct} f_{ct} - A_{sc} f_{sc}$$

If the condition is considered when steel and concrete simultaneously reach their limiting values in tension, that is $f_{st} = f_{yk}$ and $f_{ct} = f_{ct,eff}$ = the tensile strength of concrete at the time when cracking is expected to develop (usually taken as three days), then

$$A_s f_{yk} = A_{ct} f_{ct,eff} - A_{sc} f_{sc}$$

The value of f_{sc} can be calculated but is generally very small and may be taken as zero without introducing undue inaccuracy. Hence the critical value of steel area is

$$A_{s,min} = A_{ct} f_{ct,eff} / f_{yk}$$

If the steel area is less than this amount then the steel will yield in tension, resulting in a few wide cracks; however, if it is greater, then more cracks will be formed but of narrower width. In EC2 this formula is modified by the inclusion of a stress distribution coefficient (k_c) taken as 1.0 for pure tension and a further coefficient (k) that accounts for non-linear stress distribution within the section. For thermal and shrinkage effects, k can range from 1.0 for webs where $h \leq 300$ mm or flanges with width ≤ 300 mm to 0.65 for webs with $h \geq 800$ mm or flanges ≥ 800 mm interpolating accordingly. Hence the recommended minimum steel area required to control thermal and shrinkage cracking is given by

$$A_{s,min} = k_c k A_{ct} f_{ct,eff} / f_{yk}$$
$$= k A_{ct} f_{ct,eff} / f_{yk} \qquad (6.18)*$$

6.5.1 Crack width calculation

The expression for the design crack width given in EC2, and discussed in section 6.4.2 for the case of flexural cracking, can be used for the calculation of thermal and

Table 6.15 Restraint factor values

Pour configuration	R
Thin wall cast onto massive concrete base	0.6 to 0.8 at base
	0.1 to 0.2 at top
Massive pour cast onto blinding	0.1 to 0.2
Massive pour cast onto existing concrete	0.3 to 0.4 at base
	0.1 to 0.2 at top
Suspended slabs	0.2 to 0.4
Infill bays, i.e rigid restraint	0.8 to 1.0

shrinkage cracking with some minor modifications. The crack width is given in equation 6.15 by

$$w_k = s_{r,\max}(\varepsilon_{sm} - \varepsilon_{cm})$$

where w_k is the design crack width, $s_{r,\max}$ is the maximum crack spacing and ε_{sm} is the mean strain in the section.

For steel areas greater than the minimum required value as given by equation 6.18, and when the total contraction exceeds the ultimate tensile strain for the concrete, the shrinkage and thermal movement will be accommodated by controlled cracking of the concrete. Any tensile strain in the concrete between cracks, ε_{cm}, is small and the effect may be approximated for building structures by using the expression $(\varepsilon_{sm} - \varepsilon_{cm}) = 0.8R\varepsilon_{imp}$, where ε_{imp} is the sum of the free shrinkage and thermal strains. That is

$$\varepsilon_{imp} = (\varepsilon_{cs} + T\alpha_{T,c}) \tag{6.19}*$$

where ε_{cs} is the shrinkage strain, T is the fall in temperature from the hydration peak and $\alpha_{T,c}$ is the coefficient of thermal expansion of concrete – often taken as half the value for mature concrete to allow for creep effects.

The restraint factor, R, is to allow for differences in restraint according to pour configuration, and typical values are given in table 6.15.

In practice, variations in restraint conditions cause large variations within members, and between otherwise similar members, with 'full' restraint seldom occurring as indicated in table 6.15. Cracking behaviour thus depends considerably on the degree and nature of the restraint and temperatures at the time of casting. CIRIA Guide C660 (ref. 25) offers further guidance on early-age crack control.

The maximum crack spacing, $s_{r,\max}$ is given by equation 6.17 with factor k_2 taken as 1.0. Hence for ribbed bars:

$$s_{r,\max} = 3.4c + 0.425 \times 0.8 \times 1.0\phi/\rho_{p,eff} \tag{6.20}$$

Calculations of crack widths should therefore be considered as realistic 'estimates' only and engineering judgement may need to be applied in interpreting such results.

EXAMPLE 6.5

Calculation of shrinkage and thermal crack widths

A section of reinforced concrete wall is 150 mm thick and is cast onto a massive concrete base. A drying shrinkage strain of 50 microstrain (ε_{cs}) is anticipated together with a temperature drop (T) of 20°C after setting. Determine the minimum reinforcement to control cracking in the lower part of the wall and calculate the design crack width and maximum spacing for a suitable reinforcement arrangement. The following design parameters should be used:

Three-day tensile strength of concrete $(f_{c,\,eff}) = 1.5\,\text{N/mm}^2$

Effective modulus of elasticity of concrete $(E_{c,\,eff}) = 10\,\text{kN/mm}^2$

Coefficient of thermal expansion for mature concrete $(\alpha_{T,c}) = 12\,\text{microstrain/}^{\circ}\text{C}$

Characteristic yield strength of reinforcement $(f_{yk}) = 500\,\text{N/mm}^2$

Modulus of elasticity of reinforcement $= 200\,\text{kN/mm}^2$

Minimum steel area to be provided, from equation 6.18:

$$A_{s,\,min} = 1.0 A_{ct} f_{ct,\,eff}/f_{yk}$$

assuming a value of 1.0 for factor k in equation 6.18

If horizontal steel is to be placed in two layers the area of concrete within the tensile zone, A_{ct}, can be taken as the full wall thickness multiplied by a one metre height. Hence

$$A_{s,\,min} = 1.0(150 \times 1000) \times 1.5/500$$
$$= 450\,\text{mm}^2/\text{m}$$

This could be conveniently provided as 10 mm bars at 300 mm centres in each face of the member (524 mm^2/m). For this reinforcement and assuming 35 mm cover, the crack spacing is given by equation 6.20 as

$$s_{r,\,max} = 3.4c + 0.425 \times 0.8 \times 1.0\phi/\rho_{p,\,eff}$$

where

$$\phi = 10\,\text{mm}$$
$$\rho_{p,\,eff} = A_s/A_{c,c,eff} = 524/(150 \times 1000) = 0.0035$$

therefore

$$s_{r,\,max} = 3.4 \times 35 + 0.425 \times 0.8 \times 1.0 \times 10/0.0035 = 1090\,\text{mm}$$

The imposed strain in the section is given by equation 6.19:

$$\varepsilon_{imp} = (\varepsilon_{cs} + T\alpha_{T,c})$$
$$= (50 + 20\{12/2\}) \times 10^{-6}$$
$$= 170\,\text{microstrain}$$

The ultimate tensile strain for the concrete

$$\varepsilon_{ult} = f_{ct,\,eff}/E_{c,\,eff}$$
$$= 1.5/(10000)$$
$$= 150\,\text{microstrain}$$

Therefore the section can be considered as cracked. The design crack width is given as

$$w_k = s_{r,\,max} \times 0.8R\varepsilon_{imp}$$

Thus taking $R = 0.8$ (table 6.15)

$$w_k = 1090 \times 0.8 \times 0.8 \times 170 \times 10^{-6}$$
$$= 0.12\,\text{mm}$$

6.6　Other serviceability requirements

The two principal other serviceability considerations are those of durability and resistance to fire, although occasionally a situation arises in which some other factor may be of importance to ensure the proper performance of a structural member in service. This may include fatigue due to moving loads or machinery, or specific thermal and sound insulation properties. The methods of dealing with such requirements may range from the use of reduced working stresses in the materials, to the use of special concretes, for example lightweight aggregates for good thermal resistance.

6.6.1　Durability

Deterioration will generally be associated with water permeating the concrete, and the opportunities for this to occur should be minimised as far as possible by providing good architectural details with adequate drainage and protection to the concrete surface.

Permeability is the principal characteristic of the concrete which affects durability, although in some situations it is necessary to consider also physical and chemical effects which may cause the concrete to decay.

For reinforced concrete, a further important aspect of durability is the degree of protection which is given to the reinforcement. Carbonation by the atmosphere will, in time, destroy the alkalinity of the surface zone concrete, and if this reaches the level of the reinforcement will render the steel vulnerable to corrosion in the presence of moisture and oxygen.

If a concrete is made with a sound inert aggregate, deterioration will not occur in the absence of an external influence. Since concrete is a highly alkaline material, its resistance to other alkalis is good, but it is however very susceptible to attack by acids or substances which easily decompose to produce acids. Concrete made with Portland cement is thus not suitable for use in situations where it comes into contact with such materials, which include beer, milk and fats. Some neutral salts may also attack concrete, the two most notable being calcium chloride and soluble sulfates. These react with a minor constituent of the hydration products in different ways. The chloride must be in concentrated solution, when it has a solvent effect on the concrete in addition to its more widely recognised action in promoting the corrosion of the reinforcement, while sulfates need only be present in much smaller quantities to cause internal expansion of the concrete with consequent cracking and strength loss. Sulfates present the most commonly met chemical-attack problem for concrete since they may occur in groundwater and sewage. In such cases cements containing reduced proportions of the vulnerable tricalcium aluminate, such as Sulfate Resisting Portland Cement, should be used. The addition of Pulverised Fuel Ash (Pfa) or ground granulated blast furnace slag (ggbfs) may also be beneficial. Both chlorides and sulfates are present in sea water, and because of this the chemical actions are different, resulting in reduced sulfate damage, although if the concrete is of poor quality, serious damage may occur from reactions of soluble magnesium salts with the hydrated compounds. Well-constructed Portland cement structures have nevertheless been found to endure for many years in sea water.

The matter of exposure classifications related to environmental conditions is dealt with in detail in EN 206 and BS 8500 together with the provision of appropriate concrete materials. BS 8500 includes the use of a system of classification of a wide range of chemically aggressive environments based on recommendations of the UK Building Research Establishment (BRE Special Digest 1). In some cases liable to aggressive chemical attack Additional Protective Measures (APMs) such as controlled permeability formwork, surface protection, sacrificial layers or site drainage may be recommended.

Physical attack of the concrete must also be considered. This may come from abrasion or attrition as may be caused by sand or shingle, and by alternate wetting and drying. The latter effect is particularly important in the case of marine structures near the water surface, and causes stresses to develop if the movements produced are restrained. It is also possible for crystal growth to occur from drying out of sea water in cracks and pores, and this may cause further internal stresses, leading to cracking. Alternate freezing and thawing is another major cause of physical damage, particularly in road and runway slabs and other situations where water in pores and cracks can freeze and expand, thus leading to spalling. It has been found that the entrainment of a small percentage of air in the concrete in the form of small discrete bubbles offers the most effective protection against this form of attack. Although this reduces the strength of the concrete, it is recommended that between 4.0 and 6.0 per cent by volume of entrained air should be included in concrete subjected to regular wetting and drying combined with severe frost.

All these forms of attack may be minimised by the production of a dense, well-compacted and well cured concrete with low permeability, thus restricting damage to the surface zone of the member. Aggregates which are likely to react with the alkali matrix should be avoided (or the alkali levels of the cement carefully limited), as must those which exhibit unusually high shrinkage characteristics. If this is done, then permeability, and hence durability, is affected by

1. aggregate type and density;
2. water–cement ratio;
3. degree of hydration of cement;
4. degree of compaction.

A low water–cement ratio is necessary to limit the voids due to hydration, which must be well advanced with the assistance of good curing techniques. EN 206 recommends minimum curing periods taking account of exposure classification, concrete strength development rate, concrete temperature and ambient conditions. Coupled with this is the need for non-porous aggregates which are hard enough to resist any attrition, and for thorough compaction. It is essential that the mix is designed to have adequate workability for the situation in which it is to be used, thus the cement content of the mix must be reasonably high.

EN 206 specifies minimum cement contents for various exposure conditions according to cement types, as well as minimum strength and maximum water cement ratio which can also be related to minimum cover requirements as described in section 6.1.1.

The consequences of thermal effects on durability must not be overlooked, and very high cement contents should only be used in conjunction with a detailed cracking assessment. A cement content of $550 \, \text{kg/m}^3$ is often regarded as an upper limit for general use.

Provided that such measures are taken, and that adequate cover of sound concrete is given to the reinforcement, deterioration of reinforced concrete is unlikely. Thus although the surface concrete may be affected, the reinforcing steel will remain protected by an alkaline concrete matrix which has not been carbonated by the atmosphere. Once this cover breaks down and water and possibly chemicals can reach the steel, rusting and consequent expansion lead rapidly to cracking and spalling of the cover concrete and severe damage – visually and sometimes structurally.

6.6.2 Fire resistance

Depending on the type of structure under consideration, it may be necessary to consider the fire resistance of the individual concrete members. Three conditions must be examined:

1. effects on structural strength
2. flame penetration resistance } in the case of dividing members such as wall
3. heat transmission properties } and slabs

Concrete and steel in the form of reinforcement or prestressing tendons exhibit reduced strength after being subjected to high temperatures. Although concrete has low thermal conductivity, and thus good resistance to temperature rise, the strength begins to drop significantly at temperatures above 300°C and it has a tendency to spall at high temperatures. The extent of this spalling is governed by the type of aggregate, with siliceous materials being particularly susceptible while calcareous and lightweight aggregate concretes suffer very little. Reinforcement will retain about 50 per cent of its normal strength after reaching about 550°C, while for prestressing tendons the corresponding temperature is only 400°C.

Thus as the temperature rises the heat is transferred to the interior of a concrete member, with a thermal gradient established in the concrete. This gradient will be affected by the area and mass of the member in addition to the thermal properties of the concrete, and may lead to expansion and loss of strength. Dependent on the thickness and nature of cover, the steel will rise in temperature and lose strength, thus leading to deflections and eventual structural failure of the member if the steel temperature becomes excessive. Design must therefore be aimed at providing and maintaining sound cover of concrete as a protection, thus delaying the temperature rise in the steel. The presence of plaster, screeds and other non-combustible finishes assists the cover in protecting the reinforcement and may thus be allowed for in the design.

EC2 gives tabulated values of minimum dimensions and covers for various types of concrete member which are necessary to permit the member to withstand fire for a specified period of time. These values, which have been summarised in tables 6.4, 6.5 and 6.6 for siliceous aggregates may be considered adequate for most normal purposes. More detailed information concerning design for fire resistance is given in Part 1.2 of Eurocode 2 including concrete type, member type and details of finishes. The period that a member is required to survive, both in respect of strength in relation to working loads and the containment of fire, will depend upon the type and usage of the structure – and minimum requirements are generally specified by building regulations. Prestressed concrete beams must be considered separately in view of the increased vulnerability of the prestressing steel.

6.7 Limitation of damage caused by accidental loads

While it would be unreasonable to expect a structure to withstand extremes of accidental loading as may be caused by collision, explosion or similar, it is important that resulting damage should not be disproportionate to the cause. It follows therefore that a major structural collapse must not be allowed to be caused by a relatively minor mishap which may have a reasonably high probability of happening in the anticipated lifetime of the structure.

The possibilities of a structure buckling or overturning under the 'design' loads will have been considered as part of the ultimate limit state analysis. However, in some instances a structure will not have an adequate lateral strength even though it has been designed to resist the specified combinations of wind load and vertical load. This could be the case if there is an explosion or a slight earth tremor, since then the lateral loads are proportional to the mass of the structure. Therefore it is recommended that at any floor level, a structure should always be capable of resisting a minimum lateral force as detailed in section 3.4.2.

Damage and possible instability should also be guarded against wherever possible, for example vulnerable load-bearing members should be protected from collision by protective features such as banks or barriers.

6.7.1 Ties

In addition to these precautions, the general stability and robustness of a building structure can be increased by providing reinforcement acting as ties. These ties should act both vertically between roof and foundations, and horizontally around and across each floor (figure 6.15), and all external vertical load-bearing members should be anchored to the floors and beams. If a building is divided by expansion joints into structurally independent sections, then each section should have an independent tying system.

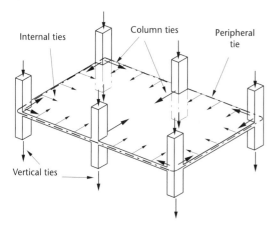

Internal ties   Column ties   Peripheral tie

Vertical ties

Figure 6.15
Tie forces

Vertical ties

Vertical ties are not generally necessary in structures of less than five storeys but in higher buildings should be provided by reinforcement, effectively continuous from roof to foundation by means of proper laps, running through all vertical load-bearing members. This steel should be capable of resisting a tensile force equal to the maximum design ultimate load carried by the column or wall from any one storey or the roof. Although the accidental load case is an ultimate limit state, the ultimate load used should reflect the loads likely to be acting at the time and the *quasi-permanent* value would normally be taken. The aim is to contribute to a bridging system in the event of loss of the member at a lower level. In *in situ* concrete, this requirement is almost invariably satisfied by a normal design, but joint detailing may be affected in precast work.

Horizontal ties

Horizontal ties should be provided for all buildings, irrespective of height, in three ways:

1. peripheral ties;
2. internal ties;
3. column and wall ties.

The resistance of these ties when stressed to their characteristic strength is given in terms of a force F_t, where $F_t = 60$ kN or $(20 + 4 \times$ number of storeys in structure) kN, whichever is less. This expression takes into account the increased risk of an accident in a large building and the seriousness of the collapse of a tall structure.

(a) Peripheral ties

The peripheral tie must be provided, by reinforcement which is effectively continuous, around the perimeter of the building at each floor and roof level. This reinforcement must lie within 1.2 m of the outer edge and at its characteristic stress be capable of resisting a force of at least F_t.

(b) Internal ties

Internal ties should also be provided at each floor in two perpendicular directions and be anchored at each end, either to the peripheral tie or to the continuous column or wall ties.

These ties must be effectively continuous and they may either be spread evenly across a floor, or grouped at beams or walls as convenient. Where walls are used, the tie reinforcement must be within 0.5 m of the top or bottom of the floor slab.

The resistance required is related to the span and loading. Internal ties must be capable of resisting a force of F_t kN per metre width or $[F_t(g_k + q_k)/7.5]l_r/5$ kN per metre width, if this is greater. In this expression, l_r is the greatest horizontal distance in the direction of the tie between the centres of vertical load-bearing members. The loading $(g_k + q_k)$ kN/m^2 is the average characteristic load on unit area of the floor considered. If the ties are grouped their maximum spacing should be limited to $1.5l_r$.

(c) Column and wall ties

Column and wall ties must be able to resist a force of at least 3 per cent of the total vertical ultimate load at that level for which the member has been designed. Additionally, the resistance provided must not be less than the smaller of $2F_t$ or $F_t l_s/2.5$ kN where l_s is the floor to ceiling height in metres. Wall ties are assessed on the basis of the above forces acting per metre length of the wall, while column ties are concentrated within 1 m of either side of the column centre line. Particular care should be taken with corner columns to ensure they are tied in two perpendicular directions.

In considering the structure subjected to accidental loading it is assumed that no other forces are acting. Thus reinforcement provided for any other purposes may also act as ties. Indeed, peripheral and internal ties may also be considered to be acting as column or wall ties.

As with vertical ties, the provision of horizontal ties for *in situ* construction will seldom affect the amount of reinforcement provided. Detailing of the reinforcement may however be affected, and particular attention must be paid to the manner in which internal ties are anchored to peripheral ties. Typical details for the anchorage of internal

Full anchorage length

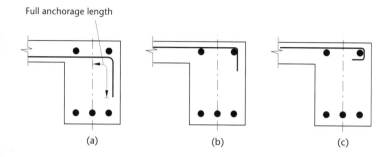

Figure 6.16
Typical anchorage details for
internal ties

ties are illustrated in figure 6.16. If full anchorage is not possible then the assumed stresses in the ties must be reduced appropriately.

Precast concrete construction, however, presents a more serious problem since the requirements of tie forces and simple easily constructed joints are not always compatible. Unless the required tie forces can be provided with the bars anchored by hooks and bends in the case of column and wall ties, an analysis of the structure must be performed to assess the remaining stability after a specified degree of structural damage.

EXAMPLE 6.6

Stability ties

Calculate the stability ties required in an eight-storey building of plan area shown in figure 6.17:

$$\text{Clear storey height under beams} = 2.9\,\text{m}$$

$$\text{Floor to ceiling height } (l_s) = 3.4\,\text{m}$$

$$\text{Characteristic permanent load } (g_k) = 6\,\text{kN/m}^2$$

$$\text{Characteristic variable load } (q_k) = 3\,\text{kN/m}^2$$

$$\text{Characteristic steel strength } (f_{yk}) = 500\,\text{N/mm}^2$$

$$F_t = (20 + 4 \times \text{number of storeys})$$

$$= 20 + 4 \times 8 = 52\,\text{kN} < 60\,\text{kN}$$

(a) Peripheral ties

$$\text{Force to be resisted} = F_t = 52\,\text{kN}$$

$$\text{Bar area required} = \frac{52 \times 10^3}{500} = 104\,\text{mm}^2$$

This could be provided by one H12 bar.

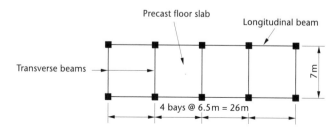

Precast floor slab

Longitudinal beam

Transverse beams

7 m

4 bays @ 6.5m = 26m

Figure 6.17
Structure layout

(b) Internal ties

$$\text{Force to be resisted} = \frac{F_t(g_k + q_k)}{7.5} \times \frac{l_r}{5} \text{ kN per metre}$$

(1) Transverse direction

$$\text{Force} = \frac{52(6+3)}{7.5} \times \frac{7}{5} = 87.4 \text{ kN/m} > F_t$$

Force per bay $= 87.4 \times 6.5$

$$= 568.1 \text{ kN}$$

Therefore, bar area required in each transverse interior beam is

$$\frac{568.1 \times 10^3}{500} = 1136 \text{ mm}^2$$

This could be provided by 4 H20 bars.

(2) Longitudinal direction

$$\text{Force} = \frac{52(6+3)}{7.5} \times \frac{6.5}{5} = 81.1 \text{ kN/m} > F_t$$

Therefore force along length of building $= 81.1 \times 7 = 567.7 \text{ kN}$, hence bar area required in each longitudinal beam is

$$\frac{567.7 \times 10^3}{2 \times 500} = 567 \text{ mm}^2$$

This could be provided by 2 H20 bars.

(3) Column ties

Force to be designed for is

$$\left(\frac{l_s}{2.5}\right) F_t = \left(\frac{3.4}{2.5}\right) 52 = 70.7 \text{ kN} < 2F_t$$

or 3 per cent of ultimate floor load on a column is

$$8\left[\frac{3}{100}(1.35 \times 6 + 1.5 \times 3) \times 6.5 \times \frac{7}{2}\right] = 69 \text{ kN at ground level}$$

To allow for 3 per cent of column self-weight, take design force to be 72 kN, say, at ground level.

$$\text{Area of ties required} = \frac{72 \times 10^3}{500} = 144 \text{mm}^2$$

This would be provided by 1 H20 bar and incorporated with the internal ties. At higher floor levels a design force of 70.7 kN would be used giving a similar practical reinforcement requirement.

(c) Vertical ties

Assume *quasi-permanent* loading with $\Psi_2 = 0.6$. Thus the ultimate design load $= 1.0 \times 6 + 0.6 \times 3 = 7.8 \text{ kN/m}^2$.

Maximum column load from one storey is approximately equal to

$$7.8 \times 3.5 \times 6.5 = 177.5\,\text{kN}$$

Therefore bar area required throughout each column is equal to

$$\frac{177.5 \times 10^3}{500} = 355\,\text{mm}^2$$

This would be provided by 4 H12 bars.

6.7.2 Analysis of 'damaged' structure

This must be undertaken when a structure has five or more storeys and does not comply with the vertical-tie requirements, or when every precast floor or roof unit does not have sufficient anchorage to resist a force equal to F_t kN per metre width acting in the direction of the span. The analysis must show that each key load-bearing member, its connections, and the horizontal members which provide lateral support, are able to withstand a specified loading from any direction. If this cannot be satisfied, then the analysis must demonstrate that the removal of any single vertical load-bearing element, other than key members, at each storey in turn will not result in collapse of a significant part of the structure.

The minimum loading that may act from any direction on a key member is recommended as 34 kN/m². The decision as to what loads should be considered acting is left to the engineer, but will generally be on the basis of permanent and realistic variable load estimates, depending on the building usage. This method is attempting, therefore, to assess quantitatively the effects of exceptional loading such as explosion. The design 'pressure' must thus be regarded as a somewhat arbitrary value.

The 'pressure' method will generally be suitable for application to columns in precast framed structures; however, where precast load-bearing panel construction is being used an approach incorporating the removal of individual elements may be more appropriate. In this case, vertical loadings should be assessed as described, and the structure investigated to determine whether it is able to remain standing by a different structural action. This action may include parts of the damaged structure behaving as a cantilever or a catenary, and it may also be necessary to consider the strength of non-load-bearing partitions or cladding.

Whichever approach is adopted, such analyses are tedious, and the provision of effective tie forces within the structure should be regarded as the preferred solution both from the point of view of design and performance.

Continuity reinforcement and good detailing will greatly enhance the overall fire resistance of a structure with respect to collapse. A fire-damaged structure with reduced member strength may even be likened to a structure subjected to accidental overload, and analysed accordingly.

6.8 Design and detailing for seismic forces

Earthquakes are caused by movement of the earth's crust along faults or slip planes. These movements result in horizontal and vertical forces with vibrations of varying frequency, amplitude and duration to act on structures within the earthquake zone.

The earth's crust is not one continuous outer layer but consists of seven major plus a number of minor tectonic plates as shown in figure 6.18. These plates bear against each other continually moving and grinding at their adjacent edges. Occasionally a large slippage takes place with a large release of energy, causing an earthquake with horizontal and vertical vibrations. These can cause destruction and damage to structures and landslides with large losses of life. Most of these earthquakes occur near the boundaries of the tectonic plates but powerful earthquakes sometimes occur in the interior of a plate.

When the earthquake occurs on the ocean floor, large waves called tsunamis are generated which spread out radially from the epicentre at a large speed of possibly 800 km per hour. The speed is proportional to the depth of water so as the waves approach the coastline they slow down and there is a large build-up of water with massive waves hitting the coast and causing extensive major destruction.

Seismic disturbances are measured according to their intensity on the Richter logarithmic scale. Intensities up to 3 in magnitude are generally considered to be moderate but higher intensities of over 6 are severe. Intensities as high as 9.5 have been measured.

In many parts of the world, such as Turkey, Japan and California, where earthquakes can be severe, resistance to seismic forces forms a critical part of the structural design. In other areas of the world, such as the British Isles, earthquakes are less common and not nearly so severe so that the design for wind loading, or the requirements for the structure to be able to resist a minimum horizontal force, plus the provision of continuity steel throughout the structure according to the requirements in section 6.7 are generally adequate. Nevertheless, with important structures, such as major dams or nuclear power stations where failure or damage can have catastrophic effects, the resistance to seismic disturbances must also be considered, even in the British Isles.

The nature of the vibrations and the forces induced by an earthquake are complex phenomena, as is the dynamic response of a highly indeterminate concrete structure. This has led to the development of computer programs to carry out the analysis, sometimes referred to as a multi-modal response spectrum analysis. A simpler approach is the equivalent static analysis in which the base shear at the foot of the structure is

Figure 6.18
Tectonic plates

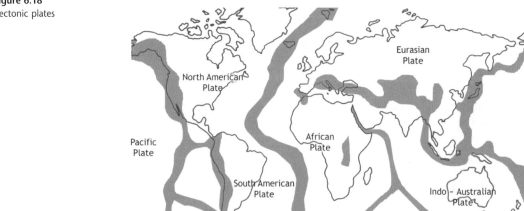

calculated and distributed as horizontal forces at each floor level according to certain defined criteria. This approach is allowed in many national codes of practice for the design of approximately regular and symmetrical structures. Eurocode 8 provides guidance relevant to countries within the European Union.

The full numerical design requirements of the codes of practice are beyond the scope of this book but it is hoped that highlighting some of the important principles and requirements for the overall design and detailing may be of some help in the design of safer structures.

6.8.1   Construction and general layout

It is particularly important that good quality materials, including high ductility reinforcing steel, are used together with rigorous testing and control procedures. Design should ensure that sudden brittle shear or compressive failure is avoided with emphasis on energy dissipation. Good construction practices, including steel fixing, compaction, curing and inspection are also essential if a structure is to perform satisfactorily under seismic loading.

Foundations should be designed to provide a regular layout in plan and elevation and to be approximately symmetrical about both orthogonal axes in plan with no sudden and major change in layout or construction. It is important that there is adequate bending and torsional resistance about both axes of the structure. Some illustrative examples of good and poor practice are shown in figures 6.19 and 6.20.

Sway effects under horizontal motions should be minimised by ensuring approximately equal loading at each floor level with no heavy loads in the higher storeys. Efforts should be made to provide a highly indeterminate structure that is well

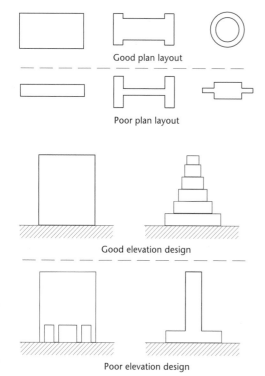

Good plan layout

Poor plan layout

Figure 6.19
Examples of good and poor plan layouts

Good elevation design

Poor elevation design

Figure 6.20
Examples of good and poor elevations

Figure 6.21
One-storey building

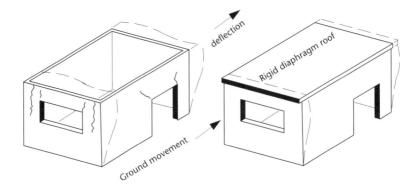

tied together with continuity reinforcement so that the loading can be redistributed and alternative structural actions may develop if necessary. The principles discussed in section 6.7 are relevant to this.

Slabs can provide rigid diaphragms to transfer loads at the roof and each floor. Figure 6.21 shows how, in a one-storey building, a rigid horizontal slab or bracing at roof level enables the structure to act as a closed box giving more rigidity and strength to resist cracking.

6.8.2 Foundations

In addition to a regular and symmetrical layout in plan as discussed above, it is preferable that only one type of foundation is used throughout a structure and that it is constructed on a level ground base and tied together with strong ground beams to limit relative movement. This is illustrated in figure 6.22.

Landslides are a common feature of earthquakes and they cause much structural damage and loss of life. Therefore structures should not be built on steep slopes, in or near gulleys or near cliffs. It must also be recognised that vibrations during an earthquake can cause liquefaction of some soils, such as sandy or silty soils, causing loss of bearing strength, excessive settlement and failure.

Figure 6.22
Examples of good and poor foundation design

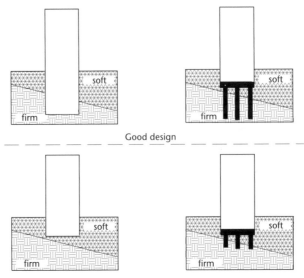

6.8.3 Shear walls

Shear walls provide a strong resistance to the lateral forces from an earthquake and they should continue down to, and be anchored into, foundations. They should never be supported on beams, slabs or columns

If coupling beams are required these should be reinforced with diagonal cages of steel bars and diagonal reinforcing bars should also be provided to resist horizontal sliding at construction joints in the wall. These bars should have at least a tension anchorage on either side of the construction joint. Some typical reinforcing steel details are given in EC 8 and a typical detail for a coupling beam is shown in figure 6.23.

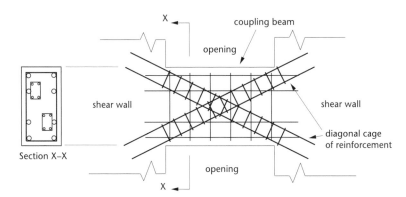

Figure 6.23
Typical reinforcement detail for a coupling beam

6.8.4 Columns

Columns and their connections to beams are critical parts of a structure. Failure of a column in a building can be catastrophic leading to a progressive collapse, and the formation of plastic hinges in columns above the base of a building should be avoided.

Horizontal hoops of helical reinforcing bars have been found to give a stronger containment to the longitudinal vertical bars than that provided by rectangular links and at a beam-to-column joint horizontal steel reinforcement hoops not less than 6 mm diameter are advisable within the depth of the beam.

At external columns the longitudinal reinforcement of beams should be well anchored within the column. This may require special measures such as the provision of beam haunches or anchorage plates and some typical examples of details are given in EC8.

6.8.5 Beams and slabs

Beams should be ductile so that plastic hinges can form, and these should be distributed throughout a structure, avoiding 'soft' storeys. This will provide a gradual type of failure and not a sudden catastrophic failure such as that associated with shear or brittle compressive failure. The formation of plastic hinges also allows the maximum moments to be redistributed to other parts of the statically indeterminate structure, thus providing more overall safety.

The first plastic hinges are likely to form in the sections of the beam near the column where the maximum moments are hogging, causing compression on the lower fibres so that the section acts effectively as a rectangular section. Plastic hinges which form later at mid-span will have compression on the upper fibres so that the section is effectively a T-section with the slab acting as the flange and there is a large area to resist the compression. Further discussion of the design of ductile sections is given in Sections 4.2, 4.4 and 4.7.

The beam sections near the support should be reinforced by closed steel links, closely spaced to resist the shear and to provide greater compressive resistance to the enclosed concrete The provision of compressive steel reinforcement also ensures a more ductile section.

The slabs in a building act as rigid horizontal diaphragms to stiffen the structure against torsion during seismic disturbances and also transfer the horizontal forces into the columns and shear walls. The slabs should be well tied into the columns, the shear walls and the perimeter beams with continuity reinforcement as indicated previously.

When precast concrete slabs are used they should have good lengths of bearings onto the supporting beams and shear walls should also be provided with continuity steel over their supports so that they can act as continuous indeterminate members. In this way they can also develop their full ultimate reserve of strength by enabling a tensile catenary action.

CHAPTER 7

Design of reinforced concrete beams

CHAPTER INTRODUCTION

Reinforced concrete beam design consists primarily of producing member details which will adequately resist the ultimate bending moments, shear forces and torsional moments. At the same time serviceability requirements must be considered to ensure that the member will behave satisfactorily under working loads. It is difficult to separate these two criteria, hence the design procedure consists of a series of interrelated steps and checks. These steps are shown in detail in the flow chart in figure 7.1, but may be condensed into three basic design stages:

1. preliminary analysis and member sizing;
2. detailed analysis and design of reinforcement;
3. serviceability calculations.

Much of the material in this chapter depends on the theory and design specification from the previous chapters. The loading and calculation of moments and shear forces should be carried out using the methods described in chapter 3. The equations used for calculating the areas of reinforcement were derived in chapters 4 and 5.

Full details of serviceability requirements and calculations are given in chapter 6, but it is normal practice to make use of simple rules which are specified

→

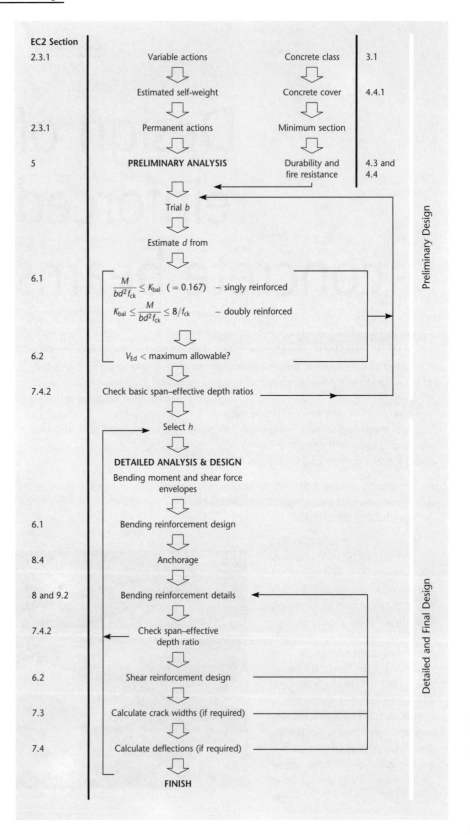

Figure 7.1
Beam design flowchart

$\longrightarrow$

in the Code of Practice and are quite adequate for most situations. Typical of these are the span–effective depth ratios to ensure acceptable deflections, and the rules for maximum bar spacings, maximum bar sizes and minimum quantities of reinforcement, which are to limit cracking, as described in chapter 6.

Design and detailing of the bending reinforcement must allow for factors such as anchorage bond between the steel and concrete. The area of the tensile bending reinforcement also affects the subsequent design of the shear and torsion reinforcement. Arrangement of reinforcement is constrained both by the requirements of the codes of practice for concrete structures and by practical considerations such as construction tolerances, clearance between bars and available bar sizes and lengths. Many of the requirements for correct detailing are illustrated in the examples which deal with the design of typical beams.

All calculations should be based on the effective span of a beam which is given by

$$l_{\text{eff}} = l_n + a_1 + a_2$$

where

l_n is the clear distance between the faces of the supports; for a cantilever l_n is its length to the face of the support

a_1, a_2 are the lesser of half the width, t, of the support, or half the overall depth, h, of the beam, at the respective ends of the span

7.1 | Preliminary analysis and member sizing

The layout and size of members are very often controlled by architectural details, and clearances for machinery and equipment. The engineer must either check that the beam sizes are adequate to carry the loading, or alternatively, decide on sizes that are adequate. The preliminary analysis need only provide the maximum moments and shears in order to ascertain reasonable dimensions. Beam dimensions required are

1. cover to the reinforcement
2. breadth (b)
3. effective depth (d)
4. overall depth (h)

Adequate concrete cover is required to ensure adequate bond and to protect the reinforcement from corrosion and damage. The necessary cover depends on the class of concrete, the exposure of the beam, and the required fire resistance. Table 6.2 gives the nominal cover that should be provided to all reinforcement, including links. This cover may need to be increased to meet the fire resistance requirements of the Code of Practice.

The strength of a beam is affected considerably more by its depth than its breadth. The span–depth ratios usually vary between say 14 and 30 but for large spans the ratios can be greater. A suitable breadth may be one-third to one-half of the depth; but it may be much less for a deep beam. At other times wide shallow beams are used to conserve headroom. The beam should not be too narrow; if it is much less than 200 mm wide there may be difficulty in providing adequate side cover and space for the reinforcing bars.

Suitable dimensions for b and d can be decided by a few trial calculations as follows:

1. For no compression reinforcement

$$K = M/bd^2 f_{ck} \leq K_{bal}$$

where

$$K_{bal} = 0.167 \text{ for } f_{ck} \leq C50$$

With compression reinforcement it can be shown that

$$M/bd^2 f_{ck} < 8/f_{ck}$$

approximately, if the area of bending reinforcement is not to be excessive.

2. The maximum design shear force $V_{Ed,max}$ should not be greater than $V_{Rd,max} = 0.18 b_w d(1 - f_{ck}/250)f_{ck}$. To avoid congested shear reinforcement, $V_{Ed,max}$ should preferably be somewhat closer to half (or less) of the maximum allowed.

3. The span–effective depth ratio for spans not exceeding 7 m should be within the basic values given in table 6.10 or figure 6.3. For spans greater than 7 m the basic ratios are multiplied by 7/span.

4. The overall depth of the beam is given by

$$h = d + \text{cover} + t$$

where t = estimated distance from the outside of the link to the centre of the tension bars (see figure 7.2). For example, with nominal sized 12 mm links and one layer of 32 mm tension bars, $t = 28$, mm approximately. It will, in fact, be slightly larger than this with deformed bars as they have a larger overall dimension than the nominal bar size.

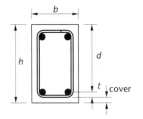

Figure 7.2
Beam dimensions

EXAMPLE 7.1

Beam sizing

A concrete lintel with an effective span of 4.0 m supports a 230 mm brick wall as shown in figure 7.3. The loads on the lintel are $G_k = 100$ kN and $Q_k = 40$ kN. Determine suitable dimensions for the lintel if class C25/30 concrete is used.

Figure 7.3
Lintel beam

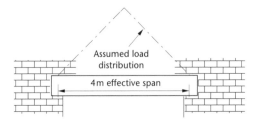

Assumed load
distribution

4m effective span

The beam breadth b will match the wall thickness so that

$b = 230\,\text{mm}$

Allowing, say, $14\,\text{kN}$ for the weight of the beam, gives the ultimate load

$F = 1.35 \times 114 + 1.5 \times 40$

$\quad = 214\,\text{kN}$

Therefore maximum design shear force

$V_{\text{Ed}} = 107\,\text{kN}$

Assuming a triangular load distribution for the preliminary analysis, we have

$$M = \frac{F \times \text{span}}{6} = \frac{214 \times 4.0}{6}$$

$$\quad = 143\,\text{kN m}$$

For such a relatively minor beam the case with no compression steel should be considered

$$K = \frac{M}{bd^2 f_{\text{ck}}} < K_{\text{bal}} = 0.167$$

therefore

$$\frac{143 \times 10^6}{230 \times d^2 \times 25} < 0.167$$

Rearranging, $d > 386\,\text{mm}$.

Assume a concrete cover of $25\,\text{mm}$ to the reinforcing steel. So for $10\,\text{mm}$ links and, say, $32\,\text{mm}$ bars

Overall beam depth $h = d + 25 + 10 + 32/2$

$$\quad = d + 51$$

Therefore make $h = 525\,\text{mm}$ as an integer number of brick courses. So that

$d = 525 - 51 = 474\,\text{mm}$

Maximum shear resistance is

$V_{\text{Rd,max}} = 0.18 b_{\text{w}} d (1 - f_{\text{ck}}/250) f_{\text{ck}}$

$\quad\quad 0.18 \times 230 \times 474 \times (1 - 25/250) \times 25 \times 10^{-3}$

$\quad\quad = 446\,\text{kN} > V_{\text{Ed}} = 107\,\text{kN}$

Basic span–effective depth $= \dfrac{4000}{479} = 8.35 < \approx 20$ (for a lightly stressed beam in C25

concrete – table 6.10)

A beam size of $230\,\text{mm}$ by $525\,\text{mm}$ deep would be suitable.

Weight of beam $= 0.23 \times 0.525 \times 4.0 \times 25$

$\quad\quad = 12.1\,\text{kN}$

which is sufficiently close to the assumed value.

7.2   Design for bending of a rectangular section with no moment redistribution

The calculation of main bending reinforcement is performed using the equations and charts derived in chapter 4. In the case of rectangular sections which require only tension steel, the lever-arm curve method is probably the simplest approach. Where compression steel is required, either design charts or a manual approach with the simplified design formulae may be used. When design charts are not applicable or not available, as in the case of non-rectangular sections, the formulae based on the equivalent rectangular stress block will simplify calculations considerably.

The grade and ductility class of reinforcing steel to be used must be decided initially since this, in conjunction with the chosen concrete class, will affect the areas required and also influence such factors as bond calculations. In most circumstances one of the available types of high-yield bars will be used. Areas of reinforcement are calculated at the critical sections with maximum moments, and suitable bar sizes selected. (Tables of bar areas are given in the appendix.) This permits anchorage calculations to be performed and details of bar arrangement to be produced, taking into account the guidance given by the Codes of Practice.

An excessive amount of reinforcement usually indicates that a member is undersized and it may also cause difficulty in fixing the bars and pouring the concrete. Therefore the code stipulates that

$100A_s/A_c \leq 4\%$ except at laps.

On the other hand too little reinforcement is also undesirable therefore

$$100A_s/b_t d \geq 26\frac{f_{ctm}}{f_{yk}} \text{ per cent and not less than 0.13 per cent}$$

where:

A_c is the area of concrete $= b \times h$ for a rectangular section

b_t is the mean width of the beam's tension zone

f_{ctm} is the concrete's mean axial tensile strength $= 0.3 \times f_{ck}^{2/3}$ for $f_{ck} \leq$ C50

Values for different concrete strengths are given in table 6.8

To avoid excessive deflections it is also necessary to check the span to effective depth ratio as outlined in chapter 6.

It should be noted that the equations derived in this chapter are for concrete classes less than or equal to C50/60. The equations for higher classes of concrete can be derived using similar procedures but using the ultimate concrete strains and constants for each class of concrete from EC2 and its National Annex.

7.2.1   Singly reinforced rectangular sections, no moment redistribution

A beam section needs reinforcement only in the tensile zone when

$$K = \frac{M}{bd^2 f_{ck}} \leq K_{bal} = 0.167$$

The singly reinforced section considered is shown in figure 7.4 and it is subjected to a sagging design moment M at the ultimate limit state. The design calculations for the longitudinal steel can be summarised as follows:

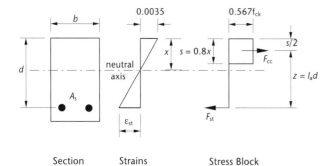

Figure 7.4
Singly reinforced section with
rectangular stress block

Section Strains Stress Block

1. Check that $K = \dfrac{M}{bd^2 f_{ck}} < K_{bal} = 0.167$

2. Determine the lever-arm, z, from the curve of figure 7.5 or from the equation

$$z = d\left[0.5 + \sqrt{(0.25 - K/1.134)}\right] \qquad (7.1)*$$

3. Calculate the area of tension steel required from

$$A_s = \frac{M}{0.87 f_{yk} z} \qquad (7.2)*$$

4. Select suitable bar sizes.

5. Check that the area of steel actually provided is within the limits required by the code, that is

$$100 \frac{A_{s,max}}{bh} \le 4.0\%$$

and

$$100 \frac{A_{s,min}}{bd} \ge 26 \frac{f_{ctm}}{f_{yk}} \% \text{ and not less than } 0.13\%$$

where $f_{ctm} = 0.3 \times f_{ck}^{2/3}$ for $f_{ck} \le C50$

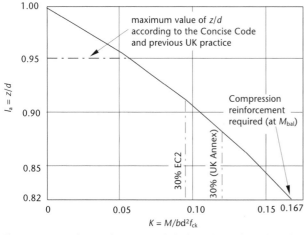

Figure 7.5
Lever-arm curve

The percentage values on the K axis mark the limits for singly reinforced
section with moment redistribution applied (see section 4.7 and table 4.2)

EXAMPLE 7.2

Design of tension reinforcement for a rectangular section, no moment redistribution

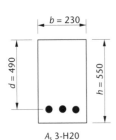

b = 230

d = 490

h = 550

A$_s$ 3-H20

Figure 7.6
Singly reinforced beam example

The beam section shown in figure 7.6 has characteristic material strengths of $f_{ck} = 25\,\text{N/mm}^2$ for the concrete and $f_{yk} = 500\,\text{N/mm}^2$ for the steel. The design moment at the ultimate limit state is 165 kN m which causes sagging of the beam.

1. $K = \dfrac{M}{bd^2 f_{ck}} = \dfrac{165 \times 10^6}{230 \times 490^2 \times 25} = 0.12$

   This is less than $K_{bal} = 0.167$ therefore compression steel is not required.

2. From the lever-arm curve of figure 7.5 $l_a = 0.88$, therefore lever arm $z = l_a d = 0.88 \times 490 = 431\,\text{mm}$ and

3. $A_s = \dfrac{M}{0.87 f_{yk} z} = \dfrac{165 \times 10^6}{0.87 \times 500 \times 431} = 880\,\text{mm}^2$

4. Provide three H20 bars, area = 943 mm^2.

5. For the steel provided $\dfrac{100 A_s}{bd} = \dfrac{100 \times 943}{230 \times 490} = 0.84 \;\; (> 0.13\%)$

   and

   $\dfrac{100 A_s}{bh} = \dfrac{100 \times 943}{230 \times 550} = 0.75 \;\; (< 4.0\%)$

   therefore the steel percentage is within the limits specified by the code.

7.2.2 Rectangular sections with tension and compression reinforcement, no moment redistribution

Compression steel is required whenever the concrete in compression, by itself, is unable to develop the necessary moment of resistance. Design charts such as the one in figure 4.9 may be used to determine the steel areas but the simplified equations based on the equivalent rectangular stress block are quick to apply. The arrangement of the reinforcement to resist a sagging moment is shown in figure 7.7

In order to have a ductile section so avoiding a sudden compressive failure of the concrete it is generally required that the maximum depth of the neutral axis is

Figure 7.7
Beam doubly reinforced to resist a sagging moment

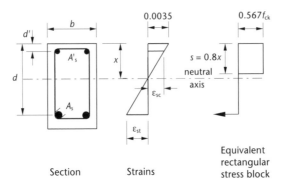

Section   Strains

Equivalent rectangular stress block

$x_{bal} = 0.45d$ and this is the value used in the design of a section with compression steel. *The design method and equations are those derived in Chapter 4 for sections subject to bending.*

The design steps are:

1. Calculate $K = \dfrac{M}{f_{ck}bd^2}$

   If $K > K_{bal} = 0.167$ compression reinforcement is required and $x = x_{bal} = 0.45d$.

2. Calculate the area of compression steel from

$$A_s' = \frac{(M - K_{bal}f_{ck}bd^2)}{f_{sc}(d - d')} \tag{7.3}*$$

   where f_{sc} is the compressive stress in the steel.

   If $d'/x \leq 0.38$ the compression steel has yielded and $f_{sc} = 0.87f_{yk}$

   If $d'/x > 0.38$ then the strain ε_{sc} in the compressive steel must be calculated from the proportions of the strain diagram and $f_{sc} = E_s\varepsilon_{sc} = 200 \times 10^3\varepsilon_{sc}$.

3. Calculate the area of tension steel required from

$$A_s = \frac{K_{bal}f_{ck}bd^2}{0.87f_{yk}z} + A_s'\frac{f_{sc}}{0.87f_{yk}} \tag{7.4}*$$

   with lever arm $z = 0.82d$.

4. Check for the areas of steel required and the areas provided that

$$(A_{s,prov}' - A_{s,req}') \geq (A_{s,prov} - A_{s,req}) \tag{7.5}$$

   This is to ensure that the depth of the neutral axis has not exceeded the maximum value of $0.45d$ by providing an over-excess of tensile reinforcement.

5. Check that the area of steel actually provided is within the limits required by the Code of Practice.

EXAMPLE 7.3

Design of tension and compression reinforcement, no moment redistribution

The beam section shown in figure 7.8 has characteristic material strengths of $f_{ck} = 25\,\text{N/mm}^2$ and $f_{yk} = 500\,\text{N/mm}^2$. The ultimate design moment is $165\,\text{kN m}$, causing hogging of the beam:

1.
$$\frac{M}{bd^2f_{ck}} = \frac{165 \times 10^6}{230 \times 330^2 \times 25}$$
$$= 0.26 > K_{bal} = 0.167$$

   so that compression steel is required.

2. $x = 0.45d = 0.45 \times 330 = 148\,\text{mm}$

   $d'/x = 50/148 = 0.34 < 0.38$

   therefore the compression steel has yielded and

   $f_{sc} = 0.87f_{yk}$

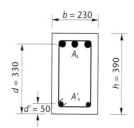

Figure 7.8
Beam doubly reinforced to resist a hogging moment

From equation 7.3

$$\text{Compression steel } A'_s = \frac{(M - 0.167f_{ck}bd^2)}{0.87f_{yk}(d - d')}$$

$$= \frac{(165 \times 10^6 - 0.167 \times 25 \times 230 \times 330^2)}{0.87 \times 500(330 - 50)}$$

$$= 496 \text{ mm}^2$$

Provide two H20 bars for A'_s, area $= 628 \text{ mm}^2$, bottom steel.

3. From equation 7.4

$$\text{Tension steel } A_s = \frac{0.167f_{ck}bd^2}{0.87f_{yk}z} + A'_s$$

$$= \frac{0.167 \times 25 \times 230 \times 330^2}{0.87 \times 500 \times 0.82 \times 330} + 496$$

$$= 888 + 496 = 1384 \text{ mm}^2$$

Provide three H25 bars for A_s, area $= 1470 \text{ mm}^2$, top steel.

4. Check equation 7.5 for the areas of steel required and provided for the compression and tension reinforcement to ensure ductility of the section

$$(A'_{s, \text{prov}} - A'_{s, \text{req}}) \geq (A_{s, \text{prov}} - A_{s, \text{req}})$$

That is

$$628 - 496 \ (= 132) > 1470 - 1384 \ (= 86) \text{ mm}^2$$

5. The bar areas provided are within the upper and lower limits specified by the code. To restrain the compression steel, at least 8 mm links at 300 mm centres should be provided.

7.3 Design for bending of a rectangular section with moment redistribution

The redistribution of the moments obtained from the elastic analysis of a concrete structure takes account of the plasticity of the reinforced concrete as it approaches the ultimate limit state. In order to achieve this plasticity the concrete sections must be designed so that plastic hinges can form with the yielding of the tensile reinforcement. This will result in a ductile structure that has a gradual failure at the ultimate limit state and not a sudden catastrophic failure of the concrete in compression. To ensure this EC2 places limits on the maximum depth of the neutral axis x_{bal} so that there are high strains in the tension steel allowing sufficient rotation of the section for plastic hinges to form. The limit to x_{bal} is set according to the amount of redistribution δ. For the EC2 code it is

$$x_{bal} \leq 0.8(\delta - 0.44)d \quad \text{for } f_{ck} \leq \text{C50} \tag{7.6a}$$

where

$$\delta = \frac{\text{moment at the section after redistribution}}{\text{moment at the section before redistribution}}$$

However the UK Annex to the EC2 modifies the limit to x_{bal} as

$$x_{bal} \leq (\delta - 0.4)d \tag{7.6b}*$$

Table 7.1   Moment redistribution factors for concrete classes $\leq$ C50/60

Redistribution (%)	δ	x_{bal}/d	z_{bal}/d	K_{bal}	d'/d
According to EC2, $k_1 = 0.44$ and $k_2 = 1.25$					
0	1.0	0.448	0.821	0.167	0.171
10	0.9	0.368	0.853	0.142	0.140
15	0.85	0.328	0.869	0.129	0.125
20[a]	0.8	0.288	0.885	0.116	0.109
25	0.75	0.248	0.900	0.101	0.094
30[b]	0.70	0.206	0.917	0.087	0.079
According to EC2, UK Annex, $k_1 = 0.4$ and $k_2 = 1.0$					
0	1.0	0.45	0.82	0.167	0.171
10	0.9	0.45	0.82	0.167	0.171
15	0.85	0.45	0.82	0.167	0.171
20[a]	0.8	0.40	0.84	0.152	0.152
25	0.75	0.35	0.86	0.137	0.133
30[b]	0.70	0.30	0.88	0.120	0.114

[a] Maximum permitted redistribution for class A normal ductility steel
[b] Maximum permitted redistribution for class B and C higher ductility steel, see section 1.6.2

In this chapter the examples will be based on the UK Annex's equation 7.6b, but, because many of the designs in the UK are for projects overseas which may require the use of the EC2 specifications, example 4.9 part (a) was based on the use of the EC2 equation 7.6a. Also table 7.1, which is a copy of table 4.2, lists all the relevant design factors such as x_{bal}, z_{bal} and K_{bal} for both the EC2 and the UK Annex equations so that the examples on moment redistribution in this chapter can be readily amended for use in terms of the EC2 equation. The ratio d'/d in table 7.1 sets the limiting upper value for the yield of the compression steel.

The moment redistribution is generally carried out on the maximum moments along a beam and these are generally the hogging moments at the beams supports. Example 3.9 on moment redistribution shows how the hogging moment may be reduced without increasing the maximum sagging moment in the bending moment envelope. Thus there is an economy on the amount of steel reinforcement required and a reduction of the congestion of steel bars at the beam–column connection.

The equations used in the design procedures that follow are based on the equations derived in section 4.7.

7.3.1   Singly reinforced rectangular sections with moment redistribution

The design procedure using the equations based on the UK Annex to EC2 is

1. Calculate $K = M/bd^2 f_{ck}$

2. Take K_{bal} from table 7.1, or alternatively calculate

$$K_{bal} = 0.454(\delta - 0.4) - 0.182(\delta - 0.4)^2 \quad \text{for } f_{ck} \leq C50$$

where δ = moment after redistribution/moment before redistribution

and check that $K < K_{bal}$. Therefore compression steel is not required.

3. Calculate $z = d\left[0.5 + \sqrt{(0.25 - K/1.134)}\right]$

4. Calculate $A_s = \dfrac{M}{0.87 f_{yk} z}$

5. Check that the area of steel provided is within the maximum and minimum limits required.

7.3.2 Rectangular sections with tension and compression reinforcement with moment redistribution applied (based on the UK Annex to EC2)

The steps in the design are :

1. Calculate $x_{bal} \leq (\delta - 0.4)d$

2. Calculate $K = M/bd^2 f_{ck}$

3. Take K_{bal} from table 7.1 or alternatively calculate

$$K_{bal} = 0.454(\delta - 0.4) - 0.182(\delta - 0.4)^2 \quad \text{for } f_{ck} \leq \text{C50}$$

If $K > K_{bal}$, compression steel is required.

4. Calculate the area of compression steel from

$$A'_s = \frac{(K - K_{bal}) f_{ck} bd^2}{f_{sc}(d - d')} \tag{7.7}*$$

where f_{sc} is the stress in the compression steel

If $d'/x \leq 0.38$ the compression steel has yielded and $f_{sc} = 0.87 f_{yk}$

If $d'/x > 0.38$ then the strain ε_{sc} in the compressive steel must be calculated from the proportions of the strain diagram and $f_{sc} = E_s \varepsilon_{sc} = 200 \times 10^3 \varepsilon_{sc}$.

5. Calculate the area of tension steel from

$$A_s = \frac{K_{bal} f_{ck} bd^2}{0.87 f_{yk} z} + A'_s \frac{f_{sc}}{0.87 f_{yk}} \tag{7.8}*$$

where $z = d - 0.8 x_{bal}/2$.

6. Check equation 7.5 for the areas of steel required and the areas provided that

$$(A'_{s, prov} - A'_{s, req}) \geq (A_{s, prov} - A_{s, req})$$

This is to ensure that the depth of the neutral axis has not exceeded the maximum value of x_{bal} by providing an over-excess of tensile reinforcement.

7. Check that the area of steel provided is within the maximum and minimum limits required.

EXAMPLE 7.4

Design of tension and compression reinforcement, with 20 per cent moment redistribution, $\delta = 0.8$ (based on the UK Annex to EC2)

The beam section shown in figure 7.9 has characteristic material strengths of $f_{ck} = 25 \, \text{N/mm}^2$ and $f_{yk} = 500 \, \text{N/mm}^2$. The ultimate moment is $370 \, \text{kN m}$, causing hogging of the beam.

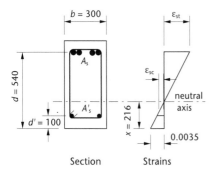

Figure 7.9
Beam doubly reinforced to
resist a hogging moment

Section   Strains

1. As the moment reduction factor $\delta = 0.80$, the limiting depth of the neutral axis is

$$x = (\delta - 0.4)d$$
$$= (0.8 - 0.4) \times 540 = 216 \, \text{mm}$$

2. $K = M/bd^2 f_{ck} = 370 \times 10^6/(300 \times 540^2 \times 25)$
$$= 0.169$$

3. $K_{bal} = 0.454(\delta - 0.4) - 0.182(\delta - 0.4)^2$
$$= 0.454(0.8 - 0.4) - 0.182(0.8 - 0.4)^2$$
$$= 0.152$$

$K > K_{bal}$ therefore compression steel is required.

4. $d'/x = 100/216 = 0.46 > 0.38$

therefore $f_{sc} < 0.87 f_{yk}$

From the proportions of the strain diagram

Steel compressive strain $\varepsilon_{sc} = \dfrac{0.0035(x - d')}{x}$

$$= \dfrac{0.0035(216 - 100)}{216} = 0.00188$$

Steel compressive stress $\quad = E_s \varepsilon_{sc}$
$$= 200\,000 \times 0.00188$$
$$= 376 \, \text{N/mm}^2$$

Compression steel $\quad A'_s = \dfrac{(K - K_{bal})f_{ck}bd^2}{f_{sc}(d - d')}$

$$= \dfrac{(0.169 - 0.152)25 \times 300 \times 540^2}{376(540 - 100)}$$

$$= 224 \, \text{mm}^2$$

Provide two H20 bars for A'_s, area $= 628 \, \text{mm}^2$, bottom steel.

5. Tension steel

$$A_s = \dfrac{K_{bal}f_{ck}bd^2}{0.87f_{yk}z} + A'_s \dfrac{f_{sc}}{0.87f_{yk}}$$

where

$$z = d - 0.8x/2 = 540 - 0.8 \times 216/2 = 454 \, \text{mm}$$

therefore

$$A_s = \frac{0.152 \times 25 \times 300 \times 540^2}{0.87 \times 500 \times 454} + 224 \times \frac{376}{0.87 \times 500}$$

$$= 1683 + 194 = 1877 \, \text{mm}^2$$

Provide four H25 bars for A_s, area $= 1960 \, \text{mm}^2$, top steel.

6. Check equation 7.5 for the areas of steel required and provided for the compression, and tension reinforcement to ensure ductility of the section

$$(A'_{s,\text{prov}} - A'_{s,\text{req}}) \geq (A_{s,\text{prov}} - A_{s,\text{req}})$$

That is

$$628 - 224 \, (= 404) > 1960 - 1877 \, (= 83) \, \text{mm}^2$$

7. These areas lie within the maximum and minimum limits specified by the code. To restrain the compression steel, at least 8 mm links at 300 mm centres should be provided.

7.4 Flanged beams

Figure 7.10 shows sections through a T-beam and an L-beam which form part of a concrete beam and slab floor with the slab spanning between the beams and the areas of the slab acting as the flanges of the beams as shown in figure 7.11. When the beams are resisting sagging moments, the slab acts as a compression flange and the members may be designed as T- or L-beams. With hogging moments the slab will be in tension and assumed to be cracked, therefore the beam must then be designed as a rectangular section of width b_w and overall depth h.

At intermediate supports of continuous beams where hogging moments occur the total area of tension reinforcement should be spread over the effective width of the flange as shown in figure 7.10. Part of the reinforcement may be concentrated over the web width.

The effective flange width b_{eff} is specified by the following equation:

$$b_{\text{eff}} = b_w + \sum b_{\text{eff},i}$$

where

$$b_{\text{eff},i} = 0.2b_i + 0.1l_0 \leq 0.2l_0 \quad \text{and also} \quad b_{\text{eff},i} \leq b_i$$

$2b_i$ is the clear distance between the webs of adjacent beams

l_0 is the distance between the points of contraflexure along the beam as shown in figure 7.11.

Figure 7.10
T-beam and L-beam

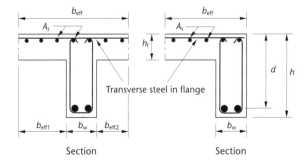

Section   Section

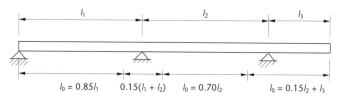

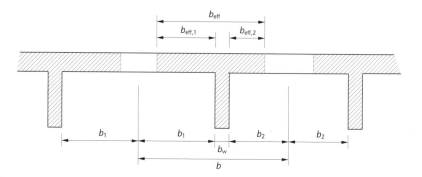

Note: (i) the length of the cantilever should be less than half the length of the adjacent span
(ii) the ratio of adjacent span lengths should be between 0.67 and 1.50

Figure 7.11
Dimensions to be used in
the calculation of effective
flange widths

So that for the interior span of a symmetrical T-beam with $b_1 = b_2 = b'$ and $l_0 = 0.7l$

$$b_{\text{eff}} = b_w + 2[0.2b' + 0.07l] \leq b_w + 2[0.14l]$$

For sagging moments the flanges act as a large compressive area. Therefore the stress block for the flanged beam section usually falls within the flange thickness. For this position of the stress block, the section may be designed as an equivalent rectangular section of breadth $b_f \, (= b_{\text{eff}})$.

Transverse reinforcement should be placed across the full width of the flange to resist the shear developed between the web and the flange, as described in section 5.1.4. Quite often this reinforcement is adequately provided for by the top steel of the bending reinforcement in the slab supported by the beam.

Design procedure for a flanged beam subject to a sagging moment

1. Calculate $\dfrac{M}{b_f d^2 f_{ck}}$ and determine l_a from the lever-arm curve of figure 7.5 or from equation 7.1

   lever arm $z = l_a d$ and the depth of the stress block $s = 2(d - z)$

   If $s \leq h$ the stress block falls within the flange depth, and the design may proceed as for a rectangular section, breadth b_f. On the very few occasions that the neutral axis does fall below the flange, reference should be made to the methods described in section 4.6.2 for a full analysis.

2. Design transverse steel in the top of the flange to resist the longitudinal shear stresses at the flange–web interface (see section 5.1.4).

   These longitudinal shear stresses are a maximum where the slopes dM/dx of the bending moment envelope are the greatest. That is (a) in the region of zero moment for the span sagging moments, and (b) the region of the maximum moments for the hogging moments at the supports.

EXAMPLE 7.5

Design of bending and transverse reinforcement for a T-section

A simply supported beam has a span $L = 6.0$ m and has the flanged cross-section shown in figure 7.12. The characteristic material strengths are $f_{ck} = 25$ N/mm^2 and $f_{yk} = 500$ N/mm^2 and the ultimate design uniformly distributed load w_u is 44 kN per metre.

Figure 7.12
T-beam

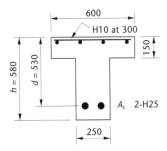

Maximum bending moment at mid-span is $M = \dfrac{44 \times 6^2}{8} = 198$ kNm

(1) Longitudinal reinforcement

$$\frac{M}{b_f d^2 f_{ck}} = \frac{198 \times 10^6}{600 \times 530^2 \times 25} = 0.047$$

From the lever-arm curve, figure 7.5, $l_a = 0.95$, therefore

lever arm $z = l_a d = 0.95 \times 530 = 503$ mm

depth of stress block $s = 2(d - z) = 2(530 - 503) = 54$ mm ($< h_f$)

Thus the stress block lies within the flange

$$A_s = \frac{M}{0.87 f_{yk} z} = \frac{198 \times 10^6}{0.87 \times 500 \times 503}$$

$$= 905 \text{ mm}^2$$

Provide two H25 bars, area $= 982$ mm^2. For these bars

$$\frac{100 A_s}{b_w d} = \frac{100 \times 982}{250 \times 530} = 0.74 \text{ per cent} > 0.13$$

Thus the steel percentage is greater than the minimum specified by the Code of Practice.

(2) Transverse steel in the flange

The design follows the procedures and equations set out in section 5.1.4

(i) Calculate the design longitudinal shear v_{Ed} at the web–flange interface

For a sagging moment the longitudinal shear stresses are the greatest over a distance of Δx measured from the point of zero moment and Δx is taken as half the distance to the maximum moment at mid-span, or $\Delta x = 0.5 \times L/2 = L/4 = 1500$ mm.

Therefore the change in moment ΔM over distance $\Delta x = L/4$ from the zero moment is

$$\Delta M = \frac{w_u \times L}{2} \times \frac{L}{4} - \frac{w_u \times L}{4} \times \frac{L}{8} = \frac{3w_u L^2}{32} = \frac{3 \times 44 \times 6^2}{32} = 149\,\text{kNm}$$

The change in longitudinal force ΔF at the flange–web interface is

$$\Delta F_d = \frac{\Delta M}{(d - h_f/2)} \times \frac{b_{fo}}{b_f}$$

where b_{fo} is the breadth of flange outstanding from the web.
Thus

$$\Delta F_d = \frac{\Delta M}{(d - h_f/2)} \times \frac{(b_f - b_w)/2}{b_f} \quad \text{as given on page 110}$$

$$= \frac{149 \times 10^3}{(530 - 150/2)} \times \frac{(600 - 250)/2}{600} = 96\,\text{kN}$$

The longitudinal shear stress v_{Ed} induced is

$$v_{Ed} = \frac{\Delta F_d}{(h_f \times \Delta x)} = \frac{96 \times 10^3}{150 \times 1500} = 0.43\,\text{N/mm}^2$$

(ii) Check the strength of the concrete strut

From equation 5.17, to prevent crushing of the compressive strut in the flange

$$v_{Ed} \leq \frac{0.6(1 - f_{ck}/250)f_{ck}}{1.5(\cot\theta_f + \tan\theta_f)}$$

The moments are sagging so the flange is in compression and the limits for θ_f are

$$26.5° \leq \theta_f \leq 45°$$

with $\theta_f = $ the minimum value of $26.5°$

$$\therefore v_{Ed(max)} = \frac{0.6(1 - 25/250) \times 25}{(2.0 + 0.5)} = 5.4 \quad (> 0.43\,\text{N/mm}^2)$$

and the concrete strut has sufficient strength with $\theta = 26.5°$ (for a flange in tension the limits on θ are $45° \geq \theta \geq 38.6°$ or $1.0 \leq \cot\theta \leq 1.25$).

(iii) Design transverse steel reinforcement

Transverse shear reinforcement is required if $v_{Ed} \geq 0.27 f_{ctk}$ where f_{ctk} is the characteristic axial tensile strength of concrete ($= 1.8\,\text{N/mm}^2$ for class 25 concrete).

The maximum allowable value of $v_{Ed} = 0.27 f_{ctk} = 0.27 \times 1.8 = 0.49\,\text{N/mm}^2$ (> 0.43) and transverse shear reinforcement is therefore not required.

A minimum area of 0.13% of transverse steel should be provided as given in table 6.8 or in table A3 in the Appendix.
Hence

$$A_{sf} = 0.13bh_f/100 = 0.13 \times 1000 \times 150/100 = 195\,\text{mm}^2/\text{m}$$

Provide H10 bars at 300 mm centres $= 262\,\text{mm}^2/\text{m}$ (see table A.3 in the Appendix).

Longitudinal reinforcement should also be provided in the flange as shown in figure 7.12.

7.5 | One-span beams

The following example describes the calculations for designing the bending reinforcement for a simply supported beam. It brings together many of the items from the previous sections. The shear reinforcement for this beam is designed later in example 7.7.

EXAMPLE 7.6

Design of a beam – bending reinforcement

The beam shown in figure 7.13 supports the following uniformly distributed loads

permanent load $g_k = 60$ kN/m, including self-weight

variable load $q_k = 18$ kN/m

The characteristic strengths of the concrete and steel are $f_{ck} = 30$ N/mm^2 and $f_{yk} = 500$ N/mm^2. Effective depth, $d = 540$ mm and breadth, $b = 300$ mm.

Figure 7.13
One-span beam-bending

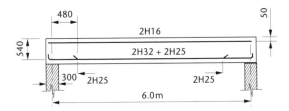

(a) Ultimate loading and maximum moment

Ultimate load $w_u = (1.35g_k + 1.5q_k)$ kN/m

$$= (1.35 \times 60 + 1.5 \times 18) = 108 \text{ kN/m}$$

therefore

maximum design moment $M = \dfrac{w_u L^2}{8} = \dfrac{108 \times 6.0^2}{8} = 486$ kN m

(b) Bending reinforcement

$$K = \frac{M}{bd^2 f_{ck}} = \frac{486 \times 10^6}{300 \times 540^2 \times 30} = 0.185 > K_{bal} = 0.167$$

Therefore compression reinforcement, A_s' is required.

$d'/d = 50/540 = 0.092 < 0.171$ in table 7.1, therefore $f_{sc} = 0.87 f_{yk}$

Compression steel $\quad A_s' = \dfrac{(K - K_{bal}) f_{ck} bd^2}{f_{sc}(d - d')}$

$$= \frac{(0.185 - 0.167) \times 30 \times 300 \times 540^2}{0.87 \times 500(540 - 50)} = 222 \text{ mm}^2$$

Provide two H16 bars, $A_s' = 402$ mm^2

Tension steel, $\quad A_s = \dfrac{0.167 f_{ck} bd^2}{0.87 f_{yk} z} + A_s'$

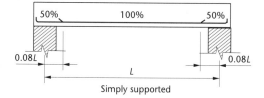

Figure 7.14
Simplified rules for curtailment of bars in beams

Simply supported

where, from the lever arm curve of figure 7.5 $l_a = 0.82$. Thus

$$A_s = \frac{0.167 \times 30 \times 300 \times 540^2}{0.87 \times 500 \times (0.82 \times 540)} + 222$$
$$= 2275 + 222 = 2497 \, \text{mm}^2$$

Provide two H32 bars and two H25 bars, area $= 2592 \, \text{mm}^2$, $100A_s/bd = 1.6 > 0.15$.

(c) Curtailment at support

The tension reinforcement should be anchored over the supports with a bend as shown in figure 7.14 which is based on past UK practice. Two bars may be curtailed near to the supports.

(d) Span–effective depth ratio

$$\rho = 100A_{s,\text{req}}/bd = (100 \times 2497)/(300 \times 540) = 1.54 \text{ per cent.}$$

From table 6.10 or figure 6.3 basic span–effective depth ratio $= 14$

Modification for steel area provided:

$$\text{Modified ratio} = 14.0 \times \frac{2592}{2497} = 14.5$$

$$\text{Span--effective depth ratio provided} = \frac{6000}{540} = 11.1$$

which is less than the allowable upper limit, thus deflection requirements are likely to be satisfied.

7.6 | Design for shear

The theory and design requirements for shear were covered in chapter 5 and the relevant design equations were derived based on the requirements of EC2 using the Variable Strut Inclination Method.

The shear reinforcement will usually take the form of vertical links or a combination of links and bent-up bars. Shear reinforcement may not be required in very minor beams such as door or window lintels with short spans of less than say 1.5 metres and light loads.

The following notation is used in the equations for the shear design

A_{sw} = the cross-sectional area of the two legs of the stirrup

s = the spacing of the stirrups

z = the lever arm between the upper and lower chord members of the analogous truss

f_{ywd} = the design yield strength of the stirrup reinforcement

f_{yk} = the characteristic strength of the stirrup reinforcement

V_{Ed} = the shear force due to the actions at the ultimate limit state

V_{wd} = the shear force in the stirrup

$V_{Rd,s}$ = the shear resistance of the stirrups

$V_{Rd,max}$ = the maximum design value of the shear which can be resisted by the concrete strut

7.6.1 Vertical stirrups or links

The procedure for designing the shear links is as follows

1. Calculate the ultimate design shear forces V_{Ed} along the beam's span.

2. Check the crushing strength $V_{Rd,max}$ of the concrete diagonal strut at the section of maximum shear, usually at the face of the beam support.

   For most cases the angle of inclination of the strut is $\theta = 22°$, with $\cot \theta = 2.5$ and $\tan \theta = 0.4$ so that from equation 5.6:

   $$V_{Rd,max} = 0.124 b_w d (1 - f_{ck}/250) f_{ck} \qquad (7.9)*$$

   and if $V_{Rd,max} \geq V_{Ed}$ then go to step (3) with $\theta = 22°$ and $\cot \theta = 2.5$

   but if $V_{Rd,max} < V_{Ed}$ then $\theta > 22°$ and θ must be calculated from equation 7.10 as

   $$\theta = 0.5 \sin^{-1} \left\{ \frac{V_{Ed}}{0.18 b_w d f_{ck} (1 - f_{ck}/250)} \right\} \leq 45° \qquad (7.10)*$$

3. The shear links required can be calculated from equation 7.11

   $$\frac{A_{sw}}{s} = \frac{V_{Ed}}{0.78 d f_{yk} \cot \theta} \qquad (7.11)*$$

   where A_{sw} is the cross-sectional area of the legs of the stirrups ($2 \times \pi \phi^2 / 4$ for single stirrups)

   For a predominately uniformly distributed load the shear V_{Ed} should be taken at a distance d from the face of the support and the shear reinforcement should continue to the face of the support.

4. Calculate the minimum links required by EC2 from

   $$\frac{A_{sw,min}}{s} = \frac{0.08 f_{ck}^{0.5} b_w}{f_{yk}} \qquad (7.12)*$$

   and the shear resistance for the links actually specified

   $$V_{min} = \frac{A_{sw}}{s} \times 0.78 d f_{yk} \cot \theta \qquad (7.13)*$$

   This value should be marked on the shear force envelope to show the extent of these links as shown in figure 7.16 of example 7.7.

5. Calculate the additional longitudinal tensile force caused by the shear force

   $$\Delta F_{td} = 0.5 V_{Ed} \cot \theta \qquad (7.14)*$$

   This additional tensile force increases the curtailment length of the tension bars as shown in section 7.9.

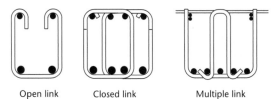

Figure 7.15
Types of shear link

Open link   Closed link   Multiple link

The minimum spacing of the links is governed by the requirements of placing and compacting the concrete and should not normally be less than about 80 mm. EC2 gives the following guidance on the maximum link spacing:

(a) Maximum longitudinal spacing between shear links in a series of links

$$s_{l, \max} = 0.75d(1 + \cot \alpha)$$

where α is the inclination of the shear reinforcement to the longitudinal axis of the beam.

(b) Maximum transverse spacing between legs in a series of shear links

$$s_{b, \max} = 0.75d \quad (\leq 600\text{mm})$$

Types of links or stirrups are shown in figure 7.15. The open links are usually used in the span of the beam with longitudinal steel consisting of top hanger bars and bottom tensile reinforcement. The closed links are used to enclose top and bottom reinforcement such as that near to the supports. Multiple links are used when there are high shear forces to be resisted.

EXAMPLE 7.7

Design of shear reinforcement for a beam

Shear reinforcement is to be designed for the one-span beam of example 7.6 as shown in figures 7.13 and 7.16. The total ultimate load is 108 kN/metre and the characteristic strengths of the concrete and steel are $f_{ck} = 30\,\text{N/mm}^2$ and $f_{yk} = 500\,\text{N/mm}^2$.

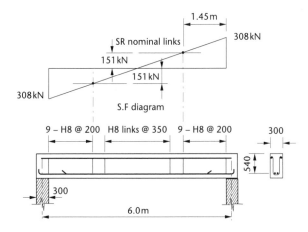

Figure 7.16
Non-continuous beam-shear reinforcement

1.45m

SR nominal links   308 kN

151 kN

151 kN

308 kN

S.F diagram

9 – H8 @ 200   H8 links @ 350   9 – H8 @ 200   300

300

540

6.0m

(a) Check maximum shear at face of support

Maximum design shear $= w_u \times$ effective span$/2 = 108 \times 6.0/2 = 324\,\text{kN}$

Design shear at face of support $V_{Ed} = 324 - 108 \times 0.15 = 308\,\text{kN}$

Crushing strength $V_{Rd,max}$ of diagonal strut, assuming angle $\theta = 22°$, $\cot\theta = 2.5$ is

$$V_{Rd,max} = 0.124 b_w d (1 - f_{ck}/250) f_{ck}$$
$$= 0.124 \times 300 \times 540 (1 - 30/250) \times 30 \times 10^{-3}$$
$$= 530\,\text{kN} \ (> V_{Ed} = 308\,\text{kN})$$

Therefore angle $\theta = 22°$ and $\cot\theta = 2.5$ as assumed.

(b) Shear links

At distance d from face of support the design shear is $V_{Ed} = 308 - w_u d = 308 - 108 \times 0.54 = 250\,\text{kN}$

$$\frac{A_{sw}}{s} = \frac{V_{Ed}}{0.78 d f_{yk} \cot\theta}$$

$$= \frac{250 \times 10^3}{0.78 \times 540 \times 500 \times 2.5} = 0.475$$

Using table A.4 in the Appendix

Provide 8 mm links at 200 mm centres, $A_{sw}/s = 0.503$.

(c) Minimum links

$$\frac{A_{sw,min}}{s} = \frac{0.08 f_{ck}^{0.5} b_w}{f_{yk}}$$

$$= \frac{0.08 \times 30^{0.5} \times 300}{500} = 0.26$$

Provide 8mm links at 350 mm centres, $A_{sw}/s = 0.287$.

The shear resistance of the links actually specified is

$$V_{min} = \frac{A_{sw}}{s} \times 0.78 d f_{yk} \cot\theta$$
$$= 0.287 \times 0.78 \times 540 \times 500 \times 2.5 \times 10^{-3} = 151\,\text{kN}$$

(d) Extent of shear links

Shear links are required at each end of the beam from the face of the support to the point where the design shear force is $V_{min} = 151\,\text{kN}$ as shown on the shear force diagram of figure 7.16.

From the face of the support

$$\text{distance } x = \frac{V_{Ed} - V_{min}}{w_u}$$

$$= \frac{308 - 151}{108} = 1.45\,\text{metres}$$

Therefore the number of H8 links at 200 mm centres required at each end of the beam is

$$1 + (x/s) = 1 + (1450/200) = 9$$

spaced over a distance of $(9 - 1)200 = 1600\,\text{mm}$.

(e) Additional longitudinal tensile force

$$\Delta F_{td} = 0.5 V_{Ed} \cot \theta$$
$$= 0.5 \times 308 \times 2.5$$
$$= 385 \, kN$$

This additional longitudinal tensile force is provided for by extending the curtailment point of the mid-span longitudinal reinforcement as discussed in section 7. 9.

7.6.2 Bent-up bars to resist shear

In regions of high shear forces it may be found that the use of links to carry the full force will cause steel congestion and lead to constructional problems. In these situations, consideration should be given to 'bending up' main reinforcement which is no longer required to resist bending forces but can be so used to resist part of the shear.

The equations for designing this type of shear reinforcement and the additional longitudinal tension force were derived in chapter 5 and are given below

$$\frac{A_{sw}}{s} = \frac{V_{Ed}}{0.78 d f_{yk} (\cot \alpha + \cot \theta) \sin \alpha}$$
$$\Delta F_{td} = 0.5 V_{Ed} (\cot \theta - \cot \alpha)$$

where α is the angle of inclination with the horizontal of the bent up bar.

Bent-up bars must be fully anchored past the point at which they are acting as tension members, as was indicated in figure 5.5.

EC2 also requires that the maximum longitudinal spacing of bent-up bars is limited to $0.6d(1 + \cot \alpha)$ and at least 50 per cent of the required shear reinforcement should be in the form of shear links.

7.7 Continuous beams

Beams, slabs and columns of a cast *in situ* structure all act together to form a continuous load-bearing structure. The reinforcement in a continuous beam must be designed and detailed to maintain this continuity by connecting adjacent spans and tying together the beam and its supporting columns. There must also be transverse reinforcement to unite the slab and the beam.

The bending-moment envelope is generally a series of sagging moments in the spans and hogging moments at the supports as in figure 7.17, but occasionally the hogging moments may extend completely over the span. Where the sagging moments occur the beam and slab act together, and the beam can be designed as a T-section. At the supports, the beam must be designed as a rectangular section because the hogging moments cause tension in the slab.

The moment of resistance of the concrete T-beam section is somewhat greater than that of the rectangular concrete section at the supports. Hence it is often advantageous to redistribute the support moments as described in chapter 3. By this means the design support moments can be reduced and the design span moments possibly increased.

Design of the beam follows the procedures and rules set out in the previous sections. Other factors which have to be considered in the detailed design are as follows:

1. At an exterior column the beam reinforcing bars which resist the design moments must have an anchorage bond length within the column.

2. In monolithic construction where a simple support has been assumed in the structural analysis, partial fixity of at least 25 per cent of the span moment should be allowed for in the design.

3. Reinforcement in the top of the slab must pass over the beam steel and still have the necessary cover. This must be considered when detailing the beam reinforcement and when deciding the effective depth of the beam at the support sections.

4. The column and beam reinforcement must be carefully detailed so that the bars can pass through the junctions without interference.

Figure 7.17 illustrates a typical arrangement of the bending reinforcement for a two-span continuous beam. The reinforcement has been arranged with reference to the bending-moment envelope and in accordance with the rules for anchorage and

Figure 7.17
Continuous beam arrangement of bending reinforcement

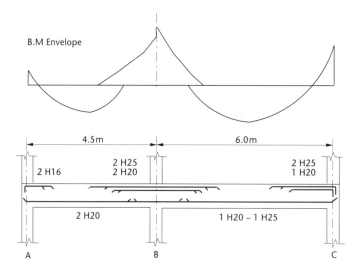

Figure 7.18
Typical arrangement of shear reinforcement

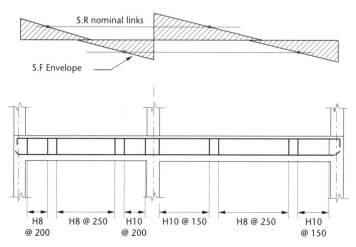

curtailment described in section 7.9. The application of these rules establishes the cut-off points beyond which the bars must extend at least a curtailment anchorage length. It should be noted that at the external columns the reinforcement has been bent to give a full anchorage bond length.

The shear-force envelope and the arrangement of the shear reinforcement for the same continuous beam are shown is figure 7.18. On the shear-force envelope the resistance of the minimum stirrups has been marked and this shows the lengths of the beam which need shear reinforcement. When designing the shear reinforcement, reference should be made to the arrangement of bending reinforcement to ensure that the longitudinal tension bars used to establish $V_{Rd,c}$ extend at least $d + l_{bd}$ beyond the section being considered.

EXAMPLE 7.8

Design of a continuous beam

The beam has a width, $b_w = 300$ mm and an overall depth, $h = 660$ mm with three equal spans, $L = 5.0$. In the transverse direction the beams spacings are $B = 4.0$ m centres with a slab thickness, $h_f = 180$ mm, as shown in figures 7.19 and 7.20. The supports have a width of 300 mm.

The uniformly distributed ultimate design load, $w_u = 190$ kN/m. The ultimate design moments and shears near mid-span and the supports are shown in figure 7.19.

The characteristic strengths of the concrete and steel are $f_{ck} = 30$ N/mm² and $f_{yk} = 500$ N/mm².

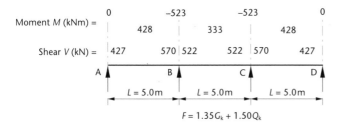

Figure 7.19
Continuous beam with ultimate design bending moments and shear forces shown

Total ultimate load on each span is

$$F = 190 \times 5.0 = 950 \, \text{kN}$$

Design for bending

(a) Mid-span of 1st and 3rd end spans – design as a T-section

Moment 428 kN m sagging

Effective width of flange

$$b_{eff} = b_w + 2[0.2b' + 0.1 \times 0.85L] \quad (\leq b_w + 2[0.2 \times 0.85L]) \text{ (see figure 7. 11)}$$

$$= 300 + 2[(0.2 \times (2000 - 300/2)) + (0.085 \times 5000)] = 1890 \, \text{mm}$$

$$b_w + 2[0.2 \times 0.85L] = 300 + 2[0.2 \times 0.85 \times 5000] = 2000 \, \text{mm}$$

Therefore $b_f = b_{eff} = 1890$ mm.

$$\frac{M}{b_f d^2 f_{ck}} = \frac{428 \times 10^6}{1890 \times 600^2 \times 30} = 0.022$$

From the lever-arm curve of figure 7.5, $l_a = 0.95$, therefore

$$z = 0.95 \times 600 = 570 \, \text{mm}$$

and

$$d - z = 600 - 570 = 30 \quad (< h_f/2)$$

so that the stress block must lie within the 180 mm thick flange and the section is designed as a rectangular section with $b = b_f$.

$$A_s = \frac{M}{0.87 f_{yk} z}$$

$$= \frac{428 \times 10^6}{0.87 \times 500 \times 570} = 1726 \, \text{mm}^2$$

Provide three H25 bars and two H16 bars, area $= 1872 \, \text{mm}^2$ (bottom steel).

(b) Interior supports – design as a rectangular section

$M = 523 \, \text{kN m hogging}$

$$\frac{M}{bd^2 f_{ck}} = \frac{523 \times 10^6}{300 \times 580^2 \times 30} = 0.173 > 0.167$$

Therefore, compression steel is required.

$$A'_s = \frac{(K - K') f_{ck} b d^2}{0.87 f_{yk} (d - d')}$$

$$= \frac{(0.173 - 0.167) \times 30 \times 300 \times 580^2}{0.87 \times 500 (580 - 50)} = 79 \, \text{mm}^2$$

This small area of reinforcement can be provided by extending the bottom span bars beyond the internal supports.

From the lever arm curve of figure 7.5 $l_a = 0.82$, therefore:

$$A_s = \frac{0.167 f_{ck} b d^2}{0.87 f_{yk} z} + A'_s$$

$$= \frac{0.167 \times 30 \times 300 \times 580^2}{0.87 \times 500 \times (0.82 \times 580)} + 79$$

$$= 2444 + 79 = 2523 \, \text{mm}^2$$

Provide four H25 bars plus two H20 bars, area $= 2588 \, \text{mm}^2$ (top steel). The arrangement of the reinforcement is shown in figure 7.20. At end support A two H25 bars have been provided as top continuity steel to meet the requirement of item (2) in section 7.7.

(c) Mid-span of interior 2nd span BC – design as a T-section

$M = 333 \, \text{kN m}$, sagging

From figure 7.11, effective flange width

$$b_{eff} = b_w + 2[0.2b' + 0.1 \times 0.70L](\leq b_w + 2[0.2 \times 0.70L])$$

$$= 300 + 2[(0.2 \times (2000 - 300/2)) + (0.07 \times 5000)] = 1740 \, \text{mm}$$

$$b_w + 2[0.2 \times 0.7L] = 300 + 2[0.2 \times 0.7 \times 5000] = 1700 \, \text{mm}$$

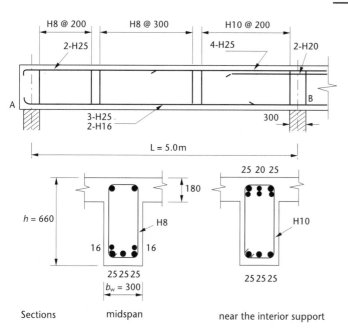

Figure 7.20
End-span reinforcement
details

Calculating $M/(b_f d^2 f_{ck})$ and using the lever-arm curve, it is found that $l_a = 0.95$

$$A_s = \frac{M}{0.87 f_{yk} z} = \frac{333 \times 10^6}{0.87 \times 500(0.95 \times 600)} = 1343 \,\text{mm}^2$$

Provide three H25 bars, area $= 1470 \,\text{mm}^2$ (bottom steel).

Design for shear

(a) Check for crushing of the concrete strut at the maximum shear force

Maximum shear is in spans AB and CD at supports B and C.
    At the face of the supports

$$V_{Ed} = 570 - w_u \times \text{support width}/2$$
$$= 570 - 190 \times 0.15 = 542 \,\text{kN}$$

Crushing strength of diagonal strut is

$$V_{Rd,\,max} = 0.124 b_w d(1 - f_{ck}/250) f_{ck} \quad \text{assuming angle } \theta = 22°, \cot\theta = 2.5$$
$$= 0.124 \times 300 \times 600(1 - 30/250) \times 30 = 589 \,\text{kN} \quad (> V_{ED} = 542 \,\text{kN})$$

Therefore angle $\theta = 22°$ and $\cot\theta = 2.5$ for all the shear calculations.

(b) Design of shear links

(i) Shear links in end spans at supports A and D

Shear distance d from face of support is $V_{Ed} = 427 - 190 \times (0.15 + 0.6) = 285 \,\text{kN}$

$$\frac{A_{sw}}{s} = \frac{V_{Ed}}{0.78 d f_{yk} \cot\theta}$$
$$= \frac{285 \times 10^3}{0.78 \times 600 \times 500 \times 2.5} = 0.49$$

Provide H8 links at 200 mm centres, $A_{sw}/s = 0.50$ (Table A4 in Appendix)

Additional longitudinal tensile force is

$$\Delta F_{td} = 0.5V_{Ed}\cot\theta$$
$$= 0.5 \times 285 \times 2.5$$
$$= 356\,\text{kN}$$

This additional longitudinal tensile force is provided for by extending the curtailment point of the longitudinal reinforcement, as discussed in section 7.9

(ii) Shear links in end spans at supports B and C

Shear V_{Ed} distance d from face of support is

$$V_{Ed} = 570 - 190(0.15 + 0.58)$$
$$= 431\,\text{kN}$$

Therefore:

$$\frac{A_{sw}}{s} = \frac{V_{Ed}}{0.78df_{yk}\cot\theta}$$
$$= \frac{431 \times 10^3}{0.78 \times 580 \times 500 \times 2.5}$$
$$= 0.762$$

Provide H10 links at 200 mm centres, $A_{sw}/s = 0.762$ (Table A4 in Appendix). Additional longitudinal tensile force is

$$\Delta F_{td} = 0.5V_{Ed}\cot\theta$$
$$= 0.5 \times 431 \times 2.5$$
$$= 539\,\text{kN}$$

This additional longitudinal tensile force is provided for by extending the curtailment point of the longitudinal reinforcement, as discussed in section 7.9.

(iii) Shear links in middle span BC at supports B and C

Shear distance d from the face of support $= 522 - 190(0.15 + 0.6) = 380\,\text{kN}$.
The calculations for the shear links would be similar to those for the other supports in sections (i) and (ii) giving 10 mm links at 225 mm centres.
The additional longitudinal tensile force, $F_{td} = 0.5 \times 380 \times 2.5 = 475\,\text{kN}$.

(iv) Minimum shear links

$$\frac{A_{sw,\min}}{s} = \frac{0.08f_{ck}^{0.5}b_w}{f_{yk}}$$
$$\frac{0.08 \times 30^{0.5} \times 300}{500}$$
$$= 0.263$$

Provide H8 links at 300 mm spacing, $A_{sw}/s = 0.335$ (Table A4 in Appendix). Shear resistance of links provided

$$V_{\min} = \frac{A_{sw}}{s} \times 0.78df_{yk}\cot\theta$$
$$= 0.335 \times 0.78 \times 600 \times 500 \times 2.5 \times 10^{-3}$$
$$= 196\,\text{kN}$$

(v) Extent of shear links

Links to resist shear are required over a distance x_i from the face of the supports to the point on the shear force diagram where the shear can be resisted by $V_{min} = 196\,kN$, as provided by the minimum links.

For the face of the end supports A and D the distance x_1 is

$$x_1 = \frac{V_{Ed} - V_{min}}{w_u} - 0.15$$

$$= \frac{427 - 196}{190} - 0.15 = 1.07\,m$$

For the interior supports B and C of the 1st and 3rd spans

$$x_2 = \frac{570 - 196}{190} - 0.15 = 1.82\,m$$

For the links at supports B and C in the middle span

$$x_3 = \frac{522 - 196}{190} - 0.15 = 1.57\,m$$

Based on these dimensions the links are arranged as shown in figure 7.20.

7.8 Cantilever beams and corbels

The effective span of a cantilever is either (a) the length to the face of the support plus half the beam's overall depth, h or (b) the distance to the centre of the support if the beam is continuous.

The moments, shears and deflections for a cantilever beam are substantially greater than those for a beam that is supported at both ends with an equivalent load. Also the moments in a cantilever can never be redistributed to other parts of the structure – the beam must always be capable of resisting the full static moment. Because of these factors and the problems that often occur with increased deflections due to creep, the design and detailing of a cantilever beam should be done with care.

Particular attention should be paid to the anchorage into the support of the top tension reinforcement. The steel should be anchored at the support by, at the very least, a full maximum anchorage length beyond the end of its effective span. Some design offices specify an anchorage length equal to the length of the cantilever, mostly to avoid steel fixing errors on site.

Loads on a cantilever can cause the adjacent interior span to be subjected to a hogging moment over all or most of its span. The critical loading pattern for this condition should be as shown in figure 7.21 where the maximum load on the cantilever together with minimum load on the interior span could cause a hogging moment to occur in the interior span.

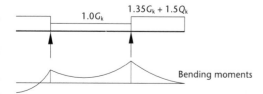

$1.0G_k$ $1.35G_k + 1.5Q_k$

Bending moments

Figure 7.21
Cantilever loading pattern

7.8.1   Design of corbels

A corbel, as shown in figure 7.22 is considered to be a short cantilever when $0.4h_c \leq a_c \leq h_c$ where h_c is the depth of the corbel at its junction with the column and a_c is the distance from the face of the column to the bearing of the vertical force, F_{Ed}.

When the vertical load has a *stiff bearing* a_c may be measured to the edge of the bearing but where a flexible bearing is used a_c is measured to the vertical force.

Corbels can be designed as a strut-and-tie system as illustrated in figure 7.22. In the figure the vertical load F_{Ed} at point B is resisted by the force F_{cd} in the inclined concrete strut CB and the force F_{td} in the horizontal steel tie AB.

The design and detailing of a corbel has the following requirements:

1. The bearing stress of the load on the corbel directly under the load should not exceed $0.48(1 - f_{ck}/250)f_{ck}$.

2. A horizontal force $H_{Ed} = 0.2F_{Ed}$ must also be resisted. This force acts at the level of the top of the bearing, a distance a_H above the horizontal tie.

3. The main tension steel, $A_{s,\,main}$ must be fully anchored into the column and the other end of these bars must be welded to an anchorage device or loops of reinforcing bars.

4. The angle of inclination, θ of the compression strut must be within the limits $22° \leq \theta \leq 45°$, or $2.5 \geq \cot\theta \geq 1.0$.

5. The design stress, f_{cd} of the concrete strut must not exceed $(\alpha_{cc}f_{ck}/\gamma_c)\nu_1$ where:

   $$\nu_1 = 0.6(1 - f_{ck}/250)$$
   $$\alpha_{cc} = 0.85$$

   $\gamma_c = 1.5$, the partial factor of safety for concrete in compression.

   Therefore f_{cd} must not exceed $0.34f_{ck}(1 - f_{ck}/250)$

6. Horizontal links of total area $A_{s,\,link}$ should be provided to confine the concrete in the compression strut and $\sum A_{s,\,link} = 0.5A_{s,\,main}$.

The strut and tie system of design

The forces on a corbel produce a complex combination of stresses due to bearing, shear, direct compression, direct tension and bending concentrated into a small area. The strut and tie system combined with good detailing is able to simplify the design to produce a workable and safe design.

Figure 7.22 shows the corbel with the inclined strut BC at an angle θ to the horizontal tie AB. The force in the strut is F_{cd} and F_{td} in the horizontal tie respectively. Point B is distance $a' = (a_c + 0.2a_H)$ from the face of the column because of the effect of the horizontal force, H_{Ed} ($= 0.2F_{Ed}$).

From the geometry of the triangle ABC, the lever-arm depth is given by $z = (a_c + 0.2a_H)\tan\theta$.

(a) Force in the concrete strut, F_{cd}

The design stress for the concrete strut is $f_{cd} = 0.34f_{ck}(1 - f_{ck}/250)$. From the geometry of figure 7.22 the width of the concrete strut measured vertically is $2(d - z)$. Hence, the width of the strut measured at right angles to its axis is given by $w_{strut} = 2(d - z)\cos\theta$.

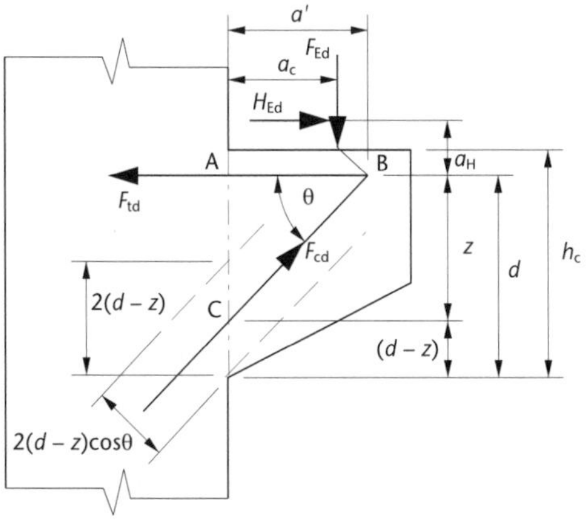

Figure 7.22
Strut and tie system in a corbel

Thus the force F_{cd} in the concrete strut is

$$
\begin{aligned}
F_{cd} &= f_{cd} \times w_{strut} \times b_w \\
&= f_{cd} \times 2(d - z) \times b_w \cos \theta
\end{aligned}
\tag{7.15}
$$

where b_w is the width of the corbel.

(b) Angle of inclination, θ of the concrete strut

Resolving vertically at point B:

$$
\begin{aligned}
F_{Ed} = F_{cd} \sin \theta &= f_{cd} \times 2(d - z) \times b_w \times \cos \theta \times \sin \theta \\
&= f_{cd} \times (d - a' \tan \theta) b_w \sin 2\theta \\
&= f_{cd} \times d \times b_w \left(1 - \left[\frac{a'}{d} \right] \tan \theta \right) \sin 2\theta
\end{aligned}
$$

Rearranging

$$
\frac{F_{Ed}}{f_{cd} d b_w} = \left(1 - \left[\frac{a'}{d} \right] \tan \theta \right) \sin 2\theta
\tag{7.16}
$$

This equation cannot be solved directly for θ but table 7.2 (overleaf), which has been developed directly from equation 7.16, can be used.

(c) Main tension steel, $A_{s,main}$

Resolving horizontally at B, the force F_{td} in the steel tie is given by

$$
F_{td} = F_{cd} \cos \theta = F_{Ed} \cos \theta / \sin \theta = F_{Ed} \cot \theta
$$

The *total* force F'_{td} in the steel tie, including the effect of the horizontal force of $0.2F_{Ed}$, is given by

$$
\begin{aligned}
F'_{td} &= F_{Ed} \cot \theta + 0.2F_{Ed} \\
&= F_{Ed} (\cot \theta + 0.2)
\end{aligned}
\tag{7.17}
$$

The area of main tension steel, $A_{s,main}$, is given by

$$
A_{s,main} = F'_{td} / 0.87 f_{yk}
\tag{7.18}
$$

Table 7.2 Values of θ to satisfy equation 7.16

θ (degs)	$\dfrac{F_{Ed}}{f_{cd}\,d\,b_w}$						
	$d'/d = 1$	$d'/d = .9$	$d'/d = .8$	$d'/d = .7$	$d'/d = .6$	$d'/d = .5$	$d'/d = .4$
22	0.429	0.458	0.487	0.516	0.545	0.574	0.603
23	0.400	0.429	0.459	0.488	0.518	0.547	0.577
24	0.371	0.401	0.431	0.461	0.490	0.520	0.550
25	0.343	0.373	0.403	0.433	0.463	0.493	0.523
26	0.315	0.345	0.375	0.405	0.435	0.466	0.496
27	0.288	0.318	0.348	0.378	0.408	0.438	0.468
28	0.262	0.292	0.321	0.351	0.381	0.411	0.440
29	0.236	0.266	0.295	0.324	0.354	0.383	0.412
30	0.211	0.240	0.269	0.298	0.327	0.356	0.385
31	0.187	0.216	0.244	0.272	0.300	0.328	0.357
32	0.164	0.192	0.219	0.247	0.274	0.301	0.329
33	0.143	0.169	0.195	0.222	0.248	0.275	0.301
34	0.122	0.147	0.172	0.198	0.223	0.248	0.274
35	0.103	0.126	0.150	0.174	0.198	0.222	0.246
36	0.085	0.107	0.129	0.152	0.174	0.197	0.219
37	0.068	0.089	0.109	0.130	0.151	0.172	0.193
38	0.053	0.072	0.091	0.110	0.129	0.147	0.166
39	0.040	0.056	0.073	0.090	0.107	0.124	0.141
40	0.028	0.043	0.057	0.072	0.086	0.101	0.115
41	0.018	0.030	0.042	0.054	0.067	0.079	0.091
42	0.010	0.020	0.029	0.039	0.048	0.057	0.067
43	0.005	0.011	0.018	0.024	0.031	0.037	0.044
44	0.001	0.005	0.008	0.011	0.015	0.018	0.021
45	0.000	0.000	0.000	0.000	0.000	0.000	0.000

EXAMPLE 7.9

Design of a corbel

Design the reinforcement for the corbel shown in figure 7.23. The corbel has a breadth $b = 350\,\text{mm}$ and supports an ultimate load of $V_{Ed} = 400\,\text{kN}$ at a distance $a_c = 200\,\text{mm}$ from the face of the column. The bearing is flexible and at a distance $a_H = 75\,\text{mm}$ above the tension tie. The bearing is 350 mm by 120 mm.

The characteristic material strengths are $f_{ck} = 30\,\text{N/mm}^2$, $f_{yk} = 500\,\text{N/mm}^2$.

Check the bearing stress

Safe bearing stress $= 0.48(1 - f_{ck}/250)f_{ck} = 0.48(1 - 30/250) \times 30 = 12.7\,\text{N/mm}^2$

Actual bearing stress $= 400 \times 10^3/(350 \times 120) = 9.6\,\text{N/mm}^2 < 12.7\,\text{N/mm}^2$

Concrete strut

The effective depth of the corbel is $d = 550\,\text{mm}$

Distance, $a' = (200 + 0.2 \times 75) = 215\,\text{mm}$

Therefore $a'/d = 215/550 = 0.40$

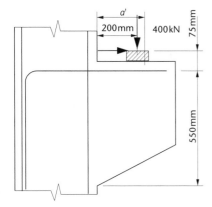

Figure 7.23
Corbel example

$$f_{cd} = 0.34 f_{ck}(1 - f_{ck}/250) = 0.34 \times 30 \times (1 - 30/250) = 8.98\,\text{N/mm}^2$$

$$\therefore \quad \frac{F_{Ed}}{f_{cd}db_w} = \frac{400 \times 1000}{8.98 \times 550 \times 350} = 0.232$$

Hence, from table 7.2, $\theta = 35.5°$.

Main tension steel

The force in the main tension steel is

$$F'_{td} = F_{Ed}(\cot\theta + 0.2) = 400(\cot 35.5° + 0.2) = 400(1.40 + 0.2) = 640\,\text{kN}$$

$$A_{s,\,main} = \frac{F'_{td}}{0.87 \times f_{yk}} = \frac{640 \times 10^3}{0.87 \times 500} = 1471\,\text{mm}^2$$

Provide two H32 bars, area $= 1610\,\text{mm}^2$.

Horizontal links

$$A_{s,\,links} > 0.5 A_{s,\,main} = 735\,\text{mm}^2$$

Provide four H16 links, $A_{s,\,links} = 804\,\text{mm}^2$.

Figure 7.24 shows the detailing of the reinforcement in the corbel.

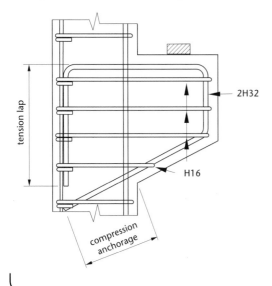

Figure 7.24
Reinforcement in corbel

7.9 | Curtailment and anchorage of reinforcing bars

As the magnitude of the bending moment on a beam decreases along its length so may the area of bending reinforcement be reduced by curtailing the bars since they are no longer required, as shown in figure 7.25. It should be recognised though that because of the approximations and assumptions made for the loading, the structural analysis and the behaviour of the reinforced concrete, the curtailment cannot be a particularly precise procedure. In addition the curtailment lengths are in many cases superseded by the requirements for serviceability, durability and detailing, such as maximum bar spacing, minimum bar numbers and curtailment beyond the critical sections for shear.

Each curtailed bar should extend a full anchorage length beyond the point at which it is no longer needed. The equations for an anchorage length were derived in section 5.2. The equation for the design anchorage length, l_{bd}, is

$$l_{bd} = \frac{f_{yk}}{4.6f_{bd}} \phi \times \alpha_n$$

where

α_n is a series of coefficients depending on the anchorage conditions

ϕ is the bar diameter

f_{bd} is the design bond strength which, for a beam, depends on the concrete strength and the bar size and whether the bar is in the top or bottom of the beam. The bar bonds better with the compacted concrete in the bottom of the beam.

For a straight bar with $\phi \leq 32$ mm, the order of anchorage lengths are $l_{bd} = 52\phi$ for a top bar and $l_{bd} = 36\phi$ for a bottom bar with class C30 concrete.

The curtailment of the tension reinforcement is based upon the envelope of tensile forces, F_s, derived from the bending moment envelope as shown in figure 7.25 such that at any location along the span

$$F_s = M_{Ed}/z + \Delta F_{td}$$

where

M_{Ed} is the design bending moment from the moment envelope

z is the lever arm

ΔF_{td} is the additional tensile force obtained from the design for shear

Figure 7.25
Curtailment of reinforcement
– envelope of tensile forces

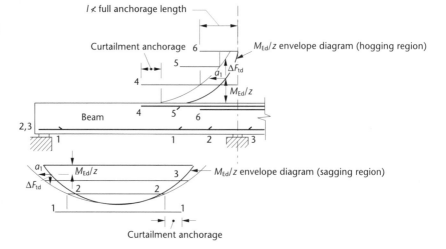

ΔF_{td} is a maximum where the shear force is a maximum at sections of zero moment, and ΔF_{td} is zero at the maximum moment near to mid-span and the interior support.

For members where shear reinforcement is not required the tensile force envelope may be estimated by simply 'shifting' the bending moment envelope diagram horizontally by a distance $a_l\ (= d)$ as shown in figure 7.25.

To determine the curtailment positions of each reinforcing bar the tensile force envelope is divided into sections as shown, in proportion to the area of each bar. In figure 7.25 the three bars provided for the sagging envelope and the four for the hogging envelope are considered to be of equal area so the envelope is divided into three equal sections for the sagging part of the envelope and four for the hogging part.

When considering the curtailment the following rules must also be applied:

1. At least one-quarter of the bottom reinforcement should extend to the supports

2. The bottom reinforcement at an end support should be anchored into the support as shown in figure 7.26.

3. At an end support where there is little or no fixity the bottom steel should be designed to resist a tensile force of $0.5V_{Ed}$ to allow for the tension induced by the shear with a minimum requirement of 25% of the reinforcement provided in the span.

4. At an end support where there is fixity but it has been analysed as a simple support, top steel should be designed and anchored to resist at least 25 per cent of the maximum span moment.

5. At internal supports the bottom steel should extend at least 10 bar diameters ϕ beyond the face of the support. To achieve continuity and resistance to such factors as accidental damage or seismic forces, splice bars should be provided across the support with a full anchorage lap on each side as shown in figure 7.27.

6. Where the loads on a beam are substantially uniformly distributed, simplified rules for curtailment may be used. These rules only apply to continuous beams if the characteristic variable load does not exceed the characteristic permanent load and the spans are approximately equal. Figure 7.28 shows the rules in diagrammatic form. However it should be noted that these rules do not appear in EC2 and are based on previous established UK practice.

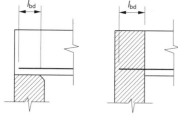

(1) Beam supported on wall or column

(2) Beam intersecting another supporting beam

Figure 7.26
Anchorage of bottom reinforcement at end supports

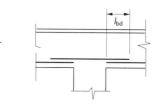

Figure 7.27
Anchorage at intermediate supports

Figure 7.28
Simplified rules for curtailment
of bars in beams

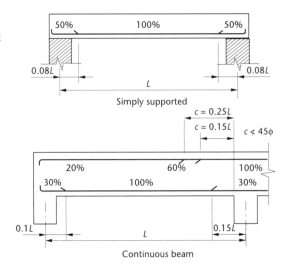

7.10　Design for torsion

The theory and design requirements for torsion were covered in section 5.4. The design procedure consists of calculations to determine additional areas of links and longitudinal reinforcement to resist the torsional moment, using an equivalent hollow box section.

Usually it is not necessary to design for torsion in statically indeterminate structures where the torsional forces are often only a secondary effect and the structure can be in equilibrium even if the torsion is neglected. When the equilibrium depends on the torsional resistance the effects of torsion must be considered.

7.2.1　Design procedure for torsion combined with shear

Notation

T_{Ed}　　Design torsion moment

$T_{Rd,max}$　Maximum torsional moment of resistance

V_{Ed}　　Design shear force

$V_{Rd,max}$　Maximum shear resistance based on crushing of the concrete

The following section outlines the procedure for designing for torsion and explains how torsional design must be considered together with the design for shear.

(1) Design for shear using the Variable Strut Inclination Method

The procedure for this is described in sections 5.1.2 and 7.6 and illustrated with examples 7.7 and 7.8.

Use the procedures previously described to determine the angle of inclination θ of the concrete compressive strut and the stirrup reinforcement to resist the shear forces. Also required is the additional horizontal tensile force ΔF_{td}.

The angle θ should range between 22° and 45° so that $\cot\theta$ is between 2.5 and 1.0. The value determined for θ should be used throughout the subsequent sections of the design.

(2) Convert the section into an equivalent hollow box section of thickness t

(See figure 7.29b.)

$$t = \frac{\text{Area of the section}}{\text{Perimeter of the section}} = \frac{A}{u}$$

so that for a rectangular section $b \times h$

$$t = \frac{bh}{2(b+h)}$$

Calculate the area A_k within the centreline of the equivalent hollow box section. For a rectangular section

$$A_k = (b - t)(h - t)$$

and the perimeter of the centreline is

$$u_k = 2(b + h - 2t)$$

(3) Check that the concrete section is adequate to resist the combined shear and torsion using the interaction condition

$$\frac{T_{Ed}}{T_{Rd, max}} + \frac{V_{Ed}}{V_{Rd, max}} \leq 1.0$$

where

$$T_{Rd, max} = \frac{1.33 v_1 f_{ck} t_{ef} A_k}{\cot \theta + \tan \theta}$$

and

$$v_1 = 0.6(1 - f_{ck}/250)$$

(4) Calculate the additional stirrup reinforcement required to resist torsion

$$\frac{A_{sw}}{s} = \frac{T_{Ed}}{2A_k 0.87 f_{yk} \cot \theta}$$

The spacing s of the stirrups should not exceed the lesser of (a) $u_k/8$, (b) $0.75d$ or (c) the least dimension of the beam's cross-section. The stirrups should be of the closed type fully anchored by means of laps.

(5) Calculate the total amount of stirrup reinforcement A_{sw}/s

This is the sum of the stirrup reinforcement for shear and torsion from steps (1) and (4).

(6) Calculate the area A_{sl} of the additional longitudinal reinforcement

$$A_{sl} = \frac{T_{Ed} u_k \cot \theta}{2A_k 0.87 f_{ylk}}$$

This reinforcement should be arranged so that there is at least one bar at each corner with the other bars distributed equally around the inner periphery of the links spaced at not more than 350 mm centres.

EXAMPLE 7.9

Design of torsional reinforcement

Torsional reinforcement is to be designed for the beam of examples 7.6 and 7.7 which is also subject to an ultimate torsional moment of $T_{Ed} = 24.0\,\text{kNm}$ in addition to the uniformly distributed loading of 108 kN/metre already considered in the previous examples.

The beam cross-section is shown in figure 7.29a. The steps in the calculations are numbered as outlined in the previous description of the design procedure. The characteristic strengths of the concrete and steel are $f_{ck} = 30\,\text{N/mm}^2$ and $f_{yk} = 500\,\text{N/mm}^2$.

1. Design for shear using the Variable Strut Inclination Method (see the design calculations of example 7.7)

   From example 7.7:

   $V_{Ed} = 308\,\text{kN}$ and $V_{Rd,max} = 530\,\text{kN}$

   The angle of inclination of the concrete strut is $\theta = 22°$ with $\cot\theta = 2.5$ and $\tan\theta = 0.4$

   For the shear links $\dfrac{A_{sw}}{s} = 0.475$ (required)

   The additional longitudinal tensile force $\Delta F_{td} = 385\,\text{kN}$

2. Convert the rectangular section to an equivalent hollow box section (see figure 7.29b)

   Thickness of box section $\quad t = \dfrac{A}{u} = \dfrac{600 \times 300}{2(600 + 300)}$

   $\qquad\qquad\qquad\qquad\qquad\qquad = 100\,\text{mm}$

   Area within centreline $\quad A_k = (b - t)(h - t)$

   $\qquad\qquad\qquad\qquad\qquad\quad = 200 \times 500$

   $\qquad\qquad\qquad\qquad\qquad\quad = 100 \times 10^3\,\text{mm}^2$

   Perimeter of centreline $\quad u_k = 2(b + h - 2t)$

   $\qquad\qquad\qquad\qquad\qquad\quad = 1400\,\text{mm}$

Figure 7.29
Torsion example

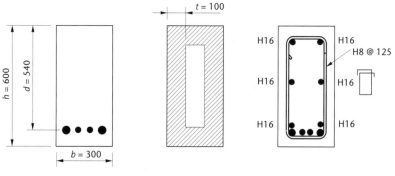

(a) Cross-section   (b) Equivalent box section   (c) Reinforcement details

3. Check if concrete section is adequate

$$\frac{T_{Ed}}{T_{Rd,max}} + \frac{V_{Ed}}{V_{Rd,max}} \le 1.0$$

where

$$T_{Rd,max} = \frac{1.33 v_1 f_{ck} t_{ef} A_k}{(\cot\theta + \tan\theta)}$$ (see equation 5.31)

with

$$v_1 = 0.6(1 - f_{ck}/250) = 0.6(1 - 30/250) = 0.528$$

Therefore

$$T_{Rd,max} = \frac{1.33 \times 0.528 \times 30 \times 100 \times 100 \times 10^{-3}}{(2.5 + 4.0)}$$

$$= 72.6\,kN \quad (> T_{Ed} = 24.0\,kN)$$

and

$$\frac{T_{Ed}}{T_{Rd,max}} = \frac{V_{Ed}}{V_{Rd,max}} = \frac{24.0}{72.6} + \frac{308}{530} = 0.33 + 0.58$$

$$= 0.91 < 1.0$$

Therefore the concrete section is adequate.

4. Calculate the additional link reinforcement required to resist torsion. (Note that A_{sw} is for one leg only)

$$\frac{A_{sw}}{s} = \frac{T_{Ed}}{2 A_k 0.87 f_{yk} \cot\theta}$$

$$= \frac{24.0 \times 10^6}{2 \times 100 \times 10^3 \times 0.87 \times 500 \times 2.5} = 0.110$$

5. Therefore for shear plus torsion and based on the area of two legs

$$\frac{A_{sw}}{s} = 0.475 + 2 \times 0.110 = 0.695$$

For 8 mm stirrups at 125 mm centres $A_{sw}/s = 0.805$ (see Appendix table A4)
Spacing $s = 125$ mm $(< u_k/8 = 175$ mm$)$

6. Calculate the area A_{sl} of the additional longitudinal reinforcement required for torsion

$$A_{sl} = \frac{T_{Ed} u_k \cot\theta}{2 A_k 0.87 f_{ylk}}$$

$$= \frac{(24 \times 10^6 \times 1400) \times 2.5}{2 \times 100 \times 10^3 \times 0.87 \times 500} = 966\,mm^2$$

This additional longitudinal steel can be provided for by six H16 bars, one in each corner and one in each of the side faces as shown in figure 7.29(c). The additional longitudinal tensile force of 385 kN resulting from the design for shear will be catered for by appropriate curtailment of the main tensile reinforcement as previously discussed in section 7.9.

7.11　Serviceability and durability requirements

The requirements for the serviceability and durability limit states have been covered extensively in Chapter 6 so this section is a short review of the factors that apply to the design and detailing of beams. Although this section is a short review at the end of the chapter it should be emphasised that the design for the serviceability and durability limit states is just as important as the design for the ultimate limit state. Failures of structures at the ULS are fortunately quite rare but can get a lot of publicity, whereas failures due to durability and serviceability are much more common during the life of a structure and they can quite easily lead eventually to a structural failure or be one of the primary causes of a failure. Also poor detailing and construction can be the cause of such problems as leaking roofs and basements and disfigurement of the structure with consequent high maintenance costs and reduced working life.

Adequate concrete cover to all the reinforcement bars is all-important to prevent ingress of moisture and corrosion of the steel bars with resultant staining and spalling of the concrete. Cover of the concrete is also required for fire resistance. The detailing and sizing of the reinforcing bars and stirrups should take account of the dimensional tolerances during bending and fabrication of the steel cages in order to maintain the required concrete cover.

The maximum and minimum spacing of the steel bars should meet the requirements of EC2 so that there is ample room for the flow and compaction of the concrete, but not be so large a gap that there is a lack of resistance to cracking of the concrete due to shrinkage, thermal movement and settlement.

For similar reasons the requirements for maximum and minimum percentages of reinforcement in concrete members must be always be checked.

The beams should be sufficiently stiff to prevent excessive deflections that could cause cracking of such features as floor finishes, glazing and partitions. This is more likely with long span beams or cantilevers. For most beams it is not necessary to carry out detailed deflection calculations. EC2 provides equations and basic span-to-depth ratios to meet this requirement. Compression steel in the compression zones of long span beams and cantilevers helps to resist the long term deflections due to creep.

Many of the more commonly used equations and tables from EC2 to meet all the above requirements are more fully described in Chapter 6 and are outlined in the Appendix at the end of the book for easy reference.

Good working practices and quality control on the construction site are also important to ensure such features as correctly designed concrete mixes, secure fixing of the formwork and reinforcing steel with adequate placement, compaction and curing of the concrete.

CHAPTER 8

Design of reinforced concrete slabs

CHAPTER INTRODUCTION

Reinforced concrete slabs are used in floors, roofs and walls of buildings and as the deck of bridges. The floor system of a structure can take many forms such as *in situ* solid slabs, ribbed slabs or precast units. Slabs may span in one direction or in two directions and they may be supported on monolithic concrete beams, steel beams, walls or directly by the structure's columns.

Continuous slabs should in principle be designed to withstand the most unfavourable arrangements of loads, in the same manner as beams. As for beams, bending moment coefficients, as given in table 8.1, may be used for one-way spanning slabs. These coefficients are comparable to those in figure 3.9 for continuous beams and are based on UK experience. If these coefficients are used the reinforcement must be of ductility class B or C and the neutral-axis depth, x, should be no greater than 0.25 of the effective depth such that the lever arm, z ($= d - 0.8x/2$), is not less than $0.9d$ to allow for moment redistribution incorporated in the values given (which may be up to 20 per cent). In addition, as for beams, table 8.1 should only be used when there are at least three spans that do not differ in length by more than 15 per cent, and Q_k should be less than or equal to $1.25G_k$ and also less than $5 \, \text{kN/m}^2$.

→

→

The moments in slabs spanning in two directions can also be determined using tabulated coefficients. Slabs which are not rectangular in plan or which support an irregular loading arrangement may be analysed by techniques such as the yield line method or the Hilleborg strip method, as described in section 8.9.

Concrete slabs are defined as members where the breadth is not less than 5 times the overall depth and behave primarily as flexural members with the design similar to that for beams, although in general it is somewhat simpler because:

1. the breadth of the slab is already fixed and a unit breadth of 1 m is used in the calculations;
2. the shear stresses are usually low in a slab except when there are heavy concentrated loads; and
3. compression reinforcement is seldom required.

Minimum thicknesses and axis distances for fire resistance are given in table 6.5 but deflection requirements will usually dominate.

Table 8.1 Ultimate bending moment and shear force coefficients in one-way spanning slabs

| | End support condition | | | | | | |
| | Pinned | | Continuous | | | | |
	Outer support	Near middle of end span	End support	End span	At first interior support	At middle of interior spans	At interior supports
Moment	0	0.086Fl	– 0.04Fl	0.075Fl	– 0.086Fl	0.063Fl	– 0.063Fl
Shear	0.4F	–	0.46F	–	0.6F	–	0.5F

l = effective span.

Area of each bay $\geq$ 30 m^2. (A bay is a strip of slab across the structure between adjacent rows of columns.)

F = total ultimate load = $1.35G_k + 1.50Q_k$.

8.1 | Shear in slabs

The shear resistance of a solid slab may be calculated by the procedures given in chapter 5. Experimental work has indicated that, compared with beams, shallow slabs fail at slightly higher shear stresses and this is incorporated into the values of the ultimate concrete shear resistance, $V_{Rd,c}$, as given by equations 5.1 and 5.2. Calculations are usually based on a strip of slab 1 m wide.

Since shear stresses in slabs subject to uniformly distributed loads are generally small, shear reinforcement will seldom be required and it would be usual to design the slab such that the design ultimate shear force, V_{Ed}, is less than the shear strength of the unreinforced section, $V_{Rd,c}$. In this case it is not necessary to provide any shear reinforcement. This can conveniently be checked using Table 8.2 which has been

Table 8.2 Shear resistance of slabs without shear reinforcement $v_{Rd,c}$ N/mm^2 (Class C30/35 concrete)

$\rho_1 = A_s/bd$	Effective depth, d (mm)								
	≤ 200	225	250	300	350	400	500	600	750
0.25%	0.54	0.52	0.50	0.47	0.45	0.43	0.40	0.38	0.36
0.50%	0.59	0.57	0.56	0.54	0.52	0.51	0.48	0.47	0.45
0.75%	0.68	0.66	0.64	0.62	0.59	0.58	0.55	0.53	0.51
1.00%	0.75	0.72	0.71	0.68	0.65	0.64	0.61	0.59	0.57
1.25%	0.80	0.78	0.76	0.73	0.71	0.69	0.66	0.63	0.61
1.50%	0.85	0.83	0.81	0.78	0.75	0.73	0.70	0.67	0.65
2.00%	0.94	0.91	0.89	0.85	0.82	0.80	0.77	0.74	0.71
k	2.000	1.943	1.894	1.816	1.756	1.707	1.632	1.577	1.516

Table 8.3 Concrete strength modification factor

f_{ck} (N/mm^2)	25	30	35	40	45	50
Modification factor	0.94	1.00	1.05	1.10	1.14	1.19

derived from Equations 5.1 and 5.2 for class C30 concrete on the basis that the allowable shear stress in the unreinforced slab is given by

$$v_{Rd,c} = \frac{V_{Rd,c}}{bd}$$

In this case, the applied ultimate shear stress

$$v_{Ed} = \frac{V_{Ed}}{bd} \le v_{Rd,c}$$

Table 8.2 also clearly illustrates the effect of increasing slab thickness on the depth related factor k, as noted above. Where different concrete strengths are used, the values in table 8.2 may be modified by the factors in table 8.3 provided $\rho_1 \ge 0.4\%$.

As for beams, the section should also be checked to ensure that V_{Ed} does not exceed the maximum permissible shear force, $V_{Rd,max}$. If shear reinforcement is required then the methods given in chapter 5 can be used, although practical difficulties concerned with bending and fixing shear reinforcement make it unlikely that shear reinforcement could be provided in slabs less than 200 mm thick.

Localised 'punching' actions due to heavy concentrated loads may, however, cause more critical conditions as shown in the following sections.

8.1.1 Punching shear – analysis

A concentrated load on a slab causes shearing stresses on a section around the load; this effect is referred to as punching shear. The critical surface for checking punching shear is shown as the perimeter in figure 8.1 which is located at $2.0d$ from the loaded area. The maximum force that can be carried by the slab without shear reinforcement ($V_{Rd,c}$) can be obtained using the values of $v_{Rd,c}$ given in table 8.2 based on equations 5.1 and 5.2 for normal shear in beams and slabs, where $\rho_1 = \sqrt{(\rho_y \rho_z)}$ where ρ_y and ρ_z are

Figure 8.1
Punching shear

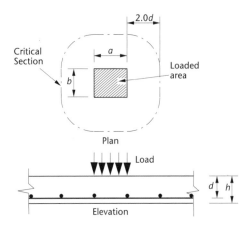

the reinforcement ratios, A_s/bd in the two mutually perpendicular directions (see table A.3 in the Appendix for A_s) then

$$V_{Rd,c} = v_{Rd,c}du \qquad (8.1)$$

where

d = effective depth of section [average of the two steel layers in perpendicular directions $= \left(\dfrac{d_y + d_z}{2}\right)$]

u = length of the punching shear perimeter.

If there are axial loads in the plane of the slab an additional term is added to $V_{Rd,c}$ to allow for the effect of these axial loads. This term is $+0.1\sigma_{cp}$ where σ_{cp} is the average of the normal compressive stresses acting in the y and z directions (from prestress or external forces). Such compressive stresses thus increase the punching shear resistance whilst conversely tensile stresses reduce the shear capacity.

Checks must also be undertaken to ensure that the maximum permissible shear force ($V_{Rd,max}$) is not exceeded at the face of the loaded area. The maximum permissible shear force is given by $V_{Rd,max} = 0.5v_1f_{cd}ud = 0.5v_1(f_{ck}/1.5)ud$ where u is the perimeter of the loaded area and v_1, the strength reduction factor $= 0.6(1 - f_{ck}/250)$.

EXAMPLE 8.1

Punching shear

A slab, 175 mm thick, average effective depth 145 mm is constructed with C25/30 concrete and reinforced with 12 mm bars at 150 mm centres one way (754 mm²/m) and 10 mm bars at 200 mm centres in the other direction (393 mm²/m). Determine the maximum ultimate load that can be carried on an area 300×400 mm.

For the unreinforced section, the first critical perimeter

$$u_1 = (2a + 2b + 2\pi \times 2d)$$
$$= 2(a + b) + 4\pi d$$
$$= 2(300 + 400) + 4\pi \times 145$$
$$= 3222 \text{ mm}$$

hence from equation 8.1

$$V_{Rd,c} = v_{Rd,c} \times 3222 \times 145$$
$$= 467\,190 v_{Rd,c}$$

Average steel ratio

$$\rho_1 = \sqrt{\rho_y \times \rho_z}$$

where

$$\rho_y = 754/(1000 \times 145) = 0.0052$$
$$\rho_z = 393/(1000 \times 145) = 0.0027$$

hence

$$\rho_1 = \sqrt{(0.0052 \times 0.0027)} = 0.0038 = 0.38\%$$

Thus from table 8.2, for a 175 mm slab, $v_{Rd,c} = 0.56\,N/mm^2$ for a class C30 concrete and from table 8.3 for class C25 concrete, as used here, modification factor = 0.94. Hence, maximum ultimate load

$$V_{Rd,c} = 0.94 \times 0.56 \times 467\,190 \times 10^{-3}$$
$$= 246\,kN$$

The maximum permissible shear force based on the face of the loaded area is given by the maximum shear resistance

$$V_{Rd,max} = 0.5ud\left[0.6\left(1 - \frac{f_{ck}}{250}\right)\right]\frac{f_{ck}}{1.5}$$
$$= 0.5 \times 2(300 + 400) \times 145 \times \left[0.6\left(1 - \frac{25}{250}\right)\right]\frac{25}{1.5} \times 10^{-3}$$
$$= 914\,kN$$

which clearly exceeds the value $V_{Rd,c}$ based on the first critical perimeter. Hence the maximum load that the slab can carry is 246 kN.

8.1.2 Punching shear – reinforcement design

If reinforcement is required to resist shear around the control perimeter indicated in Figure 8.1, it should be placed between not more than $0.5d$ from the loaded area and a distance $1.5d$ inside the outer control perimeter at which shear reinforcement is no longer required. The length of this is given by $u_{out,ef} = V_{Ed}/(v_{Rd,c}d)$ from which the necessary distance from the loaded area can be calculated. If this is less than $3d$ from the face of the loaded area, then reinforcement should be placed in the zone between $0.3d$ and $1.5d$ from this face.

Vertical links will normally be used and provided around at least two perimeters not more than $0.75d$ apart. Link spacing around a perimeter within $2d$ of the face of the loaded area should not be greater than $1.5d$, increasing to a limit of $2.0d$ at greater perimeters. Provided that the slab is greater than 200 mm thick overall then the amount of reinforcement required is given by:

$$A_{sw} \sin \alpha \geq \frac{v_{Rd,cs} - 0.75 v_{Rd,c}}{1.5 \dfrac{d \times f_{ywd,ef}}{s_r \times u_1 d}}$$

where

A_{sw} is the total area of shear reinforcement in one perimeter (mm^2)

s_r is the radial spacing of perimeters of shear reinforcement

$f_{ywd,ef}$ is the effective design strength of the shear reinforcement and is given by $f_{ywd,ef} = 250 + 0.25d \leq f_{ywd}$.

$v_{Rd,cs}$ is the punching shear resistance of the reinforced slab and α is the angle between shear reinforcement and the plane of the slab, so that $\sin \alpha = 1$ for vertical reinforcement.

This expression effectively allows for a 75 per cent contribution from the unreinforced concrete slab, and for vertical links can be expressed as:

$$A_{sw} \geq \frac{v_{Rd,cs} - 0.75 v_{Rd,c}}{1.5 \left(\dfrac{f_{ywd,ef}}{s_r u_1} \right)}$$

where the required $v_{Rd,cs}$ would be given by $\dfrac{V_{Ed}}{u_1 d}$.

A check must also be made that the calculated reinforcement satisfies the minimum requirement that:

$$A_{sw,min} \geq \frac{0.08 \sqrt{f_{ck}} (s_r . s_t)}{1.5 f_{yk}} = \frac{0.053 \sqrt{f_{ck}} (s_r . s_t)}{f_{yk}}$$

where s_t is the spacing of links around the perimeter and $A_{sw,min}$ is the area of *an individual link leg*.

Similar procedures must be applied to the regions of flat slabs which are close to supporting columns, but allowances must be made for reduced critical perimeters near slab edges and the effect of moment transfer from the columns as described in section 8.6.

EXAMPLE 8.2

Design of punching shear reinforcement

A 260 mm thick slab of class C25/30 concrete is reinforced by 12 mm high yield bars at 125 mm centres in each direction. The slab is subject to a dry environment and must be able to carry a localised concentrated ultimate load of 650 kN over a square area of 300 mm side. Determine the shear reinforcement required for $f_{yk} = 500 \, \text{N/mm}^2$.

For exposure class XC-I, cover required for a C25/30 concrete is 25 mm, thus average effective depth for the two layers of steel and allowing for 8 mm links is equal to

$$260 - (25 + 8 + 12) = 215 \, \text{mm}$$

(i) Check maximum permissible force at face of loaded area

Maximum shear resistance:

$$V_{Rd,\,max} = 0.5ud\left[0.6\left(1 - \frac{f_{ck}}{250}\right)\right]\frac{f_{ck}}{1.5}$$

$$= 0.5(4 \times 300) \times 215 \times \left[0.6\left(1 - \frac{25}{250}\right)\right]\frac{25}{1.5} \times 10^{-3}$$

$$= 1161\,\text{kN}\quad (> V_{Ed} = 650\,\text{kN})$$

(ii) Check control perimeter 2d from loaded face

Perimeter $u_1 = 2(a + b) + 4\pi d$

$$= 2(300 + 300) + 4\pi \times 215 = 3902\,\text{mm}$$

hence for concrete without shear reinforcement the shear capacity is given by:

$$V_{Rd,\,c} = v_{Rd,\,c} \times 3902 \times 215 = 838\,930v_{Rd,\,c}$$

bending steel ratio

$$\rho_1 = \frac{A_s}{bd} = \frac{905}{1000 \times 215} = 0.0042\quad (> 0.40\text{ per cent})$$

hence from table 8.2, $v_{Rd,\,c} = 0.56$ for class C30 concrete and, from table 8.3, modification factor for class C25 concrete $= 0.94$ then

$$V_{Rd,\,c} = 838\,930 \times 0.56 \times 0.94 \times 10^{-3}$$

$$= 442\,\text{kN}\quad (< V_{Ed} = 650\,\text{kN})$$

and punching shear reinforcement is required.

(iii) Check outer perimeter at which reinforcement is not required

$$u_{out,\,ef} = \frac{V_{Ed}}{v_{Rd,\,c}d} = \frac{650 \times 10^3}{0.56 \times 0.94 \times 215} = 5743\,\text{mm}$$

This will occur at a distance xd from the face of the loaded area, such that

$$5743 = 2(300 + 300) + 2\pi \times 215 \times x$$

and $x = 3.36\quad (> 3.0)$

(iv) Provision of reinforcement

Shear reinforcement should thus be provided within the zone extending from a distance not greater than $0.5d$ and less than $(3.36 - 1.5)d = 1.86d$ from the loaded face.

For perimeters $\leq 0.75d$ apart, 3 perimeters of steel will thus be adequate located at $0.4d$, $1.15d$ and $1.9d$, i.e. 85, 245 and 400 mm from the face of the loaded area (i.e. $s_r \approx 0.75d = 160$ mm apart).

Since all perimeters lie within $2d$ ($= 430$ mm) of the load and maximum link spacing, (s_t), is limited to $1.5d$ ($= 323$ mm).

The minimum link leg area is therefore given by:

$$A_{sw,\,min} = \frac{0.053\sqrt{f_{ck}}(s_r.s_t)}{f_{yk}} = \frac{0.053\sqrt{25}(160 \times 323)}{500}$$

$$= 27.3\,\text{mm}^2\text{ which is satisfied by a 6 mm diameter bar }(28.3\text{mm}^2)$$

Hence the assumed 8mm links will be adequate.

The area of steel required/perimeter is thus given by:

$$A_{sw} \geq \frac{v_{Rd,cs} - 0.75v_{Rd,c}}{1.5\frac{(f_{ywd,ef})}{s_r \times u_1}}$$

where, for the outer perimeter

$$V_{Rd,cs} = \frac{V_{Ed}}{u_1 d} = \frac{650 \times 10^3}{3902 \times 215} = 0.775 \, \text{N/mm}^2$$

$$v_{Rd,c} = 0.94 \times 0.56 = 0.526 \, \text{N/mm}^2 \quad \text{(as above)}$$

$$f_{ywd,ef} = 250 + 0.25 \times 215 = 303 \, \text{N/mm}^2 \quad (\leq 500)$$

and $\qquad s_r = 160 \, \text{mm}$

thus

$$A_{sw} \geq \frac{(0.775 - 0.75 \times 0.526) \times 160 \times 3902}{1.5 \times 303}$$

$$= 523 \, \text{mm}^2$$

(v) Number of links

The area of one leg of an 8 mm link is $50.3 \, \text{mm}^2$. Hence the number of link-legs required $= 523/50.3 = 11$ on the outer perimeter. The same number of links can conveniently be provided around each of the 3 proposed perimeters as summarised in the table below.

The table indicates the number of single-leg 8 mm diameter links (area $= 50.3 \, \text{mm}^2$) proposed for each of the three reinforcement perimeters taking account of the maximum required spacing and practical fixing considerations. Bending reinforcement is spaced at 125 mm centres in both directions; hence link spacing is set at multiples of this value.

Distance from load face (mm)	Length of perimeter (mm)	Required link spacing (mm)	Proposed link spacing (mm)	Proposed number of links
85	1734	158	125	14
245	2739	249	250	11
400	3713	323*	250	15

*maximum allowed

8.2 | Span–effective depth ratios

Excessive deflections of slabs will cause damage to the ceiling, floor finishes or other architectural finishes. To avoid this, limits are set on the span–depth ratio. These limits are exactly the same as those for beams as described in section 6.2. As a slab is usually a slender member, the restrictions on the span–depth ratio become more important and this can often control the depth of slab required. In terms of the span–effective depth ratio, the depth of slab is given by

$$\text{minimum effective depth} = \frac{\text{span}}{\text{basic ratio} \times \text{correction factors}}$$

The correction factors account for slab type and support conditions as well as cases of spans greater than 7 metres and for flat slabs greater than 8.5 metres. The basic ratio may also be corrected to account for grades of steel other than grade 500 and for when more reinforcement is provided than that required for design at the ultimate limit state. Initial values of basic ratio may be obtained from tables (e.g. table 6.10) but these are concrete strength dependent.

It may normally be assumed that, in using such tables, slabs are lightly stressed although a more exact determination can be made from figure 6.3 when the percentage of tension reinforcement is known. It can be seen that the basic ratio can be increased by reducing the stress condition in the concrete. The concrete stress may be reduced by providing an area of tension reinforcement greater than that required to resist the design moment up to a maximum of $1.5 \times$ that required.

In the case of two-way spanning slabs, the check on the span–effective depth ratio should be based on the *shorter* span length. This does not apply to flat slabs where the longer span should be checked.

8.3 | Reinforcement details

To resist cracking of the concrete, codes of practice specify details such as the minimum area of reinforcement required in a section and limits to the maximum and minimum spacing of bars. Some of these rules are as follows.

(a) Minimum areas of reinforcement

$$\text{minimum area} = 0.26 f_{ctm} b_t d / f_{yk} \geq 0.0013 b_t d$$

in both directions, where b is the mean width of the tensile zone of section. The minimum reinforcement provision for crack control, as specified in section 6.1.5 may also have to be considered where the slab depth exceeds 200 mm. Secondary transverse reinforcement should not be less than 20 per cent of the minimum main reinforcement requirement in one way slabs.

(b) Maximum areas of longitudinal and transverse reinforcement

$$\text{maximum area} = 0.04 A_c$$

where A_c is the gross cross-sectional area. This limit applies to sections away from areas of bar lapping.

(c) Maximum spacing of bars
For slabs not exceeding 200 mm thickness, bar spacing should not exceed three times the overall depth of slab or 400 mm whichever is the lesser for main reinforcement, and $3.5h$ or 450 mm for secondary reinforcement. In areas of concentrated load or maximum moment, these values are reduced to $2h \leq 250$ mm and $3h \leq 400$ mm respectively.

(d) Reinforcement in the flange of a T- or L-beam
This is described in detail in section 5.1.4.

(e) Curtailment and anchorage of reinforcement
The general rules for curtailment of bars in flexural members were discussed in section 7.9. Simplified rules for curtailment in different types of slab are illustrated in the subsequent sections of this chapter. At a simply supported end, at least half the span reinforcement should be anchored as specified in figure 7.26 and at an unsupported edge U bars with leg length at least $2h$ should be provided, anchored by top and bottom transverse bars.

8.4 | Solid slabs spanning in one direction

The slabs are designed as if they consist of a series of beams of 1 m breadth. The main steel is in the direction of the span and secondary or distribution steel is required in the transverse direction. The main steel should form the outer layer of reinforcement to give it the maximum lever arm.

The calculations for bending reinforcement follow a similar procedure to that used in beam design. The lever arm curve of figure 4.5 is used to determine the lever arm (z) and the area of tension reinforcement is then given by

$$A_s = \frac{M}{0.87 f_{yk} z}$$

For solid slabs spanning one-way the simplified rules for curtailing bars as shown in figure 8.2 may be used provided that the loads are uniformly distributed. With a continuous slab it is also necessary that the spans are approximately equal. These simplified rules are not given in EC2 but are recommended on the basis of proven satisfactory performance established in previous codes of practice.

8.4.1 Single-span solid slabs

The basic span–effective depth ratio for this type of slab is 20:1 on the basis that it is 'lightly stressed' and that grade 500 steel is used in the design. For a start-point in design a value above this can usually be estimated (unless the slab is known to be heavily loaded) and subsequently checked once the main tension reinforcement has been designed.

The effective span of the slab may be taken as the clear distance between the face of the supports plus a distance at both ends taken as the lesser of (a) the distance from the face of the support to its centreline and (b) one-half of the overall depth of the slab.

Figure 8.2
Simplified rules for curtailment of bars in slab spanning in one direction

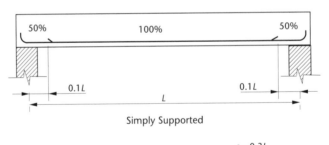

Simply Supported

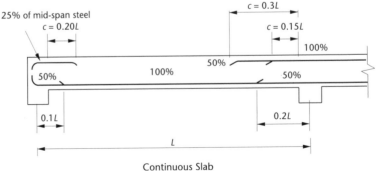

Continuous Slab

EXAMPLE 8.3

Design of a simply supported slab

The slab shown in figure 8.3 is to be designed to carry a variable load of 3.0 kN/m^2 plus floor finishes and ceiling loads of 1.0 kN/m^2. The characteristic material strengths are $f_{ck} = 25$ N/mm^2 and $f_{yk} = 500$N/mm^2. Basic span–effective depth ratio = 19 for a lightly stressed slab from Figure 6.3 for class C25/30 concrete and $\rho = 0.5\%$.

For simplicity, take the effective span to be 4.5 m between centrelines of supports.

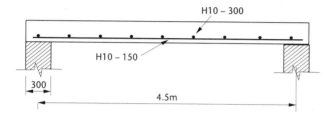

H10 – 300

H10 – 150

300

4.5m

Figure 8.3
Simply supported slab example

(1) First design solution

Estimate of slab depth

Try a basic span-depth ratio of 27 (approx. 40% above value from figure 6.3):

$$\text{minimum effective depth} = \frac{\text{span}}{27 \times \text{correction factors (c.f.)}}$$
$$= \frac{4500}{27 \times \text{c.f.}} = \frac{167}{\text{c.f.}}$$

As high yield steel is being used and the span is less than 7 m the correction factors can be taken as unity. Try an effective depth of 170 mm. For a class XC-1 exposure the cover = 25 mm. Allowing, say, 5 mm as half the bar diameter of the reinforcing bar:

overall depth of slab = $170 + 25 + 5 = 200$ mm

Slab loading

self-weight of slab = $200 \times 25 \times 10^{-3} = 5.0$ kN/m^2

total permanent load = $1.0 + 5.0 = 6.0$ kN/m^2

For a 1 m width of slab:

ultimate load = $(1.35g_k + 1.5q_k)4.5$

$= (1.35 \times 6.0 + 1.5 \times 3.0)4.5 = 56.7$ kN

$M = 56.7 \times 4.5/8 = 31.9$ kN m

Bending reinforcement

$$\frac{M}{bd^2f_{ck}} = \frac{31.9 \times 10^6}{1000 \times 170^2 \times 25} = 0.044$$

From the lever-arm curve of figure 4.5, $l_a = 0.96$. Therefore adopt upper limit of 0.95 and lever-arm $z = l_a d = 0.95 \times 170 = 161$ mm:

$$A_s = \frac{M}{0.87f_{yk}z} = \frac{31.9 \times 10^6}{0.87 \times 500 \times 161} = 455 \text{ mm}^2/\text{m}$$

Provide H10 bars at 150 mm centres, $A_s = 523$ mm^2/m (as shown in table A.3 in the Appendix).

Check span–effective depth ratio

$$\rho = \frac{100A_{s,\,req}}{bd} = \frac{100 \times 455}{1000 \times 170} = 0.268\% \quad (>0.13\% \text{ minimum requirement})$$

From figure 6.3, this corresponds to a basic span–effective depth ratio of 32. The actual ratio $= 4500/170 = 26.5$; hence the chosen effective depth is acceptable.

Shear

At the face of the support

$$\text{Shear } V_{Ed} = \frac{55.5}{2}\left(\frac{2.25 - 0.5 \times 0.3}{2.25}\right) = 25.9\,\text{kN}$$

$$\rho_1 = \frac{100 \times 523}{1000 \times 170} = 0.31$$

$V_{Rd,c} = v_{Rd,c}bd$ where $v_{Rd,c}$ from table 8.2 $= 0.55$ (note: no concrete strength adjustment since $\rho_1 < 0.4\%$). Thus:

$$V_{Rd,c} = 0.55 \times 1000 \times 170 = 93.5\,\text{kN}$$

as V_{Ed} is less than $V_{Rd,c}$ then no shear reinforcement is required.

End anchorage (figure 7.26)

From the table of anchorage lengths in the Appendix the tension anchorage length $= 40\phi = 40 \times 10 = 400\,\text{mm}$.

Distribution steel

Provide minimum $= 0.0013bd = 0.0013 \times 1000 \times 170 = 221\,\text{mm}^2/\text{m}$.

Provide H10 at 300 mm centres (262 mm^2/m) which satisfies maximum bar spacing limits.

(2) Alternative design solution

The second part of this example illustrates how a smaller depth of slab is adequate provided it is reinforced with steel in excess of that required for bending thus working at a lower stress in service. Try a thickness of slab, $h = 170\,\text{mm}$ and $d = 140\,\text{mm}$:

$$\text{self-weight of slab} = 0.17 \times 25 = 4.25\,\text{kN/m}^2$$

$$\text{total permanent load} = 1.0 + 4.25 = 5.25\,\text{kN/m}^2$$

$$\text{ultimate load} = (1.35g_k + 1.5q_k)$$

$$= (1.35 \times 5.25 + 1.5 \times 3.0)4.5 = 52.1\,\text{kN}$$

Bending reinforcement

$$M = 52.1 \times \frac{4.5}{8} = 29.3\,\text{kN m}$$

$$\frac{M}{bd^2 f_{ck}} = \frac{29.3 \times 10^6}{1000 \times 140^2 \times 25} = 0.060$$

From the lever-arm curve of figure 4.5, $l_a = 0.945$. Therefore, lever-arm $z = l_a d = 0.945 \times 140 = 132\,\text{mm}$:

$$A_s = \frac{M}{0.87f_{yk}z} = \frac{29.3 \times 10^6}{0.87 \times 500 \times 132} = 510\,\text{mm}^2/\text{m}$$

Provide H10 bars at 150 mm centres, $A_s = 523\,\text{mm}^2/\text{m}$.

Check span–effective depth ratio

$$\rho = \frac{100A_s}{bd} = \frac{100 \times 510}{1000 \times 140} = 0.364\%$$

From figure 6.3 this corresponds to a basic span–effective depth ratio of 24.0.

Actual $\quad\quad \dfrac{\text{Span}}{\text{Eff. depth}} = \dfrac{45\,000}{140} = 32.1$

This is inadequate but can be overcome by increasing the steel area.

Limiting $\quad\quad \dfrac{\text{Span}}{\text{Eff. depth}} = \text{basic ratio} \times \dfrac{A_{s,\,prov}}{A_{s,\,req}}$

Try 10mm bars at 100 mm centres, $A_{s,\,prov} = 785 \text{ mm}^2/\text{m}$.
Hence:

$$\frac{A_{s,\,prov}}{A_{s,\,req}} = \frac{785}{510} = 1.54$$

Upper limit to correction factor (UK National Annex) = 1.5.

Hence allowable $\dfrac{\text{span}}{\text{effective depth}} = 24 \times 1.5 = 36$ which is greater than that provided.

Therefore $d = 140$ mm is adequate.

8.4.2 Continuous solid slab spanning in one direction

For a continuous slab, bottom reinforcement is required within the span and top reinforcement over the supports. The effective span is the distance between the centreline of the supports and the basic span–effective depth ratio of an interior span is 30.0 for 'lightly stressed' where grade 500 steel and class C30/35 concrete are used. The corresponding limit for an end span is 26.0.

If the conditions given on page 209 are met, the bending moment and shear force coefficients given in table 8.1 may be used.

EXAMPLE 8.4

Design of a continuous solid slab

The four-span slab shown in figure 8.4 supports a variable load of 3.0 kN/m^2 plus floor finishes and a ceiling load of 1.0 kN/m^2. The characteristic material strengths are $f_{ck} = 25 \text{ N/mm}^2$ and $f_{yk} = 500 \text{ N/mm}^2$.

Estimate of slab depth

As the end span is more critical than the interior spans, try a basic span–effective depth ratio 30 per cent above the end-span limit of 26.0 (i.e. 33.0):

$$\text{minimum effective depth} = \frac{\text{span}}{33.0 \times \text{correction factor}}$$
$$= \frac{4500}{33.0 \times \text{c.f.}} = \frac{136}{\text{c.f.}}$$

Figure 8.4
Continuous slab

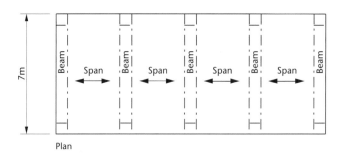

Plan

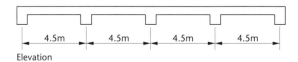

Elevation

As high yield steel is being used and the span is less than 7 m the correction factor can be taken as unity. Try an effective depth of 140 mm. For a class XC-1 exposure the cover = 25 mm. Allowing, say, 5 mm as half the bar diameter of the reinforcing bar:

$$\text{overall depth of slab} = 140 + 25 + 5$$
$$= 170\,\text{mm}$$

Slab loading

$$\text{self-weight of slab} = 170 \times 25 \times 10^{-3} = 4.25\,\text{kN/m}^2$$
$$\text{total permanent load} = 1.0 + 4.25 = 5.25\,\text{kN/m}^2$$

For a 1 m width of slab

$$\text{ultimate load, } F = (1.35g_k + 1.5q_k)4.5$$
$$= (1.35 \times 5.25 + 1.5 \times 3.0)4.5 = 52.14\,\text{kN}$$

Using the coefficients of table 8.1, assuming the end support is pinned, the moment at the middle of the end span is given by

$$M = 0.086Fl = 0.086 \times 52.14 \times 4.5 = 20.18\,\text{kN m}$$

Bending reinforcement

$$\frac{M}{bd^2f_{ck}} = \frac{20.18 \times 10^6}{1000 \times 140^2 \times 25} = 0.0412$$

From the lever-arm curve of figure 4.5, $l_a = 0.96$. Therefore, lever-arm $z = l_a d = 0.95 \times 140 = 133$ mm:

$$A_s = \frac{M}{0.87f_{yk}z} = \frac{20.18 \times 10^6}{0.87 \times 500 \times 133}$$
$$= 349\,\text{mm}^2/\text{m}$$

Provide H10 bars at 200 mm centres, $A_s = 393\,\text{mm}^2/\text{m}$.

Check span–effective depth ratio

$$\frac{100A_{s,\,req}}{bd} = \frac{100 \times 349}{1000 \times 140} = 0.249$$

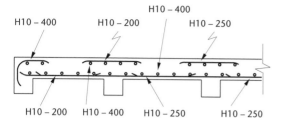

Figure 8.5
Reinforcement in a continuous slab

From figure 6.3 this corresponds to a basic span–effective depth ratio in excess of 32×1.3 (for an end span) $= 41$. The actual ratio $= 4500/140 = 32.1$; hence the chosen effective depth is acceptable.

Similar calculations for the supports and the interior span give the steel areas shown in figure 8.5.

At the end supports there is a monolithic connection between the slab and the beam, therefore top steel should be provided to resist any negative moment. The moment to be designed for is a minimum of 25 per cent of the span moment, that is 5.1 kN m. In fact, to provide a minimum of 0.13 per cent of steel, H10 bars at 400 mm centres have been specified. The layout of the reinforcement in figure 8.5 is according to the simplified rules for curtailment of bars in slabs as illustrated in figure 8.2.

$$\text{Transverse reinforcement} = 0.0013bd$$
$$= 0.0013 \times 1000 \times 140$$
$$= 182 \, \text{mm}^2/\text{m}$$

Provide H10 at 400 mm centres top and bottom, wherever there is main reinforcement $(196 \, \text{mm}^2/\text{m})$.

8.5 | Solid slabs spanning in two directions

When a slab is supported on all four of its sides it effectively spans in both directions, and it is sometimes more economical to design the slab on this basis. The amount of bending in each direction will depend on the ratio of the two spans and the conditions of restraint at each support.

If the slab is square and the restraints are similar along the four sides then the load will span equally in both directions. If the slab is rectangular then more than one-half of the load will be carried in the stiffer, shorter direction and less in the longer direction. If one span is much longer than the other, a large proportion of the load will be carried in the short direction and the slab may as well be designed as spanning in only one direction.

Moments in each direction of span are generally calculated using tabulated coefficients. Areas of reinforcement to resist the moments are determined independently for each direction of span. The slab is reinforced with bars in both directions parallel to the spans with the steel for the shorter span placed furthest from the neutral axis to give it the greater effective depth.

The span–effective depth ratios are based on the shorter span and the percentage of reinforcement in that direction.

With a uniformly distributed load the loads on the supporting beams may generally be apportioned as shown in figure 8.6.

Figure 8.6
Loads carried by supporting beams

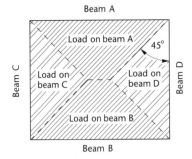

8.5.1 Simply supported slab spanning in two directions

A slab simply supported on its four sides will deflect about both axes under load and the corners will tend to lift and curl up from the supports, causing torsional moments. When no provision has been made to prevent this lifting or to resist the torsion then the moment coefficients of table 8.4 may be used and the maximum moments are given by

$$M_{sx} = a_{sx}nl_x^2 \quad \text{in direction of span } l_x$$

and

$$M_{sy} = a_{sy}nl_x^2 \quad \text{in direction of span } l_y$$

where

$\quad M_{sx}$ and M_{sy} are the moments at mid-span on strips of unit width with spans l_x and l_y respectively

$\quad n = (1.35g_k + 1.5q_k)$, that is the total ultimate load per unit area

$\quad l_y =$ the length of the longer side

$\quad l_x =$ the length of the shorter side

$\quad a_{sx}$ and a_{sy} are the moment coefficients from table 8.4.

The area of reinforcement in directions l_x and l_y respectively are

$$A_{sx} = \frac{M_{sx}}{0.87f_{yk}z} \quad \text{per metre width}$$

and

$$A_{sy} = \frac{M_{sy}}{0.87f_{yk}z} \quad \text{per metre width}$$

The slab should be reinforced uniformly across the full width, in each direction.

The effective depth d used in calculating A_{sy} should be less than that for A_{sx} because of the different depths of the two layers of reinforcement.

Table 8.4 Bending-moment coefficients for slabs spanning in two directions at right angles, simply supported on four sides

l_y/l_x	1.0	1.1	1.2	1.3	1.4	1.5	1.75	2.0
a_{sx}	0.062	0.074	0.084	0.093	0.099	0.104	0.113	0.118
a_{sy}	0.062	0.061	0.059	0.055	0.051	0.046	0.037	0.029

Established practice suggests that at least 40 per cent of the mid-span reinforcement should extend to the supports and the remaining 60 per cent should extend to within $0.1l_x$ or $0.1l_y$ of the appropriate support.

It should be noted that the above method is not specially mentioned in EC2; however, as the method was deemed acceptable in BS8110, its continued use should be an acceptable method of analysing this type of slab.

EXAMPLE 8.5

Design the reinforcement for a simply supported slab

The slab is 220 mm thick and spans in two directions. The effective span in each direction is 4.5 m and 6.3 m and the slab supports a variable load of 10 kN/m². The characteristic material strengths are $f_{ck} = 25$ N/mm² and $f_{yk} = 500$ N/mm².

$$l_y/l_x = 6.3/4.5 = 1.4$$

From table 8.4, $a_{sx} = 0.099$ and $a_{sy} = 0.051$.

$$\text{Self-weight of slab} = 220 \times 25 \times 10^{-3} = 5.5 \text{ kN/m}^2$$
$$\text{ultimate load} = 1.35g_k + 1.5q_k$$
$$= 1.35 \times 5.5 + 1.5 \times 10.0 = 22.43 \text{ kN/m}^2$$

Bending – short span

With class XC-1 exposure conditions take $d = 185$ mm.

$$M_{sx} = a_{sx}nl_x^2 = 0.099 \times 22.43 \times 4.5^2$$
$$= 45.0 \text{ kN m}$$
$$\frac{M_{sx}}{bd^2 f_{ck}} = \frac{45.0 \times 10^6}{1000 \times 185^2 \times 25} = 0.053$$

From the lever-arm curve, figure 4.5, $l_a = 0.95$. Therefore

$$\text{lever-arm } z = 0.95 \times 185 = 176 \text{ mm}$$

and

$$A_s = \frac{M_{sx}}{0.87f_{yk}z} = \frac{45.0 \times 10^6}{0.87 \times 500 \times 176}$$
$$= 588 \text{ mm}^2/\text{m}$$

Provide H12 at 175 mm centres, $A_s = 646$ mm²/m.

Span–effective depth ratio

$$\rho_1 = \frac{100A_{s,\text{req}}}{bd} = \frac{100 \times 588}{1000 \times 185} = 0.318$$

From figure 6.3, this corresponds to a basic span–effective depth ratio of 28.0:

$$\text{actual } \frac{\text{span}}{\text{effective depth}} = \frac{4500}{185} = 24.3$$

Thus $d = 185$ mm is adequate.

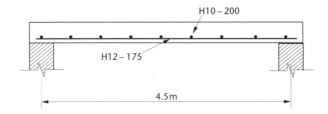

Figure 8.7
Simply supported slab
spanning in two directions

Bending – long span

$$M_{sy} = a_{sy}nl_x^2$$
$$= 0.051 \times 22.43 \times 4.5^2$$
$$= 23.16 \, \text{kN m}$$

Since the reinforcement for this span will have a reduced effective depth, take $z = 176 - 12 = 164$ mm. Therefore

$$A_s = \frac{M_{sy}}{0.87 f_{yk} z}$$
$$= \frac{23.16 \times 10^6}{0.87 \times 500 \times 164}$$
$$= 325 \, \text{mm}^2/\text{m}$$

Provide H10 at 200 mm centres, $A_s = 393 \, \text{mm}^2/\text{m}$

$$\frac{100 A_s}{bd} = \frac{100 \times 393}{1000 \times 164}$$
$$= 0.24$$

which is greater than 0.13, the minimum for transverse steel, with class C25/30 concrete. The arrangement of the reinforcement is shown in figure 8.7.

8.5.2　Restrained slab spanning in two directions

When the slabs have fixity at the supports and reinforcement is added to resist torsion and to prevent the corners of the slab from lifting then the maximum moments per unit width are given by

$$M_{sx} = \beta_{sx}nl_x^2 \quad \text{in direction of span } l_x$$

and

$$M_{sy} = \beta_{sy}nl_x^2 \quad \text{in direction of span } l_y$$

where β_{sx} and β_{sy} are the moment coefficients given in table 8.5, based on previous experience, for the specified end conditions, and $n = (1.35g_k + 1.5q_k)$, the total ultimate load per unit area.

The slab is divided into middle and edge strips as shown in figure 8.8 and reinforcement is required in the middle strips to resist M_{sx} and M_{sy}. The arrangement this reinforcement should take is illustrated in figure 8.2. In the edge strips only nominal reinforcement is necessary, such that $A_s/bd = 0.26 f_{ctm}/f_{yk} \geq 0.0013$ for high yield steel.

Table 8.5 Bending moment coefficients for two-way spanning rectangular slabs supported by beams

Type of panel and moments considered	Short span coefficients for values of l_y/l_x					Long-span coefficients for all values of l_y/l_x
	1.0	1.25	1.5	1.75	2.0	
Interior panels						
Negative moment at continuous edge	0.031	0.044	0.053	0.059	0.063	0.032
Positive moment at midspan	0.024	0.034	0.040	0.044	0.048	0.024
One short edge discontinuous						
Negative moment at continuous edge	0.039	0.050	0.058	0.063	0.067	0.037
Positive moment at midspan	0.029	0.038	0.043	0.047	0.050	0.028
One long edge discontinuous						
Negative moment at continuous edge	0.039	0.059	0.073	0.083	0.089	0.037
Positive moment at midspan	0.030	0.045	0.055	0.062	0.067	0.028
Two adjacent edges discontinuous						
Negative moment at continuous edge	0.047	0.066	0.078	0.087	0.093	0.045
Positive moment at midspan	0.036	0.049	0.059	0.065	0.070	0.034

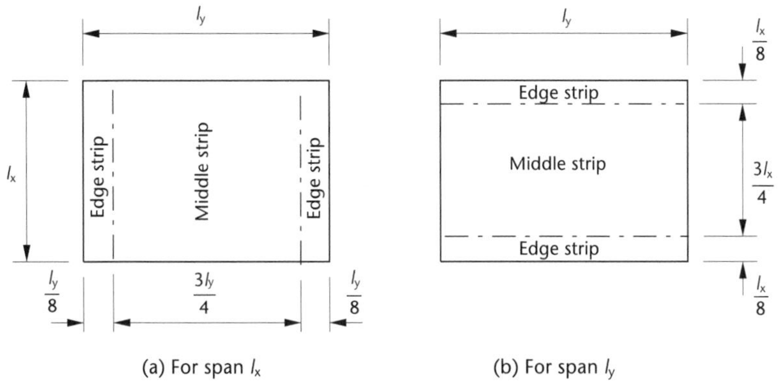

Figure 8.8
Division of slab into middle and edge strips

(a) For span l_x (b) For span l_y

In addition, torsion reinforcement is provided at discontinuous corners and it should:

1. consist of top and bottom mats, each having bars in both directions of span;
2. extend from the edges a minimum distance $l_x/5$;
3. at a corner where the slab is discontinuous in both directions have an area of steel in each of the four layers equal to three-quarters of the area required for the maximum mid-span moment;
4. at a corner where the slab is discontinuous in one direction only, have an area of torsion reinforcement only half of that specified in rule 3.

Torsion reinforcement is not, however, necessary at any corner where the slab is continuous in both directions.

Where $l_y/l_x > 2$, the slabs should be designed as spanning in one direction only.

It should be noted that the coefficients for both shear and moments can only be used if class B or C ductility reinforcement is specified and the ratio x/d is limited to 0.25.

Figure 8.9
Continuous panel spanning in
two directions

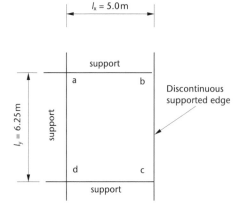

EXAMPLE 8.6

Moments in a continuous two-way slab

The panel considered is an edge panel, as shown in figure 8.9 and the uniformly distributed load, $n = (1.35g_k + 1.5q_k) = 10 \text{ kN/m}^2$.

The moment coefficients are taken from table 8.5.

$$\frac{l_y}{l_x} = \frac{6.25}{5.0} = 1.25$$

Positive moments at mid-span

$$M_{sx} = \beta_{sx}nl_x^2 = 0.045 \times 10 \times 5^2$$
$$= 11.25 \text{ kN m in direction } l_x$$
$$M_{sy} = \beta_{sy}nl_x^2 = 0.028 \times 10 \times 5^2$$
$$= 7.0 \text{ kN m in direction } l_y$$

Negative moments

$$\text{Support ad, } M_x = 0.059 \times 10 \times 5^2 = 14.75 \text{ kN m}$$
$$\text{Supports ab and dc, } M_y = 0.037 \times 10 \times 5^2 = 9.25 \text{ kN m}$$

The moments calculated are for a metre width of slab.

The design of reinforcement to resist these moments would follow the usual procedure. Torsion reinforcement, according to rule 4 is required at corners b and c. A check would also be required on the span–effective depth ratio of the slab.

8.6   Flat slab floors

A flat slab floor is a reinforced concrete slab supported directly by concrete columns without the use of intermediary beams. The slab may be of constant thickness throughout or in the area of the column it may be thickened as a drop panel. The column may also be of constant section or it may be flared to form a column head or capital. These various forms of construction are illustrated in figure 8.10.

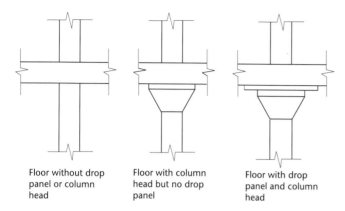

Figure 8.10
Drop panels and column heads

Floor without drop panel or column head

Floor with column head but no drop panel

Floor with drop panel and column head

The drop panels are effective in reducing the shearing stresses where the column is liable to punch through the slab, and they also provide an increased moment of resistance where the negative moments are greatest.

The flat slab floor has many advantages over the beam and slab floor. The simplified formwork and the reduced storey heights make it more economical. Windows can extend up to the underside of the slab, and there are no beams to obstruct the light and the circulation of air. The absence of sharp corners gives greater fire resistance as there is less danger of the concrete spalling and exposing the reinforcement. Deflection requirements will generally govern slab thickness which should not normally be less than 180 mm for fire resistance as indicated in table 8.6.

The analysis of a flat slab structure may be carried out by dividing the structure into a series of equivalent frames. The moments in these frames may be determined by:

(a)   a method of frame analysis such as moment distribution, or the stiffness method on a computer;

(b)   a simplified method using the moment and shear coefficients of table 8.1 subject to the following requirements:

    (i)   the lateral stability is not dependent on the slab-column connections;

    (ii)   the conditions for using table 8.1 described on page 209 are satisfied;

    (iii)   there are at least three rows of panels of approximately equal span in the direction being considered;

    (iv)   the bay size exceeds $30 \, m^2$

Table 8.6   Minimum dimensions and axis distance for flat slabs for fire resistance

Standard fire resistance	Minimum dimensions (mm)	
	Slab thickness, h_s	Axis distance, a
REI 60	180	15
REI 90	200	25
REI 120	200	35
REI 240	200	50

Note:

1. Redistribution of moments not to exceed 15%.
2. For fire resistance R90 and above, 20% of the total top reinforcement in each direction over intermediate supports should be continuous over the whole span and placed in the column strip.

Figure 8.11

Flat slab divided into strips

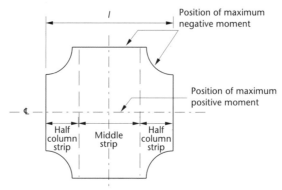

Width of half column strip = $l/4$ with no drops
or = half drop width when drops are used

Interior panels of the flat slab should be divided as shown in figure 8.11 into column and middle strips. Drop panels should be ignored if their smaller dimension is less than one-third of the smaller panel dimension l_x. If a panel is not square, strip widths in both directions are based on l_x.

Moments determined from a structural analysis or the coefficients of table 8.1 are distributed between the strips as shown in table 8.7 such that the negative and positive moments resisted by the column and middle strips total 100 per cent in each case.

Reinforcement designed to resist these slab moments may be detailed according to the simplified rules for slabs, and satisfying normal spacing limits. This should be spread across the respective strip but, in solid slabs without drops, top steel to resist negative moments in column strips should have one-half of the area located in the central quarter-strip width. If the column strip is narrower because of drops, the moments resisted by the column and middle strips should be adjusted proportionally as illustrated in example 8.7.

Column moments can be calculated from the analysis of the equivalent frame. Particular care is needed over the transfer of moments to edge columns. This is to ensure that there is adequate moment capacity within the slab adjacent to the column since moments will only be able to be transferred to the edge column by a strip of slab considerably narrower than the normal internal panel column strip width. As seen in table 8.7, a limit is placed on the negative moment transferred to an edge column, and slab reinforcement should be concentrated within width b_e as defined in figure 8.12. If exceeded the moment should be limited to this value and the positive moment increased to maintain equilibrium.

Table 8.7 Division of moments between strips

	Column strip	Middle strip
Negative moment at edge column	100% but not more than $0.17b_e d^2 f_{ck}$	0
Negative moment at internal column	60–80%	40–20%
Positive moment in span	50–70%	50–30%

b_e = width of edge strip.

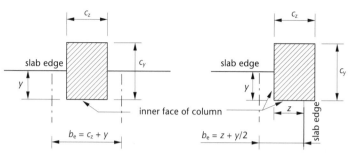

Figure 8.12
Definition of b_e

Note: All slab reinforcement perpendicular to a free edge transferring moment to the column should be concentrated within the width b_e

       (a) Edge column   (b) Corner column

The reinforcement for a flat slab should generally be arranged according to the rules illustrated in figure 8.2, but at least 2 bottom bars in each orthogonal direction should pass through internal columns to enhance robustness.

Important features in the design of the slabs are the calculations for punching shear at the head of the columns and at the change in depth of the slab, if drop panels are used. The design for shear should follow the procedure described in the previous section on punching shear except that EC2 requires that the design shear force be increased above the calculated value by 15 per cent for internal columns, up to 40 per cent for edge columns and 50 per cent for corner columns, to allow for the effects of moment transfer. These simplified rules only apply to braced structures where adjacent spans do not differ by more than 25%.

In considering punching shear, EC2 places additional requirements on the amount and distribution of reinforcement around column heads to ensure that full punching shear capacity is developed.

The usual basic span–effective depth ratios may be used but where the greater span exceeds 8.5 m the basic ratio should be multiplied by 8.5/span. For flat slabs the span–effective depth calculation should be based on the longer span.

Reference should be made to codes of practice for further detailed information describing the requirements for the analysis and design of flat slabs, including the use of bent-up bars to provide punching shear resistance.

EXAMPLE 8.7

Design of a flat slab

The columns are at 6.5 m centres in each direction and the slab supports a variable load of 5 kN/m². The characteristic material strengths are $f_{ck} = 25$ N/mm² for the concrete, and $f_{yk} = 500$ N/mm² for the reinforcement.

It is decided to use a floor slab as shown in figure 8.13 with 250 mm overall depth of slab, and drop panels 2.5 m square by 100 mm deep. The column heads are to be made 1.2 m diameter.

Permanent load

Weight of slab $= 0.25 \times 25 \times 6.5^2 = 264.1$ kN

Weight of drop $= 0.1 \times 25 \times 2.5^2 = \underline{15.6\,\text{kN}}$

                     Total $= 279.7$ kN

Figure 8.13
Flat slab example

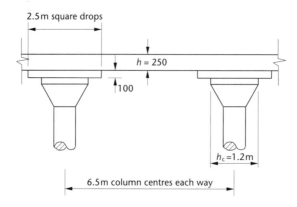

Variable load

Total $= 5 \times 6.5^2 = 211.3 \,\text{kN}$

Therefore

ultimate load on the floor, $F = 1.35 \times 279.7 + 1.5 \times 211.3$

$$= 695 \,\text{kN per panel}$$

and equivalent distributed load, $n = \dfrac{695}{6.5^2} = 16.4 \,\text{kN/m}^2$

The effective span,

$$L = \text{clear span between column heads} + \frac{\text{slab thickness}}{2} \text{ at either end}$$

$$= (6.5 - 1.2) + \frac{350}{2} \times 2 \times 10^{-3}$$

$$= 5.65 \,\text{m}$$

A concrete cover of 25 mm has been allowed, and where there are two equal layers of reinforcement the effective depth has been taken as the mean depth of the two layers in calculating the reinforcement areas. ($d = 205$ mm in span and 305 mm at supports.)

The drop dimension is greater than one-third of the panel dimension, therefore the column strip is taken as the width of the drop panel (2.5 m).

Bending reinforcement

Since the variable load is less than the permanent load and bay size $= 6.5 \times 6.5 = 42.25 \,\text{m}^2$ ($\geq 30 \,\text{m}^2$), from table 8.1:

1. Centre of interior span

Positive moment $= 0.063 Fl$

$$= 0.063 \times 695 \times 5.65 = 247 \,\text{kN m}$$

The width of the middle strip is $(6.5 - 2.5) = 4$ m which is greater than half the panel dimension, therefore the proportion of this moment taken by the middle strip can be taken as 0.45 from table 8.6 adjusted as shown:

$$0.45 \times \frac{4}{6.5/2} = 0.55$$

Thus middle strip positive moment $= 0.55 \times 247 = 136 \,\text{kN m}$.

The column strip positive moment $= (1 - 0.55) \times 247 = 111 \,\text{kN m}$.

(a) For the middle strip

$$\frac{M}{bd^2f_{ck}} = \frac{136 \times 10^6}{4000 \times 205^2 \times 25} = 0.032$$

From the lever-arm curve, figure 4.5, $l_a = 0.97$, therefore

$$A_s = \frac{M}{0.87f_{yk}z} = \frac{136 \times 10^6}{0.87 \times 500 \times 0.95 \times 205}$$

$$= 1605 \, \text{mm}^2 \text{ bottom steel}$$

Thus provide sixteen H12 bars ($A_s = 1809 \, \text{mm}^2$) each way in the span, distributed evenly across the 4 m width of the middle strip (spacing $= 250 \, \text{mm}$ = maximum allowable for a slab in an area of maximum moment).

(b) The column strip moments will require $1310 \, \text{mm}^2$ bottom steel which can be provided as twelve H12 bars ($A_s = 1356 \, \text{mm}^2$) in the span distributed evenly across the 2.5 m width of the column strip (spacing approx 210 mm).

2. Interior support

$$\text{Negative moment} = -0.063Fl$$
$$= -0.063 \times 695 \times 5.65 = 247 \, \text{kN m}$$

and this can also be divided into

$$\text{middle strip} = 0.25 \times \frac{4}{6.5/2} \times 247 = 0.31 \times 247$$
$$= 77 \, \text{kN m}$$

$$\text{and column strip} = (1 - 0.31) \times 247 = 0.69 \times 247 = 170 \, \text{kN m}$$

(a) For the middle strip

$$\frac{M}{bd^2f_{ck}} = \frac{77 \times 10^6}{4000 \times 205^2 \times 25} = 0.018$$

From the lever-arm curve, figure 4.5, $l_a = 0.98$ (> 0.95), therefore

$$A_s = \frac{M}{0.87f_{yk}z} = \frac{77 \times 10^6}{0.87 \times 500 \times 0.95 \times 205}$$

$$= 909 \, \text{mm}^2$$

Provide eleven evenly spaced H12 bars as top steel ($A_s = 1243 \, \text{mm}^2$) to satisfy 400 mm maximum spacing limit.

(b) For the column strip

$$\frac{M}{bd^2f_{ck}} = \frac{170 \times 10^6}{2500 \times 305^2 \times 25} = 0.029$$

From the lever-arm curve, figure 4.5, $l_a = 0.97$ (> 0.95), therefore

$$A_s = \frac{M}{0.87f_{yk}z} = \frac{170 \times 10^6}{0.87 \times 500 \times 0.95 \times 305}$$

$$= 1349 \, \text{mm}^2$$

Provide H12 bars as top steel at 200 centres. This is equivalent to fourteen bars ($A_s = 1582 \, \text{mm}^2$) over the full 2.5 m width of the column strip. The bending reinforcement requirements are summarised in figure 8.14.

Figure 8.14
Details of bending
reinforcement

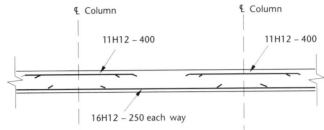

(a) Middle strip 4.0m wide

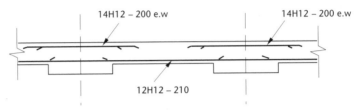

(b) Column strip 2.5m wide

Punching shear

1. At the column head

$$\text{perimeter } u_0 = \pi \times \text{diameter of column head}$$
$$= \pi \times 1200 = 3770\,\text{mm}$$
$$\text{shear force } V_{Ed} = F - \frac{\pi}{4}1.2^2 n = 695 - \frac{\pi}{4}1.2^2 \times 16.4 = 676.4\,\text{kN}$$

To allow for the effects of moment transfer, V is increased by 15 per cent for an internal column, thus

$$V_{Ed,\,eff} = 1.15 \times 676.4 = 778\,\text{kN}$$

Maximum permissible shear force,

$$V_{Rd,\,max} = 0.5ud\left[0.6\left(1 - \frac{f_{ck}}{250}\right)\right]\frac{f_{ck}}{1.5}$$
$$= 0.5 \times 3770 \times 305 \times \left[0.6\left(1 - \frac{25}{250}\right)\right]\frac{25}{1.5} \times 10^{-3} = 5174\,\text{kN}$$

thus $V_{Ed,\,eff}$ is significantly less than $V_{Rd,\,max}$.

2. The first critical section for shear is $2.0 \times$ effective depth from the face of the column head, that is, a section of diameter $1.2 + 2 \times 2.0 \times 0.305 = 2.42$ m. (i.e. within the drop panel).

Thus the length of the perimeter $u_1 = \pi \times 2420 = 7602\,\text{mm}$

Ultimate shear force, $V_{Ed} = 695 - \dfrac{\pi}{4} \times 2.42^2 \times 16.4 = 620\,\text{kN}$

$$V_{Ed,\,eff} = 1.15 \times 620 = 713\,\text{kN}$$

For the unreinforced section

$$V_{Rd,\,c} = v_{Rd,\,c}u_1 d = v_{Rd,\,c} \times 7602 \times 305$$
$$\text{with } \rho_1 = \rho_y = \rho_z = \frac{100 \times 1582}{2500 \times 305} = 0.21\%$$

thus from table 8.2, $v_{Rd,c} = 0.47\,\text{N/mm}^2$; therefore

$$V_{Rd,c} = 0.47 \times 7602 \times 305 \times 10^{-3} = 1090\,\text{kN}$$

As $V_{Ed,eff}$ is less than $V_{Rd,c}$ the section is adequate, and shear reinforcement is not needed.

3. At the dropped panel the critical section is $2.0 \times 205 = 410\,\text{mm}$ from the panel with a perimeter given by

$$u = (2a + 2b + 2\pi \times 2d)$$
$$= (4 \times 2500 + 2\pi \times 410) = 12\,576\,\text{mm}$$

The area within the perimeter is given by

$$(2.5 + 3d)^2 - (4 - \pi)(2.0 \times 0.205)^2$$
$$= (2.5 + 3 \times 0.205)^2 - (4 - \pi)(0.410)^2$$
$$= 9.559\,\text{m}^2$$

Ultimate shear force,

$$V_{Ed} = 695 - 9.559 \times 16.4 = 538\,\text{kN}$$

so

$$V_{Ed,eff} = 1.15 \times 538 = 619\,\text{kN}$$
$$V_{Rd,c} = v_{Rd,c}ud \quad \text{where } u = 12\,576\,\text{mm and } d = 205\,\text{mm}$$
$$\rho_1 = \frac{100 \times 1582}{2500 \times 205} = 0.31\%, \text{ thus from table 8.2 } v_{Rd,c} \cong 0.55\,\text{N/mm}^2$$

hence

$$V_{Rd,c} = 0.55 \times 12,576 \times 205$$
$$= 1418\,\text{kN}$$

As $V_{Ed,eff}$ is less than $V_{Rd,c}$ the section is adequate.

[*Note* – in the above calculation ρ_1 has been based on column strip reinforcement at the support. Since the critical zone will lie partially in the middle strip, this value will be a minor over-estimate but is not significant in this case.]

Span–effective depth ratios

At the centre of the span

$$\frac{100A_{s,req}}{bd} = \frac{100 \times 1605}{4000 \times 205} = 0.20$$

From figure 6.3 the limiting basic span–effective depth ratio is 32 for class C25 concrete and this is multiplied by a K factor of 1.2 for a flat slab (see table 6.10) giving $32 \times 1.2 = 38.4$.

actual span − effective depth ratio $= 6500/205 = 31.7$

Hence the slab effective depth is acceptable. To take care of stability requirements, extra reinforcement may be necessary in the column strips to act as a tie between each pair of columns – see section 6.7, and the requirement for at least two bottom bars to pass through each column will be satisfied by the spacings calculated above and shown in figure 8.14.

8.7 | Ribbed and hollow block floors

Cross-sections through a ribbed and hollow block floor slab are shown in figure 8.15. The ribbed floor is formed using temporary or permanent shuttering while the hollow block floor is generally constructed with blocks made of clay tile or with concrete containing a lightweight aggregate. If the blocks are suitably manufactured and have adequate strength they can be considered to contribute to the strength of the slab in the design calculations, but in many designs no such allowance is made.

The principal advantage of these floors is the reduction in weight achieved by removing part of the concrete below the neutral axis and, in the case of the hollow block floor, replacing it with a lighter form of construction. Ribbed and hollow block floors are economical for buildings where there are long spans, over about 5 m, and light or moderate live loads, such as in hospital wards or apartment buildings. They would not be suitable for structures having a heavy loading, such as warehouses and garages.

Figure 8.15
Sections through ribbed and hollow block floors, and waffle slab

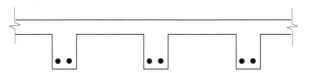

(a) Section through a ribbed floor

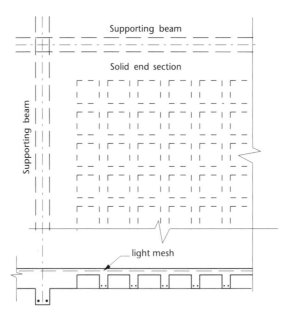

(b) Partial plan of and section through a waffle slab

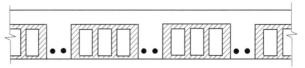

(c) Section through a hollow block floor

Near to the supports the hollow blocks are stopped off and the slab is made solid. This is done to achieve a greater shear strength, and if the slab is supported by a monolithic concrete beam the solid section acts as the flange of a T-section. The ribs should be checked for shear at their junction with the solid slab. It is good practice to stagger the joints of the hollow blocks in adjacent rows so that, as they are stopped off, there is no abrupt change in cross-section extending across the slab. The slabs are usually made solid under partitions and concentrated loads.

During construction the hollow tiles should be well soaked in water prior to placing the concrete, otherwise shrinkage cracking of the top concrete flange is liable to occur.

The thickness of the concrete flange should not be less than:

1. 40 mm or one-tenth of the clear distance between ribs, whichever is the greater, for slabs with permanent blocks;
2. 50 mm or one-tenth of the clear distance between ribs, whichever is the greater, for slabs without permanent blocks.

If these requirements are not met, than a check of longitudinal shear between web and flange should be made to see if additional transverse steel is needed.

The rib width will be governed by cover, bar-spacing and fire resistance (section 6.1). The ribs should be spaced no further apart than 1.5 m and their depth below the flange should not be greater than four times their width. Transverse ribs should be provided at spacings no greater than ten times the overall slab depth.

Provided that the above dimensional requirements are met, ribbed slabs can be treated for analysis as solid slabs and the design requirements can be based on those of a solid slab. Calculations of reinforcement will require evaluation of effective flange breadths using the procedures described for T-beams in Chapter 7.

Ribbed slabs will be designed for shear using the approach described previously with b_w taken as the breadth of the rib. Although no specific guidance is given in EC2, previous practice suggests that, where hollow blocks are used, the rib width may be increased by the wall thickness of the block on one side of the rib.

Span–effective depth ratios will be based on the shorter span with the basic values given in figure 6.3 multiplied by 0.8 where the ratio of the flange width to the rib width exceeds 3. Again, no specific guidance is given in the Code but previous practice suggests that the thickness of the rib width may include the thickness of the two adjacent block-walls.

At least 50 per cent of the tensile reinforcement in the span should continue to the supports and be anchored. In some instances the slabs are supported by steel beams and are designed as simply supported even though the topping is continuous. Reinforcement should be provided over the supports to prevent cracking in these cases. This top steel should be determined on the basis of 25 per cent of the mid-span moment and should extend at least 0.15 of the clear span into the adjoining span.

A light reinforcing mesh in the topping flange can give added strength and durability to the slab, particularly if there are concentrated or moving loads, or if cracking due to shrinkage or thermal movement is likely. The minimum area of reinforcement required to control shrinkage and thermal cracking can be calculated, as given in chapter 6, but established practice suggests that an area of mesh equivalent to 0.13 per cent of the topping flange will be adequate.

Waffle slabs are designed as ribbed slabs and their design moments each way are obtained from the moment coefficients tabulated in table 8.5 for two-way spanning slabs.

EXAMPLE 8.8

Design of a ribbed floor

The ribbed floor is constructed with permanent fibreglass moulds; it is continuous over several spans of 5.0 m. The characteristic material strengths are $f_{ck} = 25\,\text{N/mm}^2$ and $f_{yk} = 400\,\text{N/mm}^2$.

An effective section, as shown in figure 8.16, which satisfies requirements for a 60 minute fire resistance (see table 6.5) is to be tried. The characteristic permanent load including self-weight and finishes is $4.5\,\text{kN/m}^2$ and the characteristic variable load is $2.5\,\text{kN/m}^2$.

The calculations are for an end span (which will be most critical) for which the moments and shears can be determined from the coefficients in table 8.1.

Considering a 0.4 m width of floor as supported by each rib:

$$\text{Ultimate load} = 0.4(1.35g_k + 1.5q_k)$$

$$= 0.4(1.35 \times 4.5 + 1.5 \times 2.5)$$

$$= 3.93\,\text{kN/m}$$

Ultimate load on the span, $F = 3.93 \times 5.0 = 19.65\,\text{kN}$

Bending

(1) At mid-span design as a T-section:

$$M = 0.086Fl = 0.086 \times 19.65 \times 5.0 = 8.45\,\text{kN m}$$

The effective breadth of flange $= b_w + b_{eff1} + b_{eff2}$ (see section 7.4) where

$$b_{eff1} = b_{eff2} = 0.2b_1 + 0.1l_0 \le 0.2l_0 \le b_1$$

with $b_1 = (400 - 125)/2 = 137\,\text{mm}$ and $l_0 = 0.85 \times 5000 = 4250\,\text{mm}$

thus

$$b_w + b_{eff1} + b_{eff2} = 125 + 2(0.2 \times 137 + 0.1 \times 4250) = 1030\,\text{mm}$$
$$\text{or } 0.2 \times 4250 = 850\,\text{mm}$$

which both exceed the rib spacing of 400mm, which governs

$$\frac{M}{bd^2 f_{ck}} = \frac{8.45 \times 10^6}{400 \times 160^2 \times 25} = 0.033$$

Figure 8.16
Ribbed slab

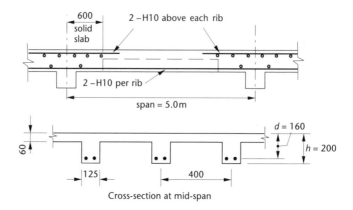

Cross-section at mid-span

From the lever-arm curve, figure 4.5, $l_a = 0.98$ (> 0.95). Thus the neutral axis depth lies within the flange and

$$A_s = \frac{M}{0.87 f_{yk} l_a d} = \frac{8.45 \times 10^6}{0.87 \times 500 \times 0.95 \times 160} = 128\,\text{mm}^2$$

Provide two H10 bars in the ribs, $A_s = 157\,\text{mm}^2$.

2. At the end interior support design as a rectangular section for the solid slab:

$$M = 0.086Fl = 0.086 \times 19.65 \times 5.0 = 8.45\,\text{kN m as in (1)}$$

and $A_s = 128\,\text{mm}^2$ as at mid-span

Provide two H10 bars in each 0.4 m width of slab, $A_s = 157\,\text{mm}^2$.

3. At the section where the ribs terminate: this occurs 0.6 m from the centreline of the support and the moment may be hogging so that 125 mm ribs must provide the concrete area in compression to resist the design moment. The maximum moment of resistance of the concrete is

$$M = 0.167 f_{ck} b d^2 = 0.167 \times 25 \times 125 \times 160^2 \times 10^{-6}$$
$$= 13.36\,\text{kN m}$$

which must be greater than the moment at this section, therefore compression steel is not required.

Span–effective depth ratio

At the centre of the span

$$\rho = \frac{100 A_{s,\,req}}{bd} = \frac{100 \times 128}{400 \times 160} = 0.20\%$$

From figure 6.3 and table 6.10 the limiting basic span–effective depth ratio ($\rho = 0.3\%$) for an end span is $32 \times 1.3 = 41.6$.

For a T-section with a flange width greater than three times the web width this should be multiplied by 0.8 to give a limiting ratio of $0.8 \times 41.6 = 33.2$.

actual span$-$effective depth ratio $= 5000/160 = 31.3$

Thus $d = 160\,\text{mm}$ is adequate.

Shear

Maximum shear in the rib 0.6 m from the support centreline (end span)

$$V_{Ed} = 0.6F - 0.6 \times 3.93 = 0.6 \times 19.65 - 0.6 \times 3.93 = 9.43\,\text{kN}$$
$$\rho_1 = \frac{A_s}{bd} = \frac{157}{125 \times 160} = 0.0079$$

From table 8.2, the shear resistance without reinforcement $V_{Rd,c} = v_{Rd,c} bd$ where $v_{Rd,c} = 0.68\,\text{N/mm}^2$ and, from table 8.3, the strength modification factor $= 0.94$. Hence:

$$V_{Rd,c} = 0.94 \times 0.68 \times 125 \times 160 = 12.78\,\text{kN}$$

As $V_{Rd,c}$ is greater than V_{Ed} then no shear reinforcement is required provided that the bars in the ribs are securely located during construction.

EXAMPLE 8.9

Design of a waffle slab

Design a waffle slab for an internal panel of a floor system, each panel spanning 6.0 m in each direction. The characteristic material strengths are $f_{ck} = 25 \, \text{N/mm}^2$ and $f_{yk} = 500 \, \text{N/mm}^2$. The section as used in example 8.8, figure 8.16 is to be tried with characteristic permanent load including self-weight of 6.0 kN/m^2 and characteristic variable load of 2.5 kN/m^2.

Design ultimate load $= (1.35g_k + 1.5q_k)$

$$= (1.35 \times 6.0) + (1.5 \times 2.5) = 11.85 \, \text{kN/m}^2$$

As the slab has the same span in each direction the moment coefficients, β_{sx}, β_{sy} are taken from table 8.5 with $l_y/l_x = 1.0$. Calculations are given for a single 0.4 m wide beam section and in both directions of span.

Bending

1. At mid-span: design as a T-section.

Positive moment at mid-span $= m_{sx} = \beta_{sx}nl_x^2$

$$= 0.024 \times 11.85 \times 6^2 = 10.24 \, \text{kN m/m}$$

Moment carried by each rib $= 0.4 \times 10.24 = 4.10 \, \text{kN m}$

$$\frac{M}{bd^2 f_{ck}} = \frac{4.10 \times 10^6}{400 \times 160^2 \times 25} = 0.016$$

where the effective breadth is 400 mm as in the previous example.

From the lever-arm curve, figure 4.5, $l_a = 0.95$. Thus the neutral axis lies within the flange and

$$A_s = \frac{M}{0.87 f_{yk} z} = \frac{4.10 \times 10^6}{0.87 \times 500 \times 0.95 \times 160} = 62 \, \text{mm}^2$$

Provide two H10 bars in each rib at the bottom of the beam, $A_s = 157 \, \text{mm}^2$ to satisfy minimum requirement of $0.13bd\% = 0.0013 \times 400 \times 160 = 83 \, \text{mm}^2/\text{rib}$. Note that since the service stress in the steel will be reduced, this will lead to a higher span–effective depth ratio thus ensuring that the span–effective depth ratio of the slab is kept within acceptable limits.

2. At the support: design as a rectangular section for the solid slab.

Negative moment at support $m_{sx} = \beta_{sx}nl_x^2 = 0.031 \times 11.85 \times 6^2$

$$= 13.22 \, \text{kN m/m}$$

Moment carried by each 0.4 m width $= 0.4 \times 13.22 = 5.29 \, \text{kN m}$

$$\frac{M}{bd^2 f_{ck}} = \frac{5.29 \times 10^6}{400 \times 160^2 \times 25} = 0.021$$

From the lever-arm curve, figure 4.5 $l_a = 0.95$. Thus

$$A_s = \frac{M}{0.87 f_{yk} z} = \frac{5.29 \times 10^6}{0.87 \times 500 \times 0.95 \times 160} = 80 \, \text{mm}^2$$

Provide two H10 bars in each 0.4 m width of slab, $A_s = 157 \, \text{mm}^2$.

3. At the section where the ribs terminate: the maximum hogging moment of resistance of the concrete ribs is 13.36 kN m, as in the previous example. This is greater than the moment at this section, therefore compression steel is not required.

Span–effective depth ratio

At mid-span $\rho = \dfrac{100A_{s,\,req}}{bd} = \dfrac{100 \times 62}{400 \times 160} = 0.096\%$

hence from figure 6.3, limiting basic span depth ratio $= 32 \times 1.5$ (for interior span) $\times 0.8$ (for flange $> 3 \times$ web thickness) when $\rho < 0.3\%$.

Thus allowable ratio $= 32 \times 1.5 \times 0.8 = 38.4$

actual $\dfrac{\text{span}}{\text{effective depth}} = \dfrac{6000}{160} = 37.5$

Thus $d = 160$ mm is just adequate. It has not been necessary here to allow for the increased span/effective depth resulting from providing an increased steel area, thus consideration could be given to reducing the rib reinforcement to two H8 bars (101 mm^2) which still satisfies nominal requirements.

Shear

From Table 3.6 in the *Designers Guide* (ref. 20) the shear force coefficient for a continuous edge support is 0.33. Hence, for one rib, the shear at the support

$V_{sx} = \beta_{vx} n l_x \times b = 0.33 \times 11.85 \times 6 \times 0.4 = 9.38\,\text{kN}$

Maximum shear in the rib 0.6 m from the centre-line is

$V_{Ed} = 9.38 - 0.6 \times 11.85 \times 0.4 = 6.54\,\text{kN}$

At this position, $V_{Rd,c} = v_{Rd,c} \times 125 \times 160$ and

$\rho_1 = \dfrac{100A_s}{bd} = \dfrac{100 \times 157}{125 \times 160} = 0.79\%$

hence from table 8.2, $v_{Rd,c} = 0.68\,\text{N/mm}^2$ and from table 8.3 $K = 0.94$

$\therefore V_{Rd,c} = 0.68 \times 0.94 \times 125 \times 160 \times 10^{-3} = 12.8\,\text{kN}$

Therefore the unreinforced section is adequate in shear, and no links are required provided that the bars in the ribs are securely located during construction.

Reinforcement in the topping flange

Light reinforcing mesh should be provided in the top of the flange.

Area required $= 0.13 \times b \times h/100 = 0.13 \times 1000 \times 60/100 = 78\,\text{mm}^2/\text{m}$

Provide D98 mesh (see table A.5), $A_s = 98\,\text{mm}^2/\text{m}$.

8.8 ▎ Stair slabs

The usual form of stairs can be classified into two types: (1) those spanning horizontally in the transverse direction, and (2) those spanning longitudinally.

8.8.1 Stairs spanning horizontally

Stairs of this type may be supported on both sides or they may be cantilevered from a supporting wall.

Figure 8.17 shows a stair supported on one side by a wall and on the other by a stringer beam. Each step is usually designed as having a breadth b and an effective depth of $d = D/2$ as shown in the figure; a more rigorous analysis of the section is rarely justified. Distribution steel in the longitudinal direction is placed above the main reinforcement.

Details of a cantilever stair are shown in figure 8.18. The effective depth of the member is taken as the mean effective depth of the section and the main reinforcement must be placed in the top of the stairs and anchored into the support. A light mesh of reinforcement is placed in the bottom face to resist shrinkage cracking.

Figure 8.17
Stairs spanning horizontally

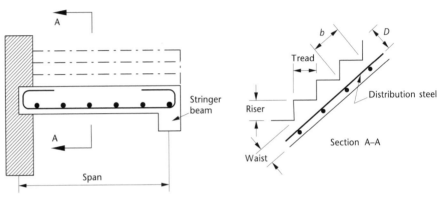

Figure 8.18
Cantilevered stairs

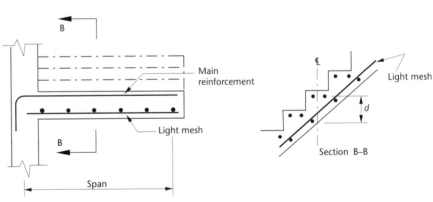

8.8.2 Stair slab spanning longitudinally

The stair slab may span into landings which span at right angles to the stairs as in figure 8.19 or it may span between supporting beams as in figure 8.20 of example 8.10.

The permanent load is calculated along the slope length of the stairs but the variable load is based on the plan area. Loads common on two spans which intersect at right angles and surround an open well may be assumed to be divided equally between the spans. The effective span (l) is measured horizontally between the centres of the supports and the thickness of the waist (h) is taken as the slab thickness.

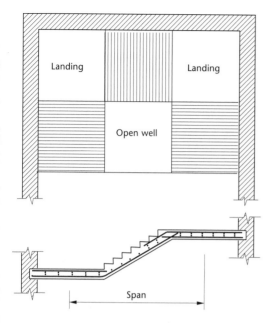

Figure 8.19
Stairs spanning into landings

Stair slabs which are continuous and constructed monolithically with their supporting slabs or beams can be designed for a bending moment of say $Fl/10$, where F is the total ultimate load. However, in many instances the stairs are precast or constructed after the main structure, pockets with dowels being left in the supporting beams to receive the stairs, and with no appreciable end restraint the design moment should be $Fl/8$.

EXAMPLE 8.10

Design of a stair slab

The stairs are of the type shown in figure 8.20 spanning longitudinally and set into pockets in the two supporting beams. The effective span is 3 m and the rise of the stairs is 1.5 m with 260 mm treads and 150 mm risers. The variable load is 3.0 kN/m² and the characteristic material strengths are $f_{ck} = 30$ N/mm² and $f_{yk} = 500$ N/mm².

Try a 140 mm thick waist, effective depth, $d = 115$ mm. This would give an initial estimate of the span–effective depth ratio of 26.1 (3000/115) which, from table 6.10, lies a little above the basic value for a 'lightly stressed' simply supported slab.

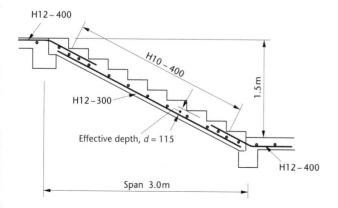

Figure 8.20
Stairs supported by beams

Slope length of stairs $= \sqrt{(3^2 + 5^2)} = 3.35 \, \text{m}$

Consider a 1 m width of stairs:

$$\text{Weight of waist plus steps} = (0.14 \times 3.35 + 0.26 \times 1.5/2)25$$
$$= 16.60 \, \text{kN}$$

$$\text{Variable load} = 3.0 \times 3 = 9.0 \, \text{kN}$$

$$\text{Ultimate load, } F = 1.35 \times 16.60 + 1.5 \times 9.0 = 35.91 \, \text{kN}$$

With no effective end restraint:

$$M = \frac{Fl}{8} = \frac{35.91 \times 3.0}{8} = 13.46 \, \text{kN m}$$

Bending reinforcement

$$\frac{M}{bd^2 f_{ck}} = \frac{13.46 \times 10^6}{1000 \times 115^2 \times 30} = 0.034$$

From the lever-arm curve, figure 4.5, $l_a = 0.95$ (the maximum normally adopted in practice), therefore

$$A_s = \frac{M}{0.87 f_{yk} z} = \frac{13.46 \times 10^6}{0.87 \times 500 \times 0.95 \times 115} = 283 \, \text{mm}^2/\text{m}$$

Maximum allowable spacing is $3h = 3 \times 140 = 420 \, \text{mm}$ with an upper limit of 400 mm. Provide H12 bars at 300 mm centres, area $= 377 \, \text{mm}^2/\text{m}$.

Span–effective depth ratio

At the centre of the span

$$\frac{100 A_{s,\,\text{prov}}}{bd} = \frac{100 \times 377}{1000 \times 115} = 0.33$$

which is greater than the minimum requirement of 0.15 for class C30 concrete (see Table 6.8).

From table 6.10 the basic span–effective depth ratio for a simply supported span with $\rho_{\text{req}} = 0.5\%$ is 20. Allowing for the actual steel area provided:

$$\text{limiting span–effective depth ratio} = 20 \times A_{s,\,\text{prov}}/A_{s,\,\text{req}}$$
$$= 20 \times 377/283 = 26.6$$

$$\text{actual span–effective depth ratio} = 3000/115 = 26.09$$

Hence the slab effective depth is acceptable. (Note that the allowable ratio will actually be greater than estimated above since the required steel ratio is less than the 0.5% used with table 6.10.)

Secondary reinforcement

Transverse distribution steel $\geq 0.2 A_{s,\,\text{min}} = 0.2 \times 377 = 75.4 \, \text{mm}^2/\text{m}$

This is very small, and adequately covered by H10 bars at the maximum allowable spacing of 400 mm centres, area $= 174 \, \text{mm}^2/\text{m}$.

Continuity bars at the top and bottom of the span should be provided and, whereas about 50 per cent of the main steel would be reasonable, the maximum spacing is limited to 400mm. Hence provide, say, H12 bars at 400 mm centres as continuity steel.

8.9   Yield line and strip methods

For cases which are more complex as a result of shape, support conditions, the presence of openings, or loading conditions it may be worthwhile adopting an ultimate analysis method. The two principal approaches are the yield line method, which is particularly suitable for slabs with a complex shape or concentrated loading, and the strip method which is valuable where the slab contains openings.

These methods have been the subject of research, and are well documented although they are of a relatively specialised nature. A brief introduction is included here to illustrate the general principles and features of the methods, which are particularly valuable in assisting an understanding of failure mechanisms. In practical design situations care must be taken to allow for the effects of tie-down forces at corners and torsion at free edges of slabs.

8.9.1   Yield line method

The capacity of reinforced concrete to sustain plastic deformation has been described in section 3.6. For an under-reinforced section the capacity to develop curvatures between the first yield of reinforcement and failure due to crushing of concrete is considerable. For a slab which is subjected to increasing load, cracking and reinforcement yield will first occur in the most highly stressed zone. This will then act as a plastic hinge as subsequent loads are distributed to other regions of the slab. Cracks will develop to form a pattern of 'yield lines' until a mechanism is formed and collapse is indicated by increasing deflections under constant load. To ensure that adequate plastic deformation can take place the Code specifies that slabs designed by the yield line method must be reinforced with Class B or C (medium or high) ductility steel and the ratio x/d should not exceed 0.25 for concrete up to Class C50/60.

For continuous slabs, the intermediate support moment should also lie between half and twice the magnitude of the span moments.

It is assumed that a pattern of yield lines can be superimposed on the slab, which will cause a collapse mechanism, and that the regions between yield lines remain rigid and uncracked. Figure 8.21 shows the yield line mechanism which will occur for the simple case of a fixed ended slab spanning in one direction with a uniform load. Rotation along the yield lines will occur at a constant moment equal to the ultimate moment of

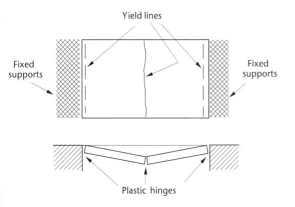

Figure 8.21
Development of yield lines

resistance of the section, and will absorb energy. This can be equated to the energy expended by the applied load undergoing a compatible displacement and is known as the virtual work method.

Considerable care must be taken over the selection of likely yield line patterns, since the method will give an 'upper bound' solution, that is, either a correct or unsafe solution. Yield lines will form at right angles to bending moments which have reached the ultimate moment of resistance of the slab, and the following rules may be helpful:

1. Yield lines are usually straight and end at a slab boundary.

2. Yield lines will lie along axes of rotation, or pass through their points of intersection.

3. Axes of rotation lie along supported edges, pass over columns or cut unsupported edges.

In simple cases the alternative patterns to be considered will be readily determined on the basis of common sense, while for more complex cases differential calculus may be used. The danger of missing the critical layout of yield lines, and thus obtaining an incorrect solution, means that the method can only be used with confidence by experienced designers.

A number of typical patterns are shown in figure 8.22.

Figure 8.22
Examples of yield line patterns

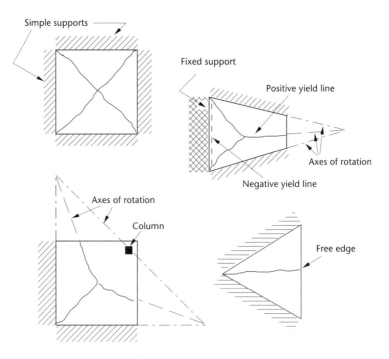

A yield line caused by a sagging moment is generally referred to as a 'positive' yield line and is represented by a full line, while a hogging moment causing cracking on the top surface of the slab causes a 'negative' yield line shown by a broken line.

The basic approach of the method is illustrated for the simple case of a one-way spanning slab in example 8.11

EXAMPLE 8.11

Simply supported, one-way spanning rectangular slab

The slab shown in figure 8.23 is subjected to a uniformly distributed load w per unit area. Longitudinal reinforcement is provided as indicated giving a uniform ultimate moment of resistance m per unit width.

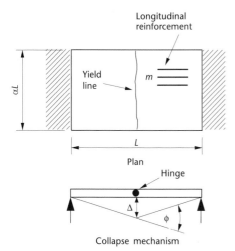

Figure 8.23
One-way spanning slab

The maximum moment will occur at midspan and a positive yield line can thus be superimposed as shown. If this is considered to be subject to a small displacement Δ, then

external work done = area × load × average distance moved for each rigid half of
$$\text{the slab}$$
$$= \left(\alpha L \times \frac{L}{2} \right) \times w \times \frac{\Delta}{2}$$

therefore

$$\text{total} = \frac{1}{2} \alpha L^2 w \Delta$$

Internal energy absorbed by rotation along the yield line is

$$\text{moment} \times \text{rotation} \times \text{length} = m \phi \alpha L$$

where

$$\phi \approx 2 \left(\frac{\Delta}{0.5L} \right) = \frac{4\Delta}{L}$$

hence

$$\text{internal energy} = 4 m \alpha \Delta$$

thus equating internal energy absorbed with external work done

$$4 m \alpha \Delta = \frac{1}{2} \alpha L^2 w \Delta \quad \text{or} \quad m = \frac{wL^2}{8} \quad \text{as anticipated}$$

Since the displacement Δ is eliminated, this will generally be set to unity in calculations of this type.

In the simple case of example 8.11, the yield line crossed the reinforcement at right angles and transverse steel was not involved in bending calculations. Generally, a yield line will lie at an angle θ to the orthogonal to the main reinforcement and will thus also cross transverse steel. The ultimate moment of resistance developed is not easy to define, but Johansen's stepped yield criteria is the most popular approach. This assumes that an inclined yield line consists of a number of steps, each orthogonal to a reinforcing bar as shown in figure 8.24.

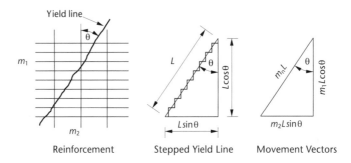

Figure 8.24
Stepped yield line

If the ultimate moments of resistance provided by main and transverse steel are m_1 and m_2 per unit width, it follows that for equilibrium of the vectors shown, the ultimate moment of resistance normal to the yield line m_n per unit length is given by

$$m_n L = m_1 L \cos\theta \times \cos\theta + m_2 L \sin\theta \times \sin\theta$$

hence

$$m_n = m_1 \cos^2\theta + m_2 \sin^2\theta$$

In the extreme case of $\theta = 0$, this reduces to $m_n = m_1$, and when $m_1 = m_2 = m$, then $m_n = m$ for any value of θ. This latter case of an orthotropically reinforced slab (reinforcement mutually perpendicular) with equal moments of resistance is said to be isotropically reinforced.

When applying this approach to complex situations it is often difficult to calculate the lengths and rotations of the yield lines, and a simple vector notation can be used. The total moment component m_n can be resolved vectorially in the x and y directions and since internal energy dissipation along a yield line is given by moment $\times$ rotation $\times$ length it follows that the energy dissipated by rotation of yield lines bounding any rigid area is given by

$$m_x l_x \phi_x + m_y l_y \phi_y$$

where m_x and m_y are yield moments in directions x and y, l_x and l_y are projections of yield lines along each axis, and ϕ_x and ϕ_y are rotations about the axes. This is illustrated in example 8.12.

EXAMPLE 8.12

Slab simply supported on three sides

The slab shown in figure 8.25 supports a uniformly distributed load (u.d.l.) of w per unit area.

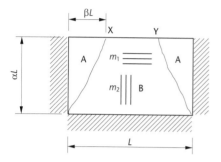

Figure 8.25
Slab supported on three sides

Internal energy absorbed (E) for unit displacement at points X and Y

Area A

$$E_A = m_x l_x \phi_x + m_y l_y \phi_y$$

where $\phi_x = 0$; hence

$$E_A = m_1 \alpha L \times \frac{1}{\beta L} = m_1 \frac{\alpha}{\beta}$$

Area B

$$E_B = m_x l_x \phi_x + m_y l_y \phi_y$$

where $\phi_y = 0$; hence

$$E_B = 2m_2 \beta L \times \frac{1}{\alpha L} = 2m_2 \frac{\beta}{\alpha}$$

hence total for all rigid areas is

$$2E_A + E_B = \frac{2}{\alpha \beta}(m_1 \alpha^2 + m_2 \beta^2)$$

External work done

This can also be calculated for each region separately

$$W_A = \frac{1}{2}(\alpha L \times \beta L)w \times \frac{1}{3} = \frac{1}{6}w\alpha\beta L^2$$

$$W_B = \left[\frac{1}{6}w\alpha\beta L^2 + \alpha L\left(\frac{L}{2} - \beta L \right)w \times \frac{1}{2} \right] \times 2$$

therefore

$$\text{total} = 2W_A + W_B$$

$$= \frac{1}{6}\alpha(3 - 2\beta)L^2 w$$

Hence equating internal and external work, the maximum u.d.l. that the slab can sustain is given by

$$w_{max} = \frac{2}{\alpha\beta}(m_1\alpha^2 + m_2\beta^2) \times \frac{6}{\alpha(3-2\beta)L^2} = \frac{12(m_1\alpha^2 + m_2\beta^2)}{\alpha^2 L^2(3\beta - 2\beta^2)}$$

It is clear that the result will vary according to the value of β. The maximum value of w may be obtained by trial and error using several values of β, or alternatively, by differentiation, let $m_2 = \mu m_1$, then

$$w = \frac{12m_1(\alpha^2 + \mu\beta^2)}{\alpha^2 L^2(3\beta - 2\beta^2)}$$

and

$$\frac{\mathrm{d}(m_1/w)}{\mathrm{d}\beta} = 0 \text{ will give the critical value of } \beta$$

hence

$$3\mu\beta^2 + 4\alpha^2\beta - 3\alpha^2 = 0$$

and

$$\beta = \frac{\alpha^2}{\mu}\left[\pm\sqrt{\left(\frac{4}{9} + \frac{\mu}{\alpha^2}\right) - \frac{2}{3}}\right]$$

A negative value is impossible, hence the critical value of β for use in the analysis is given by the positive root.

8.9.2 Hilleborg strip method

This is based on the 'lower bound' concept of plastic theory which suggests that if a stress distribution throughout a structure can be found which satisfies all equilibrium conditions without violating yield criteria, then the structure is safe for the corresponding system of external loads. Although safe, the structure will not necessarily be serviceable or economic, hence considerable skill is required on the part of the engineer in selecting a suitable distribution of bending moments on which the design can be based. Detailed analysis of a slab designed on this basis is not necessary, but the designer's structural sense and 'feel' for the way loads are transmitted to the supports are of prime importance.

Although this method for design of slabs was proposed by Hilleborg in the 1950s, developments by Wood and Armer in the 1960s have produced its currently used form. The method can be applied to slabs of any shape, and assumes that at failure the load will be carried by bending in either the x or y direction separately with no twisting action. Hence the title of 'strip method'.

Considering a rectangular slab simply supported on four sides and carrying a uniformly distributed load, the load may be expected to be distributed to the supports in the manner shown in figure 8.26.

Judgement will be required to determine the angle α, but it can be seen that if $\alpha = 90°$ the slab will be assumed to be one-way spanning and, although safe, is unlikely to be serviceable because of cracking near the supports along the y axis. Hilleborg suggests that for such a slab, α should be 45°. The load diagrams causing bending moments along typical strips spanning each direction are also shown. It will be seen that

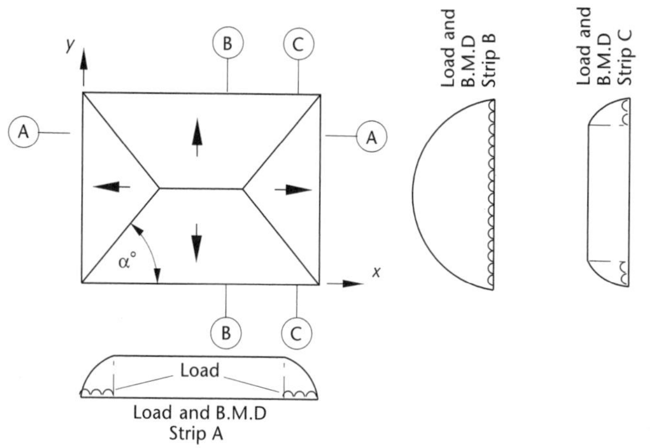

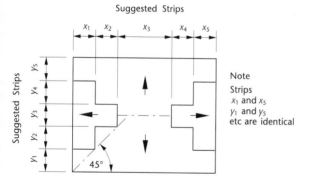

Figure 8.26
Assumed load distributions

the alternative pattern, suggested by Wood and Armer, in figure 8.27 will simplify the design, and in this case five strips in each direction may be conveniently used as shown. Each of these will be designed in bending for its particular loading, as if it were one-way spanning using the methods of section 8.4. Reinforcement will be arranged uniformly across each strip, to produce an overall pattern of reinforcement bands in two directions.

Support reactions can also be obtained very simply from each strip.

The approach is particularly suitable for slabs with openings, in which case strengthened bands can be provided round the openings with the remainder of the slab divided into strips as appropriate. A typical pattern of this type is shown in figure 8.28.

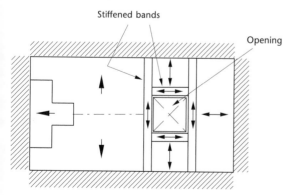

Figure 8.27
Load distribution according to Wood and Armer

Figure 8.28
Strong bands around openings

CHAPTER 9

Column
design

CHAPTER INTRODUCTION
....

The columns in a structure carry the loads from the beams and slabs down to the foundations, and therefore they are primarily compression members, although they may also have to resist bending forces due to the continuity of the structure. The analysis of a section subjected to an axial load plus bending is dealt with in chapter 4, where it is noted that a direct solution of the equations that determine the areas of reinforcement can be very laborious and impractical. Therefore, design charts or computers are often employed to facilitate the routine design of column sections.

Design of columns is governed by the ultimate limit state; deflections and cracking during service conditions are not usually a problem, but nevertheless correct detailing of the reinforcement and adequate cover are important.

Many of the principles used in this chapter for the design of a column can also be applied in a similar manner to other types of members that also resist an axial load plus a bending moment.

252

9.1 Loading and moments

The loading arrangements and the analysis of a structural frame have been described with examples in chapter 3. In the analysis it was necessary to classify the structure into one of the following types:

1. braced – where the lateral loads are resisted by shear walls or other forms of bracing capable of transmitting all horizontal loading to the foundations, and

2. unbraced – where horizontal loads are resisted by the frame action of rigidly connected columns, beams and slabs.

With a braced structure the axial forces and moments in the columns are caused by the vertical permanent and variable actions only, whereas with an unbraced structure the loading arrangements which include the effects of the lateral loads must also be considered.

Both braced and unbraced structures can be further classified as sway or non-sway. In a sway structure sidesway is likely to significantly increase the magnitude of the bending moments in the columns whereas in a non-sway structure this effect is less significant. This increase of moments due to sway, known as a 'second order' effect, is not considered to be significant if there is less than a 10 per cent increase in the normal ('first order') design moments as a result of the sidesway displacements of the structure. Substantially braced structures can normally be considered to be non-sway. EC2 gives further guidance concerning the classification of unbraced structures. In this chapter only the design of braced non-sway structures will be considered.

For a braced structure the critical arrangement of the ultimate load is usually that which causes the largest moment in the column, together with a large axial load. As an example, figure 9.1 shows a building frame with the critical loading arrangement for the design of its centre column at the first-floor level and also the left-hand column at all floor levels. When the moments in columns are large and particularly with unbraced columns, it may also be necessary to check the case of maximum moment combined with the minimum axial load.

In the case of braced frames, the axial column forces due to the vertical loading may be calculated as though the beams and slabs are simply supported, provided that the spans on either side of the column differ by no more than 30 per cent and there is not a cantilever span. In some structures it is unlikely that all the floors of a building will carry the full imposed load at the same instant, therefore a reduction is allowed in the total imposed load when designing columns or foundations in buildings which are greater than two storeys in height. Further guidance on this can be found in BS EN1991-1-1 (Actions on structures)

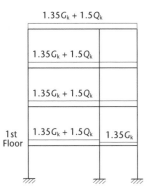

Figure 9.1
A critical loading arrangement

9.2 | Column classification and failure modes

(1) Slenderness ratio of a column

The slenderness ratio λ of a column bent about an axis is given by

$$\lambda = \frac{l_0}{i} = \frac{l_0}{\sqrt{(I/A)}} \tag{9.1}$$

where

l_0 is the effective height of the column

i is the *radius of gyration* about the axis considered

I is the second moment of area of the section about the axis

A is the cross-sectional area of the column

(2) Effective height l_0 of a column

The effective height of a column, l_0, is the height of a theoretical column of equivalent section but pinned at both ends. This depends on the degree of fixity at each end of the column, which itself depends on the relative stiffnesses of the columns and beams connected to either end of the column under consideration.

EC2 gives two formulae for calculating the effective height:

For braced members:

$$l_0 = 0.5l\sqrt{\left(1 + \frac{k_1}{0.45 + k_1}\right)\left(1 + \frac{k_2}{0.45 + k_2}\right)} \tag{9.2}$$

For unbraced members the larger of:

$$l_0 = l\sqrt{\left(1 + 10\frac{k_1 \times k_2}{k_1 + k_2}\right)} \tag{9.3.a}$$

and

$$l_0 = l\left(1 + \frac{k_1}{1 + k_1}\right)\left(1 + \frac{k_2}{1 + k_2}\right) \tag{9.3.b}$$

In the above formulae, k_1 and k_2 are the relative flexibilities of the rotational restraints at ends '1' and '2' of the column respectively. At each end k_1 and k_2 can be taken as:

$$k = \frac{\text{column stiffness}}{\sum \text{beam stiffness}} = \frac{(EI/l)_{\text{column}}}{\sum 2(EI/l)_{\text{beam}}} = \frac{(I/l)_{\text{column}}}{\sum 2(I/l)_{\text{beam}}}$$

It is assumed that any column above or below the column under consideration does not contribute anything to the rotational restraint of the joint and that the stiffness of each connecting beam is taken as $2EI/l$ to allow for cracking effects in the beam.

Hence, for a typical column in a symmetrical frame with spans of approximately equal length, as shown in figure 9.2, k_1 and k_2 can be calculated as:

$$k_1 = k_2 = k = \frac{\text{column stiffness}}{\sum \text{beam stiffness}} = \frac{(I/l)_{\text{column}}}{\sum 2(I/l)_{\text{beam}}} = \frac{(I/l)_{\text{column}}}{2 \times 2(I/l)_{\text{beam}}} = \frac{1}{4} \frac{(I/l)_{\text{column}}}{(I/l)_{\text{beam}}}$$

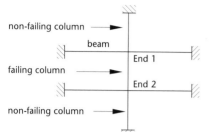

Figure 9.2
Effective length calculation
for a column in a
symmetrical frame

Note: the effective contribution of the non-failing
column to the joint stiffness may be ignored

Table 9.1 Column effective lengths

$\dfrac{1}{4}\dfrac{(I/l_{column})}{(I/l_{beam})} = k$	0 (fixed end)	0.0625	0.125	0.25	0.50	1.0	1.5	2.0
l_0 – braced (equation 9.2) $\{xl\}$	0.5	0.56	0.61	0.68	0.76	0.84	0.88	0.91
l_0 – unbraced (equation 9.3(a) and 9.3(b)). Use	1.0	1.14	1.27	1.50	1.87	2.45	2.92	3.32
greater value $\{xl\}$	1.0	1.12	1.13	1.44	1.78	2.25	2.56	2.78

Thus, for this situation typical values of column effective lengths can be tabulated using equations 9.2 and 9.3 as shown in table 9.1.

(3) Limiting slenderness ratio – short or slender columns

EC2 places an upper limit on the slenderness ratio of a single member below which second order effects may be ignored. This limit is given by:

$$\lambda_{lim} = 20 \times A \times B \times C / \sqrt{n} \qquad (9.4)$$

where:

$A = 1/(1 + 0.2\phi_{ef})$

$B = \sqrt{1 + 2w}$

$C = 1.7 - r_m$

ϕ_{ef} = effective creep ratio (if not known A can be taken as 0.7)

$w = A_s f_{yd}/(A_c f_{cd})$ (if not known B can be taken as 1.1)

f_{yd} = the design yield strength of the reinforcement

f_{cd} = the design compressive strength of the concrete

A_s = the total area of longitudinal reinforcement

$n = N_{Ed}/(A_c f_{cd})$

N_{Ed} = the design ultimate axial load in the column

$r_m = M_{01}/M_{02}$ (if r_m not known then C can be taken as 0.7)

M_{01}, M_{02} are the first order moments at the end of the column with $|M_{02}| \geq |M_{01}|$

The following conditions apply to the value of C:

(a) If the end moments, M_{01} and M_{02}, give rise to tension on the same side of the column r_m should be taken as positive from which it follows that $C \leq 1.7$.

(b) If the converse to (a) is true, i.e the column is in a state of double curvature, then r_m should be taken as negative from which it follows that $C > 1.7$.

(c) For braced members in which the first order moments arise only from transverse loads or imperfections; C can be taken as 0.7.

(d) For unbraced members; C can be taken as 0.7.

For an *unbraced* column an approximation to the limiting value of λ will be given by:

$$\lambda_{\text{lim}} = 20 \times A \times B \times C/\sqrt{n} = 20 \times 0.7 \times 1.1 \times 0.7/\sqrt{N_{\text{Ed}}/(A_c f_{cd})}$$
$$= 10.8/\sqrt{N_{\text{Ed}}/(A_c f_{cd})}$$

The limiting value of λ for a *braced* column will depend on the relative value of the column's end moments that will normally act in the same clockwise or anti-clockwise direction as in case (b) above. If these moments are of approximately equal value then $r_m = -1$, $C = 1.7 + 1 = 2.7$ and a *typical*, approximate limit on λ will be given by:

$$\lambda_{\text{lim}} = 20 \times A \times B \times C/\sqrt{n} = 20 \times 0.7 \times 1.1 \times 2.7/\sqrt{N_{\text{Ed}}/(A_c f_{cd})}$$
$$= 41.6/\sqrt{N_{\text{Ed}}/(A_c f_{cd})}$$

Alternatively for a *braced* column the *minimum* limiting value of λ will be given by taking $C = 1.7$. Hence:

$$\lambda_{\text{lim}} = 20 \times A \times B \times C/\sqrt{n} = 20 \times 0.7 \times 1.1 \times 1.7/\sqrt{N_{\text{Ed}}/(A_c f_{cd})}$$
$$= 26.2/\sqrt{N_{\text{Ed}}/(A_c f_{cd})}$$

If the actual slenderness ratio is less than the calculated value of λ_{lim} then the column can be treated as short. Otherwise the column must be treated as slender and second order effects must be accounted for in the design of the column.

EXAMPLE 9.1

Short or slender column

Determine if the column in the braced frame shown in figure 9.3 is short or slender. The concrete strength $f_{ck} = 25 \text{ N/mm}^2$, and the ultimate axial load $= 1280 \text{ kN}$.

It can be seen that the column will have the highest slenderness ratio for bending about axes YY where $h = 300 \text{ mm}$ and also the end restraints are the less stiff 300×500 beams.

Effective column height l_0

$$I_{\text{col}} = 400 \times 300^3/12 = 900 \times 10^6 \text{ mm}^4$$
$$I_{\text{beam}} = 300 \times 500^3/12 = 3125 \times 10^6 \text{ mm}^4$$
$$k_1 = k_2 = \frac{I_{\text{col}}/l_{\text{col}}}{\sum(2I_{\text{beam}}/l_{\text{beam}})} = \frac{900 \times 10^6/3.0 \times 10^3}{2(2 \times 3125 \times 10^6/4.0 \times 10^3)}$$
$$= 0.096$$

From table 9.1 and by interpolation; effective column height $l_0 = 0.59 \times 3.0 = 1.77 \text{ m}$.

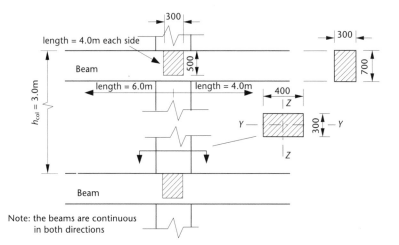

Figure 9.3
Column end support details

Slenderness ratio λ

Radius of gyration, $i = \sqrt{\left(\dfrac{I_{col}}{A_{col}}\right)} = \sqrt{\left(\dfrac{bh^3/12}{bh}\right)} = \dfrac{h}{3.46} = 86.6\,\text{mm}$

Slenderness ratio $\lambda = l_0/i = 1.77 \times 10^3/86.6 = 20.4$

For a *braced* column the *minimum* limiting value of λ will be given by

$$\lambda_{lim} = 26.2/\sqrt{N_{Ed}/(A_c f_{cd})}$$

where:

$$N_{Ed}/(A_c f_{cd}) = 1280 \times 10^3/(400 \times 300 \times 25/1.5) = 0.64$$

thus

$$\lambda_{lim} = 26.2/\sqrt{0.64} = 32.7 \qquad (> 20.4)$$

Hence, compared with the *minimum* limiting value of λ the column is short and second order moment effects would not have to be taken into account.

(4) Failure modes

Short columns usually fail by crushing but a slender column is liable to fail by buckling. The end moments on a slender column cause it to deflect sideways and thus bring into play an additional moment Ne_{add} as illustrated in figure 9.4. The moment Ne_{add} causes a further lateral deflection and if the axial load (N) exceeds a critical value this deflection, and the additional moment become self-propagating until the column buckles. Euler derived the critical load for a pin-ended strut as

$$N_{crit} = \frac{\pi^2 EI}{l^2}$$

The crushing load N_{ud} of a truly axially loaded column may be taken as

$$N_{ud} = 0.567 f_{ck} A_c + 0.87 A_s f_{yk}$$

where A_c is the area of the concrete and A_s is the area of the longitudinal steel.

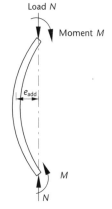

Figure 9.4
Slender column with lateral deflection

Figure 9.5
Column failure modes

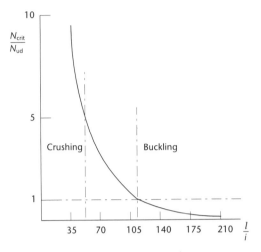

Values of N_{crit}/N_{ud} and l/i have been calculated and plotted in figure 9.5 for a typical column cross-section.

The ratio of N_{crit}/N_{ud} in figure 9.5 determines the type of failure of the column. With l/i less than, say, 50 the load will probably cause crushing, N_{ud} is much less than N_{crit}, the load that causes buckling – and therefore a buckling failure will not occur. This is not true with higher value of l/i and so a buckling failure is possible, depending on such factors as the initial curvature of the column and the actual eccentricity of the load. When l/i is greater than 110 then N_{crit} is less than N_{ud} and in this case a buckling failure will occur for the column considered.

The mode of failure of a column can be one of the following:

1. Material failure with negligible lateral deflection, which usually occurs with short columns but can also occur when there are large end moments on a column with an intermediate slenderness ratio.

2. Material failure intensified by the lateral deflection and the additional moment. This type of failure is typical of intermediate columns.

3. Instability failure which occurs with slender columns and is liable to be preceded by excessive deflections.

9.3 Reinforcement details

The rules governing the minimum and maximum amounts of reinforcement in a load bearing column are as follows.

Longitudinal steel

1. A minimum of four bars is required in a rectangular column (one bar in each corner) and six bars in a circular column. Bar diameter should not be less than 12 mm.

2. The minimum area of steel is given by

$$A_s = \frac{0.10N_{Ed}}{0.87f_{yk}} \geq 0.002A_c$$

3. The maximum area of steel, at laps is given by

$$\frac{A_{s, max}}{A_c} < 0.08$$

where A_s is the total area of longitudinal steel and A_c is the cross-sectional area of the column.

Otherwise, in regions away from laps: $\dfrac{A_{s, max}}{A_c} < 0.04$.

Links

1. Minimum size $= \frac{1}{4} \times$ size of the compression bar but not less than 6 mm.

2. Maximum spacing should not exceed the lesser of $20 \times$ size of the smallest compression bar or the least lateral dimension of the column or 400 mm. This spacing should be reduced by a factor of 0.60.

  (a) for a distance equal to the larger lateral dimension of the column above and below a beam or slab, and

  (b) at lapped joints of longitudinal bars > 14 mm diameter.

3. Where the direction of the longitudinal reinforcement changes, the spacing of the links should be calculated, while taking account of the lateral forces involved. If the change in direction is less than or equal to 1 in 12 no calculation is necessary.

4. Every longitudinal bar placed in a corner should be held by transverse reinforcement.

5. No compression bar should be further than 150 mm from a restrained bar.

Although links are popular in the United Kingdom, helical reinforcement is popular in some parts of the world and provides added strength in addition to added protection against seismic loading. Sizing and spacing of helical reinforcement should be similar to links.

Figure 9.6 shows possible arrangements of reinforcing bars at the junction of two columns and a floor. In figure 9.6a the reinforcement in the lower column is cranked so that it will fit within the smaller column above. The crank in the reinforcement should, if possible, commence above the soffit of a beam so that the moment of resistance of the column is not reduced. For the same reason, the bars in the upper column should be the

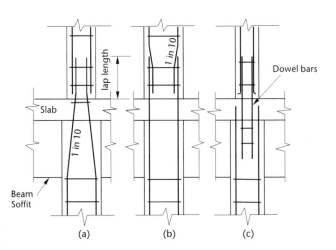

Figure 9.6
Details of splices in column reinforcement

ones cranked when both columns are of the same sizes as in figure 9.6b. Links should be provided at the points where the bars are cranked in order to resist buckling due to horizontal components of force in the inclined lengths of bar. Separate dowel bars as in figure 9.6c may also be used to provide continuity between the two lengths of column. The column–beam junction should be detailed so that there is adequate space for both the column steel and the beam steel. Careful attention to detail on this point will greatly assist the fixing of the steel during construction.

9.4 | Short columns resisting moments and axial forces

The area of longitudinal steel for these columns is determined by:

1. using design charts or constructing M–N interaction diagrams as in chapter 4.
2. a solution of the basic design equations, or
3. an approximate method

Design charts are usually used for columns having a rectangular or circular cross-section and a symmetrical arrangement of reinforcement, but interaction diagrams can be constructed for any arrangement of cross-section as illustrated in examples 4.10 and 4.11. The basic equations or the approximate method can be used when an unsymmetrical arrangement of reinforcement is required, or when the cross-section is non-rectangular as described in section 9.5.

Whichever design method is used, a column should not be designed for a moment less than $N_{Ed} \times e_{min}$, where e_{min} has the greater value of $h/30$ or 20 mm. This is to allow for tolerances in construction. The dimension h is the overall size of the column cross-section in the plane of bending. Note that UK practice is to limit the design moment to $h/20$ not $h/30$.

9.4.1 Design charts and interaction diagrams

The design of a section subjected to bending plus axial load should be in accordance with the principles described in section 4.8, which deals with the analysis of the cross-section. The basic equations derived for a rectangular section as shown in figure 9.7 and with a rectangular stress block are:

$$N_{Ed} = F_{cc} + F_{sc} + F_{s} \tag{9.5}$$
$$= 0.567 f_{ck} b s + f_{sc} A'_{s} + f_{s} A_{s}$$

$$M_{Ed} = F_{cc}\left(\frac{h}{2} - \frac{s}{2}\right) + F_{sc}\left(\frac{h}{2} - d'\right) - F_{s}\left(d - \frac{h}{2}\right) \tag{9.6}$$

N_{Ed} = design ultimate axial load

M_{Ed} = design ultimate moment

s = the depth of the stress block = $0.8x$

A'_{s} = the area of longitudinal reinforcement in the more highly compressed face

A_{s} = the area of reinforcement in the other face

f_{sc} = the stress in reinforcement A'_{s}

f_{s} = the stress in reinforcement A_{s}, negative when tensile.

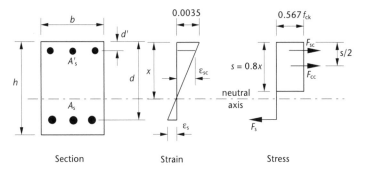

Figure 9.7
Column section

Section Strain Stress

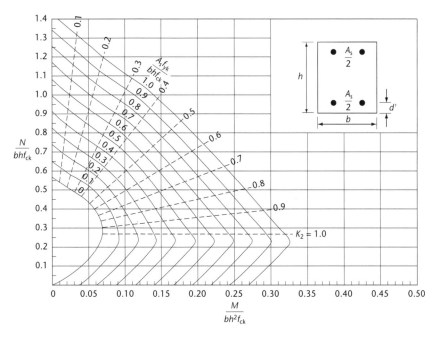

Figure 9.8
Rectangular column
($d'/h = 0.20$)

These equations are not suitable for direct solution and the design of a column with symmetrical reinforcement in each face is best carried out using design charts as illustrated in figure 9.8. Sets of these charts can be found in the Concise Eurocode (ref. 21), the Manual for the Design of Concrete Structures (ref. 23) and the website www.eurocode2.info.

EXAMPLE 9.2

Column design using design charts

Figure 9.9 shows a frame of a heavily loaded industrial structure for which the centre columns along line PQ are to be designed in this example. The frames at 4 m centres, are braced against lateral forces, and support the following floor loads:

permanent action $g_k = 10 \, \text{kN/m}^2$

variable action $q_k = 15 \, \text{kN/m}^2$

Characteristic material strengths are $f_{ck} = 25 \, \text{N/mm}^2$ for the concrete and $f_{yk} = 500$ N/mm^2 for the steel.

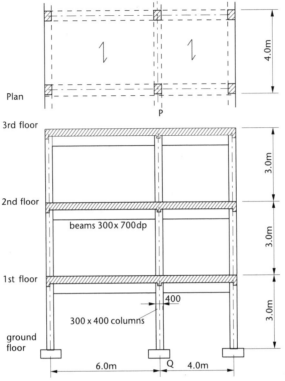

Figure 9.9
Columns in an industrial structure

Plan

P

3rd floor

2nd floor

beams 300 x 700dp

1st floor

400

300 x 400 columns

ground floor

6.0m   Q   4.0m

4.0m

3.0m

3.0m

3.0m

Section through the frame

$$\text{Maximum ultimate load at each floor} = 4.0(1.35g_k + 1.5q_k) \text{ per metre length of beam}$$
$$= 4(1.35 \times 10 + 1.5 \times 15)$$
$$= 144 \text{ kN/m}$$
$$\text{Minimum ultimate load at each floor} = 4.0 \times 1.35g_k$$
$$= 4.0 \times 1.35 \times 10$$
$$= 54 \text{ kN per metre length of beam}$$

Consider first the design of the centre column at the underside (u.s.) of the first floor. The critical arrangement of load that will cause the maximum moment in the column is shown in figure 9.10a.

Column loads

$$\text{Second and third floors} = 2 \times 144 \times 10/2 \qquad = 1440 \text{ kN}$$
$$\text{first floor} = 144 \times 6/2 + 54 \times 4/2 \ = 540$$
$$\text{Column self-weight, say } 2 \times 14 = 28$$
$$N_{Ed} = 2008 \text{ kN}$$

Similar arrangements of load will give the axial load in the column at the underside (u.s.) and top side (t.s.) of each floor level and these values of N_{Ed} are shown in table 9.2.

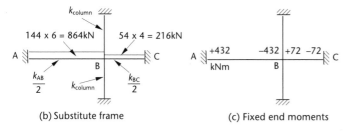

Figure 9.10
Substitute frame for column
design example

(a) Critical loading arrangement for centre columns at 1st floor

(b) Substitute frame

(c) Fixed end moments

Table 9.2

Floor	N_{Ed} (kN)	M_{Ed} (kN m)	$\dfrac{N_{Ed}}{bhf_{ck}}$	$\dfrac{M_{Ed}}{bh^2 f_{ck}}$	$\dfrac{A_s f_{yk}}{bhf_{ck}}$	A_s (mm²)
3rd u.s.	540	82.6	0.18	0.07	0	240
2nd t.s.	734 + 540	68.4	0.24	0.06	0	240
2nd u.s.	1274	68.4	0.42	0.06	0	240
1st t.s.	1468 + 540	68.4	0.49	0.06	0.10	600
1st u.s.	2008	68.4	0.67	0.06	0.30	1800

Column moments

The loading arrangement and the substitute frame for determining the column moments at the first and second floors are shown in figure 9.10(c).

Member stiffnesses are

$$\frac{k_{AB}}{2} = \frac{1}{2} \times \frac{bh^3}{12L_{AB}} = \frac{1}{2} \times \frac{0.3 \times 10.7^3}{12 \times 6} = 0.71 \times 10^{-3}$$

$$\frac{k_{BC}}{2} = \frac{1}{2} \times \frac{0.3 \times 0.7^3}{12 \times 4} = 1.07 \times 10^{-3}$$

$$k_{col} = \frac{0.3 \times 0.4^3}{12 \times 3.0} = 0.53 \times 10^{-3}$$

therefore

$$\sum k = (0.71 + 1.07 + 2 \times 0.53)10^{-3} = 2.84 \times 10^{-3}$$

and

$$\text{distribution factor for the column} = \frac{k_{col}}{\sum k} = \frac{0.53}{2.84} = 0.19$$

Fixed end moments at B are

$$F.E.M._{BA} = \frac{144 \times 6^2}{12} = 432 \, \text{kN m}$$

$$F.E.M._{BC} = \frac{54 \times 4^2}{12} = 72 \, \text{kN m}$$

Thus

$$\text{column moment } M_{Ed} = 0.19(432 - 72) = 68.4 \, \text{kN m}$$

At the 3rd floor

$$\sum k = (0.71 + 1.07 + 0.53)10^{-3}$$
$$= 2.31 \times 10^{-3}$$

and

$$\text{column moment } M_{Ed} = \frac{0.53}{2.31}(432 - 72) = 82.6 \, \text{kN m}$$

The areas of reinforcement in table 9.2 are determined by using the design chart of figure 9.8. Sections through the column are shown in figure 9.11.

Figure 9.11
Column sections in design example

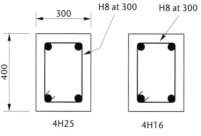

(a) Ground to 1st Floor   (b) 1st to 3rd Floor

Note: the link spacing is reduced to 0.60 x these values for 400mm above and below each floor level and at laps below 1st floor level

Cover for the reinforcement is taken as 50 mm and $d'/h = 80/400 = 0.2$. The minimum area of reinforcement allowed in the section is given by:

$$A_s = 0.002bh = 0.002 \times 300 \times 400 = 240 \, \text{mm}^2$$

and the maximum area is

$$A_s = 0.08 \times 300 \times 400 = 9600 \, \text{mm}^2$$

and the reinforcement provided is within these limits.

Although EC2 permits the use of 12 mm main steel, 16 mm bars have been used to ensure adequate rigidity of the reinforcing cage. A smaller column section could have been used above the first floor but this would have involved changes in formwork and possibly also increased areas of reinforcement.

9.4.2 Design equations for a non-symmetrical section

The symmetrical arrangement of the reinforcement with $A'_s = A_s$ is justifiable for the columns of a building where the axial loads are the dominant forces and where any moments due to the wind can be acting in either direction. But some members are required to resist axial forces combined with large bending moments so that it is not economical to have equal areas of steel in both faces, and in these cases the usual design charts cannot be applied. A rigorous design for a rectangular section as shown in figure 9.12 involves the following iterative procedure:

1. Select a depth of neutral axis, x (for this design method where the moments are relatively large, x would generally be less than h).
2. Determine the steel strains ε_{sc} and ε_s from the strain distribution.
3. Determine the steel stresses f_{sc} and f_s from the equations relating to the stress-strain curve for the reinforcing bars (see section 4.1.2).
4. Taking moments about the centroid of A_s

$$N_{Ed}\left(e + \frac{h}{2} - d_2\right) = 0.567f_{ck}bs(d - s/2) + f_{sc}A'_s(d - d') \tag{9.7}$$

   where $s = 0.8x$.

   This equation can be solved to give a value for A'_s
5. A_s is then determined from the equilibrium of the axial forces, that is

$$N_{Ed} = 0.567f_{ck}bs + f_{sc}A'_s + f_sA_s \tag{9.8}$$

6. Further values of x may be selected and steps (1) to (5) repeated until a minimum value for $A'_s + A_s$ is obtained.

The term f_{sc} in the equations may be modified to $(f_{sc} - 0.567f_{ck})$ to allow for the area of concrete displaced by the reinforcement A'_s. Stress f_s has a negative sign whenever it is tensile.

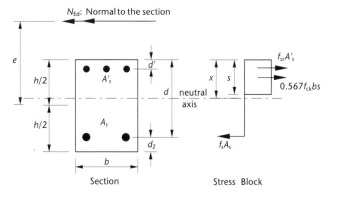

Figure 9.12
Column with a non-symmetrical arrangement of reinforcement

EXAMPLE 9.3

Column section with an unsymmetrical arrangement of reinforcement

The column section shown in figure 9.13 resists an axial load of 1100 kN and a moment of 230 kNm at the ultimate limit state. Determine the areas of reinforcement required if the characteristic material strengths are $f_{yk} = 500\,\text{N/mm}^2$ and $f_{ck} = 25\,\text{N/mm}^2$.

Figure 9.13
Unsymmetrical column design
example

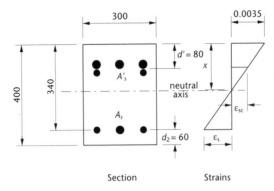

Section   Strains

1. Select a depth of neutral axis, $x = 190\,\text{mm}$.

2. From the strain diagram

$$\text{steel strain } \varepsilon_{sc} = \frac{0.0035}{x}(x - d')$$

$$= \frac{0.0035}{190}(190 - 80) = 0.00203$$

   and

$$\text{steel strain } \varepsilon_s = \frac{0.0035}{x}(d - x)$$

$$= \frac{0.0035}{190}(340 - 190) = 0.00276$$

3. From the stress–strain curve and the relevant equations of section 4.1.2 yield strain, $\varepsilon_y = 0.00217$ for grade 500 steel

$$\varepsilon_s > 0.00217; \text{ therefore } f_{sc} = 500/1.15 = 435\,\text{N/mm}^2$$

   and

$$\varepsilon_{sc} < 0.00217; \quad \text{therefore } f_{sc} = E_s\varepsilon_{sc} = 200 \times 10^3 \times 0.00203$$
$$= 406\,\text{N/mm}^2, \text{ compression.}$$

4. In equation 9.7

$$N_{Ed}\left(e + \frac{h}{2} - d_2\right) = 0.567f_{ck}bs(d - s/2) + f_{sc}A'_s(d - d')$$

$$e = \frac{M_{Ed}}{N_{Ed}} = \frac{230 \times 10^6}{1100 \times 10^3} = 209\,\text{mm}$$

$$s = 0.8x = 0.8 \times 190 = 152\,\text{mm}$$

   To allow for the area of concrete displaced

$$f_{sc} \text{ becomes } 406 - 0.567f_{ck} = 406 - 0.567 \times 25$$
$$= 392\,\text{N/mm}^2$$

   and from equation 9.7

$$A'_s = \frac{1100 \times 10^3(209 + 140) - 0.567 \times 25 \times 300 \times 152(340 - 152/2)}{392(340 - 80)}$$

$$= 2093\,\text{mm}^2$$

5. From equation 9.8

$$N_{Ed} = 0.567 f_{ck} bs + f_{sc} A'_s + f_s A_s$$

$$A_s = \frac{(0.567 \times 25 \times 300 \times 152) + (392 \times 2093) - (1100 \times 10^3)}{435}$$

$$= 843 \text{ mm}^2$$

Thus

$$A'_s + A_s = 2936 \text{ mm}^2 \text{ for } x = 190 \text{ mm}$$

6. Values of $A'_s + A_s$ calculated for other depths of neutral axis, x, are plotted in figure 9.14. From this figure the minimum area of reinforcement required occurs with $x \approx 210$ mm. Using this depth of neutral axis, steps 2 to 5 are repeated giving

$$\varepsilon_{sc} = 0.00217, \quad \varepsilon_s = 0.00217$$

$$f_{sc} = f_{yk}/\gamma_m = 435 \text{ N/mm}^2 \text{ and } f_s = 435 \text{ N/mm}^2 \text{ tension}$$

so that

$$A'_s = 1837 \text{ mm}^2 \text{ and } A_s = 891 \text{ mm}^2$$

(Alternatively separate values of A'_s and A_s as calculated for each value of x could have also have been plotted against x and their values read from the graph at $x = 210$ mm.) This area would be provided with

$$A'_s = \text{three H25 plus two H20 bars}$$
$$= 2098 \text{ mm}^2$$

and

$$A_s = \text{one H25 plus two H20 bars}$$
$$= 1119 \text{ mm}^2$$

With a symmetrical arrangement of reinforcement the area from the design chart of figure 9.8 would be $A'_s + A_s \approx 3120 \text{ mm}^2$ or 14 per cent greater than the area with an unsymmetrical arrangement, and including no allowance for the area of concrete displaced by the steel.

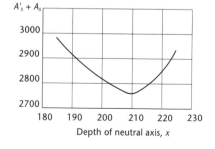

Figure 9.14
Design chart for unsymmetrical column example

These types of iterative calculations are readily programmed for solution by computer or using spreadsheets that could find the optimum steel areas without the necessity of plotting a graph.

9.4.3 Simplified design method

As an alternative to the previous rigorous method of design an approximate method may be used when the eccentricity of loading, e is not less than $(h/2 - d_2)$.

Figure 9.15
Simplified design method

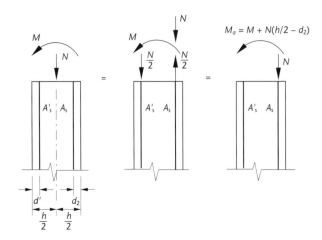

The moment M_{Ed} and the axial force N_{Ed} are replaced by an increased moment M_a where

$$M_a = M_{Ed} + N_{Ed}\left(\frac{h}{2} - d_2\right) \tag{9.9}$$

plus a compressive force N_{Ed} acting through the tensile steel A_s as shown in figure 9.15. Hence the design of the reinforcement is carried out in two parts.

1. The member is designed as a doubly reinforced section to resist M_a acting by itself. The equations for calculating the areas of reinforcement to resist M_a for grades C50 concrete (or below) are given in section 4.5 as:

$$M_a = 0.167f_{ck}bd^2 + 0.87f_{yk}A'_s(d - d') \tag{9.10}$$
$$0.87f_{yk}A_s = 0.204f_{ck}bd + 0.87f_{yk}A'_s \tag{9.11}$$

2. The area of A_s calculated in the first part is reduced by the amount $N_{Ed}/0.87f_{yk}$.

This preliminary design method is probably most useful for non-rectangular column sections as shown in example 9.5, but the procedure is first demonstrated with a rectangular cross-section in the following example.

EXAMPLE 9.4

Column design by the simplified method

Calculate the area of steel required in the 300×400 column of figure 9.13. $N_{Ed} = 1100\,\text{kN}$, $M_{Ed} = 230\,\text{kN m}$, $f_{ck} = 25\,\text{N/mm}^2$ and $f_{yk} = 500\,\text{N/mm}^2$.

$$\text{Eccentricity } e = \frac{230 \times 10^6}{1100 \times 10^3}$$

$$= 209\,\text{mm} > \left(\frac{h}{2} - d_2\right)$$

1. Increased moment

$$M_a = M_{Ed} + N_{Ed}\left(\frac{h}{2} - d_2\right)$$

$$= 230 + 1100(200 - 60)10^{-3} = 384 \, \text{kN m}$$

The area of steel to resist this moment can be calculated using formulae 9.10 and 9.11 for the design of a beam with compressive reinforcement, that is

$$M_a = 0.167 f_{ck} b d^2 + 0.87 f_{yk} A'_s (d - d')$$

and

$$0.87 f_{yk} A_s = 0.204 f_{ck} b d + 0.87 f_{yk} A'_s$$

therefore

$$384 \times 10^6 = 0.167 \times 25 \times 300 \times 340^2 + 0.87 \times 500 A'_s (340 - 80)$$

so that

$$A'_s = 2115 \, \text{mm}^2$$

and

$$0.87 \times 500 \times A_s = 0.204 \times 25 \times 300 \times 340 + 0.87 \times 500 \times 2115$$

$$A_s = 3311 \, \text{mm}^2$$

2. Reducing this area by $N_{Ed}/0.87 f_{yk}$

$$A_s = 3311 - \frac{1100 \times 10^3}{0.87 \times 500}$$

$$= 782 \, \text{mm}^2$$

This compares with $A'_s = 1837 \, \text{mm}^2$ and $A_s = 891 \, \text{mm}^2$ with the design method of example 9.3. (To give a truer comparison the stress in the compressive reinforcement should have been modified to allow for the area of concrete displaced, as was done in example 9.3.)

9.5 Non-rectangular sections

Design charts are not usually available for columns of other than a rectangular or a circular cross-section. Therefore the design of a non-rectangular section entails either (1) an iterative solution of design equations, (2) a simplified form of design, or (3) construction of *M–N* interaction diagrams.

9.5.1 Design equations

For a non-rectangular section it is much simpler to consider the equivalent rectangular stress block. Determination of the reinforcement areas follows the same procedure as described for a rectangular column in section 9.4.2, namely

1. Select a depth of neutral axis.
2. Determine the corresponding steel strains.
3. Determine the steel stresses.

Figure 9.16
Non-rectangular column
section

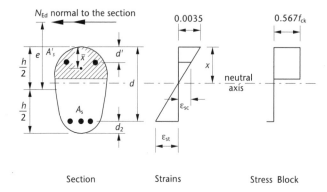

4. Take moments about A_s so that with reference to figure 9.16:

$$N_{Ed}\left(e + \frac{h}{2} - d_2\right) = 0.567f_{ck}A_{cc}(d - \bar{x}) + f_{sc}A'_s(d - d')$$

Solve this equation to give A'_s

5. For no resultant force on the section

$$N_{Ed} = 0.567f_{ck}A_{cc} + f_{sc}A'_s + f_sA_s$$

Solve this equation to give A_s.

6. Repeat the previous steps for different values of x to find a minimum $(A'_s + A_s)$.

In steps (4) and (5)

A_{cc} is the area of concrete in compression shown shaded

$\bar{x}$ is the distance from the centroid of A_{cc} to the extreme fibre in compression

f_s is the stress in reinforcement A_s, negative if tensile.

The calculation for a particular cross-section would be very similar to that described in example 9.3 except when using the design equations it would be necessary to determine A_{cc} and $\bar{x}$ for each position of a neutral axis.

9.5.2 Simplified preliminary design method

The procedure is similar to that described for a column with a rectangular section as described in section 9.4.3 and figure 9.15.

The column is designed to resist a moment M_a only, where

$$M_a = M_{Ed} + N_{Ed}\left(\frac{h}{2} - d_2\right) \tag{9.12}$$

The steel area required to resist this moment can be calculated from

$$M_a = 0.567f_{ck}A_{cc}(d - \bar{x}) + 0.87f_{ck}A'_s(d - d') \tag{9.13}$$

and

$$0.87f_{yk}A_s = 0.567f_{ck}A_{cc} + 0.87f_{yk}A'_s \tag{9.14}$$

where A_{cc} is the area of concrete in compression with $x = 0.45d$ for concrete grades C50 and below and $\bar{x}$ is the distance from the centroid of A_{cc} to the extreme fibre in compression.

The area of tension reinforcement, A_s, as given by equation 9.14 is then reduced by an amount equal to $N_{Ed}/0.87f_{yk}$.

This method should not be used if the eccentricity, e, is less than $(h/2 - d_2)$.

9.5.3 M–N interaction diagram

These diagrams can be constructed using the method described in section 4.8 with examples 4.10 and 4.11. They are particularly useful for a column in a multi-storey building where the moments and associated axial forces change at each storey. The diagrams can be constructed after carrying out the approximate design procedure in section 9.5.2 to obtain suitable arrangements of reinforcing bars.

EXAMPLE 9.5

Design of a non-rectangular column section

Design the reinforcement for the non-rectangular section shown in figure 9.17 given $M_{Ed} = 320\,\text{kN m}$, $N_{Ed} = 1200\,\text{kN}$ at the ultimate limit state and the characteristic material strengths are $f_{ck} = 25\,\text{N/mm}^2$ and $f_{yk} = 500\,\text{N/mm}^2$.

$$e = \frac{M_{Ed}}{N_{Ed}} = \frac{320 \times 10^6}{1200 \times 10^3} = 267\,\text{mm} > \left(\frac{h}{2} - d_2\right)$$

Increased moment $M_a = M_{Ed} + N_{Ed}\left(\frac{h}{2} - d_2\right)$

$$= 320 + 1200(200 - 80)10^{-3}$$

$$= 464\,\text{kN m}$$

With $x = 0.45d = 144\,\text{mm}$, $s = 0.8x = 115\,\text{mm}$ and the width (b_1) of the section at the limit of the stress block

$$b_1 = 300 + \frac{200(400 - 115)}{400}$$

$$= 443\,\text{mm}$$

$$A_{cc} = \frac{x(b + b_1)}{2}$$

$$= \frac{115(500 + 443)}{2}$$

$$= 54\,223\,\text{mm}^2$$

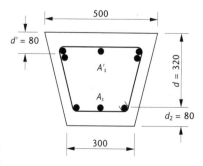

Figure 9.17
Non-rectangular section example

The depth of the centroid of the trapezium is given by

$$\bar{x} = \frac{s(b + 2b_1)}{3(b + b_1)}$$

$$= 115\frac{(500 + 2 \times 443)}{3(500 + 443)} = 56.3 \, \text{mm}$$

Therefore substituting in equation 9.13

$$464 \times 10^6 = 0.567 \times 25 \times 54223(320 - 56.3) + 0.87 \times 500A'_s(320 - 80)$$

hence

$$A'_s = 2503 \, \text{mm}^2$$

Provide three H32 plus two H16 bars, area = 2812 mm².
From equation 9.14

$$0.87f_{yk}A_s = 0.567 \times 25 \times 54\,223 + 0.87 \times 500 \times 2503$$

therefore

$$A_s = 4269 \, \text{mm}^2$$

Reducing A_s by $N_{Ed}/0.87f_{yk}$ gives

$$A_s = 4269 - \frac{1200 \times 10^3}{0.87 \times 500}$$

$$= 1510 \, \text{mm}^2$$

Provide one H16 plus two H32 bars, area = 1811 mm².
The total area of reinforcement provided = 4623 mm² which is less than the 8 per cent allowed.

An M–N interaction diagram could now be constructed for this steel arrangement, as in section 4.8, to provide a more rigorous design.

9.6 Biaxial bending of short columns

For most columns, biaxial bending will not govern the design. The loading patterns necessary to cause biaxial bending in a building's internal and edge columns will not usually cause large moments in both directions. Corner columns may have to resist significant bending about both axes, but the axial loads are usually small and a design similar to the adjacent edge columns is generally adequate.

A design for biaxial bending based on a rigorous analysis of the cross-section and the strain and stress distributions would be done according to the fundamental principles of chapter 4. For members with a rectangular cross-section, separate checks in the two principal planes are permissible if the ratio of the corresponding eccentricities satisfies one of the following conditions:

either $\quad \dfrac{e_z}{h} \Big/ \dfrac{e_y}{b} \leq 0.2$

or $\quad \dfrac{e_y}{b} \Big/ \dfrac{e_z}{b} \leq 0.2$

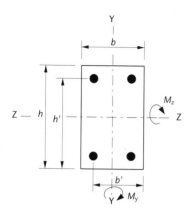

Figure 9.18
Section with biaxial bending

where e_y and e_z are the first-order eccentricities in the direction of the section dimensions b and h respectively. Where these conditions are not fulfilled biaxial bending must be accounted for and EC2 presents an interaction equation, relating the moments about the two axes to the moment of resistance about the two axes, which must be satisfied. However, the given formula cannot be used directly to design a column subject to biaxial bending but rather to check it once designed. In the absence of specific design guidance it would be acceptable in the UK that the column be designed using the method previously presented in BS 8110.

This approximate method specifies that a column subjected to an ultimate load N_{Ed} and moments M_z and M_y in the direction of the ZZ and YY axes respectively (see figure 9.18) may be designed for a single axis bending but with an increased moment and subject to the following conditions:

(a) if $\dfrac{M_z}{h'} \ge \dfrac{M_y}{b'}$

then the increased single axis design moment is

$$M'_z = M_z + \beta \frac{h'}{b'} \times M_y$$

(b) if $\dfrac{M_z}{h'} < \dfrac{M_y}{b'}$

then the increased single axis design moment is

$$M'_y = M_y + \beta \frac{b'}{h'} \times M_z$$

The dimensions h' and b' are defined in figure 9.18 and the coefficient β is specified in table 9.3. The coefficients in table 9.3 are obtained from the equation

$$\beta = 1 - \frac{N_{Ed}}{bhf_{ck}}$$

Table 9.3 Values of coefficient β for biaxial bending

$\dfrac{N_{Ed}}{bhf_{ck}}$	0	0.1	0.2	0.3	0.4	0.5	0.6	≥ 0.7
β	1.0	0.9	0.8	0.7	0.6	0.5	0.4	0.3

EXAMPLE 9.6

Design of a column for biaxial bending

The column section shown in figure 9.19 is to be designed to resist an ultimate axial load of 1200 kN plus moments of $M_z = 75\,\text{kN m}$ and $M_y = 80\,\text{kN m}$. The characteristic material strengths are $f_{ck} = 25\,\text{N/mm}^2$ and $f_{yk} = 500\,\text{N/mm}^2$.

Figure 9.19
Biaxial bending example

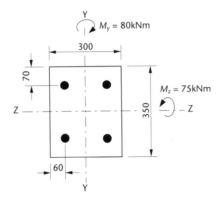

$$e_z = \frac{M_z}{N_{Ed}} = \frac{75 \times 10^6}{1200 \times 10^3} = 62.5\,\text{mm}$$

$$e_y = \frac{M_y}{N_{Ed}} = \frac{80 \times 10^6}{1200 \times 10^3} = 66.7\,\text{mm}$$

thus

$$\frac{e_z}{h} \bigg/ \frac{e_y}{b} = \frac{62.5}{350} \bigg/ \frac{66.7}{300} = 0.8 > 0.2$$

and

$$\frac{e_y}{b} \bigg/ \frac{e_z}{h} = \frac{66.7}{300} \bigg/ \frac{62.5}{350} = 1.24 > 0.2$$

Hence the column must be designed for biaxial bending.

$$\frac{M_z}{h'} = \frac{75}{(350 - 70)} = 0.268$$

$$\frac{M_y}{b'} = \frac{80}{(300 - 60)} = 0.333$$

$$\frac{M_z}{h'} < \frac{M_y}{b'}$$

therefore the increased single axis design moment is

$$M'_y = M_y + \beta \frac{b'}{h'} \times M_z$$

$$N_{Ed}/bhf_{ck} = 1200 \times 10^3 / (300 \times 350 \times 25) = 0.46$$

From table 9.3, $\beta = 0.54$

$$M'_y = 80 + 0.54 \times \frac{240}{280} \times 75 = 114.7\,\text{kN m}$$

thus

$$\frac{M_{Ed}}{bh^2 f_{ck}} = \frac{114.7 \times 10^6}{350 \times (300)^2 \times 25} = 0.15$$

From the design chart of figure 9.8

$$\frac{A_s f_{yk}}{bh f_{ck}} = 0.47$$

Therefore required $A_s = 2467\,\text{mm}^2$.

So provide four H32 bars.

9.7 Design of slender columns

As specified in section 9.2, a column is classified as slender if the slenderness ratio about either axis exceeds the value of λ_{lim}. If $\lambda \leq \lambda_{lim}$ then the column may be classified as short and the slenderness effect may be neglected.

A slender column with $\lambda > \lambda_{lim}$ must be designed for an additional moment caused by its curvature at ultimate conditions. EC2 identifies four different approaches to designing slender columns:

1. A general method based on a non-linear analysis of the structure and allowing for second-order effects that necessitates the use of computer analysis.

2. A second-order analysis based on nominal stiffness values of the beams and columns that, again, requires computer analysis using a process of iterative analysis.

3. The 'moment magnification' method where the design moments are obtained by factoring the first-order moments.

4. The 'nominal curvature' method where second-order moments are determined from an estimation of the column curvature. These second-order moments are added to the first-order moments to give the total column design moment.

Only the fourth method, as given above, will be detailed here as this method is not greatly dissimilar to the approach in the previous British Standard for concrete design, BS 8110. Further information on the other methods can be found in specialist literature.

The expressions given in EC2 for the additional moments were derived by studying the moment/curvature behaviour for a member subject to bending plus axial load. The equations for calculating the design moments are only applicable to columns of a rectangular or circular section with symmetrical reinforcement.

A slender column should be designed for an ultimate axial load (N_{Ed}) plus an increased moment given by

$$M_t = N_{Ed} e_{tot}$$

where

$e_{tot} = e_0 + e_a + e_2$

e_0 is an equivalent first-order eccentricity

e_a is an accidental eccentricity which accounts for geometric imperfections in the column

e_2 is the second-order eccentricity.

The equivalent eccentricity e_0 is given by the greater of

$$0.6e_{02} + 0.4e_{01} \quad \text{or} \quad 0.4e_{02}$$

where e_{01} and e_{02} are the first-order eccentricities at the two ends of the column as described above, and $|e_{02}|$ is greater than $|e_{01}|$.

The accidental eccentricity is given by the equation

$$e_a = v\frac{l_0}{2}$$

where l_0 is the effective column height about the axis considered and

$$v = \frac{1}{100\sqrt{l}} > \frac{1}{200}$$

where l is the height of the column in metres. A conservative estimate of e_a can be given by:

$$e_a = v\frac{l_0}{2} = \frac{1}{200} \times \frac{l_0}{2} = \frac{l_0}{400}$$

The second-order eccentricity e_2 is an estimate of the deflection of the column at failure and is given by the equation

$$e_2 = K_1 K_2 \frac{l_0^2}{\pi^2}\left(\frac{\varepsilon_{yd}}{0.45d}\right)$$

where

$$K_1 = 1 + \left(0.35 + \frac{f_{ck}}{200} - \frac{\lambda}{150}\right)\phi_{ef} \geq 1$$

$$\lambda = \text{slenderness ratio}$$

$$\phi_{ef} = \text{effective creep ratio} = \phi(\infty, t_0) \times M_{0Eqp}/M_{0Ed}$$

$$\phi(\infty, t_0) = \text{final creep coefficient}$$

$M_{0Eqp} = $ the bending moment in the quasi-permanent load combination at the SL

$M_{0Ed} = $ the bending moment in the design load combination at the ULS

ϕ_{ef} may be taken as zero if $\phi(\infty, t_0) \leq 2$ and $\lambda \leq 75$ and $M_{0Ed}/N_{Ed} \geq h$

In most practical cases the above equation may be simplified to

$$e_2 = \frac{K_1 K_2 l_0^2 f_{yk}}{\pi^2 \times 103\,500d}$$

where π^2 is sometimes approximated to a value of 10.

The coefficient K_2 is a reduction factor to allow for the fact that the deflection must be less when there is a large proportion of the column section in compression. The value for K_2 is given by the equation

$$K_2 = \frac{N_{ud} - N_{Ed}}{N_{ud} - N_{bal}} \leq 1.0 \tag{9.15}$$

where N_{ud} is the ultimate axial load such that

$$N_{ud} = 0.567f_{ck}A_c + 0.87f_{yk}A_{sc}$$

and N_{bal} is the axial load at balanced failure defined in section 4.8 and may be taken as approximately $N_{bal} = 0.29f_{ck}A_c$ for symmetrical reinforcement.

In order to calculate K_2, the area A_s of the column reinforcement must be known and hence a trial-and-error approach is necessary, taking an initial conservative value of $K_2 = 1.0$. Values of K_2 are also marked on the column design charts as shown in figure 9.8.

EXAMPLE 9.7

Design of a slender column

A non-sway column of 300×450 cross-section resists, at the ultimate limit state, an axial load of $1700\,\text{kN}$ and end moments of $70\,\text{kN m}$ and $10\,\text{kN m}$ causing double curvature about the minor axis YY as shown in figure 9.20. The column's effective heights are $l_{ey} = 6.75\,\text{m}$ and $l_{ez} = 8.0\,\text{m}$ and the characteristic material strengths $f_{ck} = 25\,\text{N/mm}^2$ and $f_{yk} = 500\,\text{N/mm}^2$. The effective creep ratio $\phi_{ef} = 0.87$.

Eccentricities are

$$e_{01} = \frac{M_1}{N_{Ed}} = \frac{10 \times 10^3}{1700} = 5.9\,\text{mm}$$

$$e_{02} = \frac{M_2}{N_{Ed}} = \frac{-70 \times 10^3}{1700} = -41.2\,\text{mm}$$

where e_{02} is negative since the column is bent in double curvature.

The limiting slenderness ratio can be calculated from equation 9.4 where:

$$A = 1/(1 + 0.2\phi_{ef}) = 1/(1 + (0.2 \times 0.87)) = 0.85$$

$$B = \text{the default value of } 1.1$$

$$C = 1.7 - M_{01}/M_{02} = 1.7 - (-10/70) = 1.84$$

$$\therefore \lambda_{lim} = 20 \times A \times B \times C/\sqrt{n} = 20 \times 0.85 \times 1.1 \times 1.84/\sqrt{n} = \frac{34.41}{\sqrt{n}}$$

$$n = \frac{N_{Ed}}{A_c f_{cd}} = \frac{1700 \times 10^3}{(300 \times 450) \times 0.567 \times 25} = 0.89$$

$$\therefore \lambda_{lim} = \frac{34.41}{\sqrt{0.89}} = 36.47$$

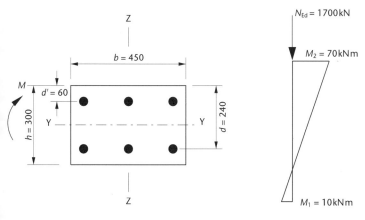

(a) Section

(b) Axial load and initial moments

Figure 9.20
Slender column example

Actual slenderness ratios are

$$\lambda_y = \frac{l_{ey}}{i_y} = \frac{6.75}{0.3} \times 3.46 = 77.85 > 36.47$$

$$\lambda_z = \frac{l_{ez}}{i_z} = \frac{8.0}{0.45} \times 3.46 = 61.55 > 36.47$$

Therefore the column is slender, and λ_y is critical.

Equivalent eccentricity $= 0.6e_{02} + 0.4e_{01} \geq 0.4e_{02}$

$$0.6e_{02} + 0.4e_{01} = 0.6 \times 41.2 + 0.4 \times (-5.9) = 22.35 \, \text{mm}$$

$$0.4e_{02} = 0.4 \times 41.2 = 16.47 \, \text{mm}$$

Therefore the equivalent eccentricity $e_e = 22.35 \, \text{mm}$.

Taking v as $1/200$ the accidental eccentricity is

$$e_a = v\frac{l_{ey}}{2} = \frac{1}{200} \times \frac{6750}{2} = 16.88 \, \text{mm}$$

The second-order eccentricity is

$$e_2 = \frac{K_1 K_2 l_0^2 f_{yk}}{\pi^2 \times 103\,500 d}$$

where

$$K_1 = 1 + \left(0.35 + \frac{f_{ck}}{200} - \frac{\lambda}{150}\right)\phi_{ef} = 1 + \left(0.35 + \frac{25}{200} - \frac{77.85}{150}\right) \times 0.87$$

$$= 0.96 \quad (\geq 1)$$

$$\therefore e_2 = \frac{K_1 K_2 l_0^2 f_{yk}}{\pi^2 \times 103\,500 d} = \frac{1 \times 1 \times 6750^2 \times 500}{\pi^2 \times 103\,500 \times 240}$$

$$= 92.92 \, \text{mm}$$

with $K_2 = 1.0$ for the initial value.

For the first iteration the total eccentricity is

$$e_{tot} = e_0 + e_a + e_2$$

$$= 22.35 + 16.88 + 92.92 = 132.15 \, \text{mm}$$

and the total moment is

$$M_t = N_{Ed}e_{tot} = 1700 \times 132.15 \times 10^{-3} = 225 \, \text{kN\,m}$$

$$\frac{N_{Ed}}{bhf_{ck}} = \frac{1700 \times 10^3}{450 \times 300 \times 25} = 0.504$$

$$\frac{M_t}{bh^2 f_{ck}} = \frac{225 \times 10^6}{450 \times 300^2 \times 25} = 0.222$$

From the design chart of figure 9.8

$$\frac{A_s f_{yk}}{bhf_{ck}} = 0.80 \quad \text{and} \quad K_2 = 0.78$$

This new value of K_2 is used to calculate e_2 and hence M_t for the second iteration. The design chart is again used to determine $A_s f_{yk}/bhf_{ck}$ and a new value of K_2 as shown in table 9.4. The iterations are continued until the value of K_2 in columns (1) and (5) of the

Table 9.4

(1) K_2	(2) M_t	(3) $\dfrac{M_t}{bh^2 f_{ck}}$	(4) $\dfrac{A_s f_{yk}}{bh f_{ck}}$	(5) K_2
1.0	225	0.222	0.80	0.78
0.78	190	0.187	0.6	0.73

table are in reasonable agreement, which in this design occurs after two iterations. So that the steel area required is

$$A_s = \frac{0.6bhf_{ck}}{f_{yk}} = \frac{0.6 \times 450 \times 300 \times 25}{500} = 4050 \,\text{mm}^2$$

and $K_2 = 0.74$.

As a check on the final value of K_2 interpolated from the design chart:

$$N_{bal} = 0.29 f_{ck} A_c$$
$$= 0.29 \times 25 \times 300 \times 450 \times 10^{-3}$$
$$= 978 \,\text{kN}$$

$$N_{Ed} = 0.567 f_{ck} A_c + 0.87 f_{yk} A_s$$
$$= (0.567 \times 25 \times 300 \times 450 + 0.87 \times 500 \times 4050) 10^{-3}$$
$$= 3675 \,\text{kN}$$

$$K_2 = \frac{N_{ud} - N_{Ed}}{N_{ud} - N_{bal}} = \frac{3675 - 1700}{3675 - 978} = 0.73$$

which agrees with the final value in column 5 of table 9.4.

9.8 Walls

Walls may take the form of non-structural dividing elements in which case their thickness will often reflect sound insulation and fire resistance requirements. Nominal reinforcement will be used to control cracking in such cases. More commonly, reinforced concrete walls will form part of a structural frame and will be designed for vertical and horizontal forces and moments obtained by normal analysis methods. In this situation a wall is defined as being a vertical load-bearing member whose length is not less than four times its thickness.

Where several walls are connected monolithically so that they behave as a unit, they are described as a wall system. Sometimes horizontal forces on a structure are resisted by more than one wall or system of walls, in which case the distribution of forces between the walls or systems will be assumed to be in proportion to their stiffnesses.

It is normal practice to consider a wall as a series of vertical strips when designing vertical reinforcement. Each strip is then designed as a column subject to the appropriate vertical load and transverse moments at its top and bottom. Slenderness effects must be considered where necessary, as for columns. If a wall is subject predominantly to lateral bending, the design and detailing will be undertaken as if it were a slab, but the wall thickness will usually be governed by slenderness limitations, fire resistance requirements and construction practicalities.

Reinforcement detailing

For a wall designed either as a series of columns or as a slab, the area of vertical reinforcement should lie between $0.002A_c$ and $0.04A_c$ and this will normally be equally divided between each face. Bar spacing along the length of the wall should not exceed the lesser of 400 mm or three times the wall thickness.

Horizontal bars should have a diameter of not less than one-quarter of the vertical bars, and with a total area of not less than 25% of the vertical bars or $0.001A_c$ whichever is greater. The horizontal bars should lie between the vertical bars and the concrete surface, with a spacing which is not greater than 400 mm.

If the area of vertical steel exceeds $0.02A_c$, then the bars should be enclosed by links designed according to the rules for columns.

CHAPTER 10

Foundations and retaining walls

CHAPTER INTRODUCTION

A building is generally composed of a superstructure above the ground and a substructure which forms the foundations below ground. The foundations transfer and spread the loads from a structure's columns and walls into the ground. The safe bearing capacity of the soil must not be exceeded otherwise excessive settlement may occur, resulting in damage to the building and its service facilities, such as the water or gas mains. Foundation failure can also affect the overall stability of a structure so that it is liable to slide, to lift vertically or even overturn.

The earth under the foundations is the most variable of all the materials that are considered in the design and construction of an engineering structure. Under one small building the soil may vary from a soft clay to a dense rock. Also the nature and properties of the soil will change with the seasons and the weather. For example Keuper Marl, a relatively common soil, is hard like rock when dry but when wet it can change into an almost liquid state.

It is important to have an engineering survey made of the soil under a proposed structure so that variations in the strata and the soil properties can be determined. Drill holes or trial pits should be sunk, *in situ* tests such as the penetration test performed and samples of the soil taken to be tested in the laboratory. From the information gained it is possible to recommend safe bearing pressures and, if necessary, calculate possible settlements of the structure.

The structural design of any foundation or retaining wall will be based on the general principles outlined in previous chapters of this book. However where the foundation interacts with the ground the geotechnical

→

→

design of the foundation must be considered i.e. the ability of the ground to resist the loading transferred by the structure.

Geotechnical design is in accordance with BS EN 1997: Eurocode 7. This code classifies design situations into three types: (i) category 1 – small and simple structures (ii) category 2 – conventional with no difficult ground or complicated loading conditions and (iii) category 3 – all other types of structures where there may be a high risk of geotechnical failure. The expectation is that structural engineers will be responsible for the design of category 1 structures, geotechnical engineers for category 3 and either type of engineer could be responsible for category 2.

This chapter will only consider foundation types that are likely to fall within the first two categories.

General design approach

Although EC7 presents three alternative design approaches the UK National Annex allows for only the first of these. In this design approach, two sets of load combinations (referred to as combinations 1 and 2 in table 10.1) must be considered at the ultimate limit state. These two combinations will be used for consideration of both structural failure, STR (excessive deformation, cracking or failure of the structure), and geotechnical failure, GEO (excessive deformation or complete failure of the supporting mass of earth).

A third combination must be taken when considering possible loss of equilibrium (EQU) of the structure such as overturning. The partial safety factors to be used for these three combinations are given in table 10.1.

Table 10.1 Partial safety factors at the ultimate limit state

Persistent or transient design situation	Permanent actions (G_k)		Leading variable action $(Q_{k,1})$		Accompanying variable action $(Q_{k,i})$	
	Unfavourable	Favourable	Unfavourable	Favourable	Unfavourable	Favourable
(a) for consideration of structural or geotechnical failure: combination 1 (STR) & (GEO)	1.35	1.00*	1.50	0	1.50	0
(b) for consideration of structural or geotechnical failure: combination 2 (STR) & (GEO)	1.00	1.00*	1.30	0	1.30	0
(c) for checking static equilibrium (EQU)	1.1	0.9	1.50	0	1.50	0

* To be applied to bearing, sliding and earth resistance forces.

In determining the design values of actions to be used at the ultimate limit state the characteristic loads should be multiplied by a partial safety factor. Appropriate values of partial safety factors can be obtained from table 10.1. In the case of the accompanying variable actions they should be further multiplied by the factor ψ_0 where appropriate values of ψ_0 can be obtained from table 2.4 in chapter 2.

In table 10.1 it should be noted that combination 1 will usually be relevant to the structural design of the foundation, whilst combination 2 will be most likely to govern the sizing of the foundation to ensure that settlement is not excessive, but this will depend on the circumstances of the particular situation.

The third combination of actions shown in the final row of table 10.1 is relevant to the design of structures such as the type shown in figure 10.1, where it may be necessary to check the possibility of uplift to the foundations and the stability of the structure when it is subjected to lateral loads. The critical loading arrangement is usually the combination of maximum lateral load with minimum permanent load and no variable load, that is $1.5W_k + 0.9G_k$. Minimum permanent load can sometimes occur during erection when many of the interior finishes and fixtures may not have been installed.

At the same time as the design values of actions are determined, as above, the soil parameters used in the geotechnical aspects of the design are multiplied by the partial factors of safety, appropriate to the load combination under consideration, as given in table 10.2. The detailed use of these factors will not be developed further in this text but are given for completeness.

For simple spread foundations such as strip and pad footings EC7 gives three alternative methods of design:

1. The 'Direct Method' where calculations are required for each limit state using the partial factors of safety as appropriate from tables 10.1 and 10.2

2. The 'Indirect Method' which allows for a simultaneous blending of ultimate limit state and serviceability limit state procedures

3. The 'Prescriptive Method' where an assumed safe bearing pressure is used to size the foundations based on the serviceability limit state followed by detailed structural design based on the ultimate limit state

In the *Prescriptive Method* the traditional UK approach to the sizing of foundations is effectively retained such that a suitable base size may be determined based on the serviceability limit state values for actions and an assumed allowable safe bearing pressure (see table 10.3). In this way settlement will be controlled, with the exception that for foundations on soft clay full settlement calculations must be carried out.

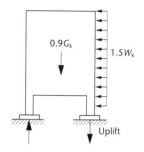

Figure 10.1
Uplift on footing

Table 10.2 Partial safety factors applied to geotechnical material properties

	Angle of shearing resistance	Effective cohesion	Undrained shear strength	Unconfined strength	Bulk density
	γ_ϕ	γ_c	γ_{cu}	γ_{qu}	γ_γ
Combination 1	1.0	1.0	1.0	1.0	1.0
Combination 2	1.25	1.25	1.4	1.4	1.0

Table 10.3 Typical allowable bearing values

Rock or soil	Typical bearing value (kN/m²)
Massive igneous bedrock	10 000
Sandstone	2000 to 4000
Shales and mudstone	600 to 2000
Gravel, sand and gravel, compact	600
Medium dense sand	100 to 300
Loose fine sand	Less than 100
Very stiff clay	300 to 600
Stiff clay	150 to 300
Firm clay	75 to 150
Soft clay	Less than 75

Where the foundations are subject to both vertical and horizontal loads the following rule can be applied:

$$\frac{V}{P_v} + \frac{H}{P_h} < 1.0$$

where

$V =$ the vertical load

$H =$ the horizontal load

$P_v =$ the allowable vertical load

$P_h =$ the allowable horizontal load.

The allowable horizontal load would take account of the passive resistance of the ground in contact with the vertical face of the foundation plus the friction and cohesion along the base.

The calculations to determine the structural strength of the foundations, that is the thickness of the bases and the areas of reinforcement, should be based on the loadings and the resultant ground pressures corresponding to the ultimate limit state and considering the worst of the combinations 1 and 2 for the actions (table 10.1) although, as previously noted, combination 1 will usually govern the structural design.

For most designs a linear distribution of soil pressure across the base of the footing is assumed as shown in figure 10.2(a). This assumption must be based on the soil acting as an elastic material and the footing having infinite rigidity. In fact, not only do most soils exhibit some plastic behaviour and all footings have a finite stiffness, but also the distribution of soil pressure varies with time. The actual distribution of bearing pressure at any moment may take the form shown in figure 10.2(b) or (c), depending on the type of soil and the stiffness of the base and the structure. But as the behaviour of foundations involves many uncertainties regarding the action of the ground and the loading, it is usually unrealistic to consider an analysis that is too sophisticated.

Figure 10.2
Pressure distributions
under footings

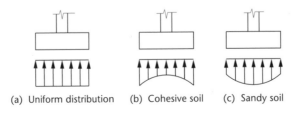

(a) Uniform distribution   (b) Cohesive soil   (c) Sandy soil

Foundations should be constructed so that the undersides of the bases are below frost level. As the concrete is subjected to more severe exposure conditions a larger nominal cover to the reinforcement is required. Despite the values suggested in tables 6.1 and 6.2 established practice in the UK would be to recommend that the minimum cover should be not less than 75 mm when the concrete is cast against the ground, or less than 50 mm when the concrete is cast against a layer of blinding concrete. A concrete class of at least C30/37 is required to meet durability requirements.

10.1 Pad footings

The footing for a single column may be made square in plan, but where there is a large moment acting about one axis it may be more economical to have a rectangular base.

Assuming there is a linear distribution the bearing pressures across the base will take one of the three forms shown in figure 10.3, according to the relative magnitudes of the axial load N and the moment M acting on the base.

1. In figure 10.3(a) there is no moment and the pressure is uniform

$$p = \frac{N}{BD} \qquad (10.1)*$$

2. With a moment M acting as shown, the pressures are given by the equation for axial load plus bending. This is provided there is positive contact between the base and the ground along the complete length D of the footing, as in figure 10.3(b) so that

$$p = \frac{N}{BD} \pm \frac{My}{I}$$

where I is the second moment area of the base about the axis of bending and y is the distance from the axis to where the pressure is being calculated.

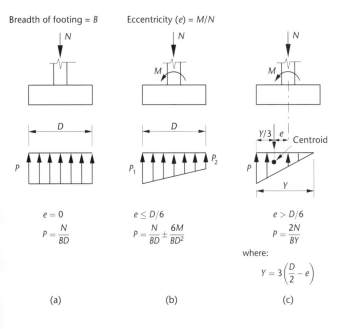

Breadth of footing = B   Eccentricity (e) = M/N

$e = 0$

$$P = \frac{N}{BD}$$

(a)

$e \leq D/6$

$$P = \frac{N}{BD} \pm \frac{6M}{BD^2}$$

(b)

$e > D/6$

$$P = \frac{2N}{BY}$$

where:

$$Y = 3\left(\frac{D}{2} - e\right)$$

(c)

Figure 10.3
Pad-footing – pressure distributions

Substituting for $I = BD^3/12$ and $y = D/2$, the maximum pressure is

$$p_1 = \frac{N}{BD} + \frac{6M}{BD^2} \qquad\qquad (10.2)*$$

and the minimum pressure is

$$p_2 = \frac{N}{BD} - \frac{6M}{BD^2} \qquad\qquad (10.3)*$$

There is positive contact along the base if p_2 from equation 10.3 is positive. When pressure p_2 just equals zero

$$\frac{N}{BD} - \frac{6M}{BD^2} = 0$$

or

$$\frac{M}{N} = \frac{D}{6}$$

So that for p_2 always to be positive, M/N – or the effective eccentricity, e – must never be greater than $D/6$. In these cases the eccentricity of loading is said to lie within the 'middle third' of the base.

3. When the eccentricity, e is greater than $D/6$ there is no longer a positive pressure along the length D and the pressure diagram is triangular as shown in figure 10.3(c). Balancing the downward load and the upward pressures

$$\frac{1}{2}pBY = N$$

therefore

$$\text{maximum pressure } p = \frac{2N}{BY}$$

where Y is the length of positive contact. The centroid of the pressure diagram must coincide with the eccentricity of loading in order for the load and reaction to be equal and opposite. Thus

$$\frac{Y}{3} = \frac{D}{2} - e$$

or

$$Y = 3\left(\frac{D}{2} - e\right)$$

therefore in the case of $e > D/6$

$$\text{maximum pressure } p = \frac{2N}{3B(D/2 - e)} \qquad\qquad (10.4)*$$

A typical arrangement of the reinforcement in a pad footing is shown in figure 10.4. With a square base the reinforcement to resist bending should be distributed uniformly across the full width of the footing. For a rectangular base the reinforcement in the short direction should be distributed with a closer spacing in the region under and near the column, to allow for the fact that the transverse moments must be greater nearer the column. It is recommended that at least two-thirds of the reinforcement in the short direction should be concentrated in a band width of $(c + 3d)$ where c is the column dimension in the long direction and d is the effective depth. If the footing should be

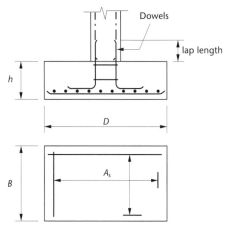

Figure 10.4
Pad footing reinforcement
details

subjected to a large overturning moment so that there is only partial bearing, or if there is a resultant uplift force, then reinforcement may also be required in the top face.

Dowels or starter bars should extend from the footing into the column in order to provide continuity to the reinforcement. These dowels should be embedded into the footing and extend into the columns a full lap length. Sometimes a 75 mm length of the column is constructed into the same concrete pour as the footing so as to form a 'kicker' or support for the column's shutters. In these cases the dowel's lap length should be measured from the top of the kicker.

The critical sections through the base for checking shear, punching shear and bending are shown in figure 10.5. The shearing force and bending moments are caused by the ultimate loads from the column and the weight of the base should not be included in these calculations.

The thickness of the base is often governed by the requirements for shear resistance.

Following the *Prescriptive Method* the principal steps in the design calculations are as follows:

1. Calculate the plan size of the footing using the permissible bearing pressure and the critical loading arrangement for the serviceability limit state.

2. Calculate the bearing pressures associated with the critical loading arrangement at the ultimate limit state.

3. Assume a suitable value for the thickness (h) and effective depth (d). Check that the shear force at the column face is less than $0.5v_1 f_{cd}ud = 0.5v_1(f_{ck}/1.5)ud$ where u is the perimeter of the column and v_1 is the strength reduction factor $= 0.6(1 - f_{ck}/250)$.

4. Carry out a preliminary check for punching shear to ensure that the footing thickness gives a punching shear stress which is within the likely range of acceptable performance.

5. Determine the reinforcement required to resist bending.

6. Make a final check for the punching shear.

7. Check the shear force at the critical sections.

8. Where applicable, both foundations and the structure should be checked for overall stability at the ultimate limit state.

9. Reinforcement to resist bending in the bottom of the base should extend at least a full tension anchorage length beyond the critical section of bending.

Figure 10.5
Critical sections for design

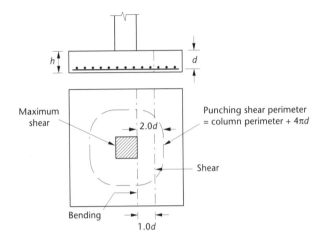

EXAMPLE 10.1

Design of a pad footing

The footing (figure 10.6) is required to resist characteristic axial loads of 1000 kN permanent and 350 kN variable from a 400 mm square column. The safe bearing pressure on the soil is 200 kN/m² and the characteristic material strengths are $f_{ck} = 30$ N/mm² and $f_{yk} = 500$ N/mm².

Assume a footing weight of 150 kN so that the total permanent load is 1150 kN and base the design on the *Prescriptive Method*.

1. For the serviceability limit state

$$\text{Total design axial load} = 1.0G_k + 1.0Q_k = 1150 + 350 = 1500 \text{ kN}$$

$$\text{Required base area} = \frac{1500}{200} = 7.5 \text{ m}^2$$

Provide a base 2.8 m square $= 7.8 \text{ m}^2$.

2. For the ultimate limit state

From table 10.1 it is apparent load combination 1 will give the largest set of actions for this simple structure. Hence, using the partial safety factors for load combination 1:

$$\text{Column design axial load, } N_{Ed} = 1.35G_k + 1.5Q_k$$

$$= 1.35 \times 1000 + 1.5 \times 350 = 1875 \text{ kN}$$

$$\text{Earth pressure} = \frac{1875}{2.8^2} = 239 \text{ kN/m}^2$$

Figure 10.6
Pad footing example

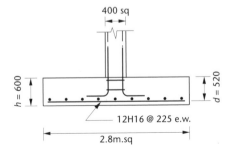

400 sq

$h = 600$

$d = 520$

12H16 @ 225 e.w.

2.8m.sq

3. Assume a 600 mm thick footing and with the footing constructed on a blinding layer of concrete the minimum cover is taken as 50 mm. Therefore take mean effective depth $= d = 520$ mm.

At the column face

Maximum shear resistance, $V_{Rd, max}$

$$= 0.5ud\left[0.6\left(1 - \frac{f_{ck}}{250}\right)\right]\frac{f_{ck}}{1.5}$$

$$= 0.5(4 \times 400) \times 520 \times \left[0.6\left(1 - \frac{30}{250}\right)\right]\frac{30}{1.5} \times 10^{-3}$$

$$= 4393\,kN \quad (> N_{Ed} = 1875\,kN)$$

4. Punching shear

The critical section for checking punching shear is at a distance $2d$ as shown in figure 10.5

$$\text{Critical perimeter} = \text{column perimeter} + 4\pi d$$
$$= 4 \times 400 + 4\pi \times 520 = 8134\,mm$$
$$\text{Area within perimeter} = (400 + 4d)^2 - (4 - \pi)(2.0d)^2$$
$$= (400 + 2080)^2 - (4 - \pi)1040^2$$
$$= 5.22 \times 10^6\,mm^2$$

therefore

$$\text{Punching shear force } V_{Ed} = 239(2.8^2 - 5.22) = 626\,kN$$

$$\text{Punching shear stress } v_{Ed} = \frac{V_{Ed}}{\text{Perimeter} \times d}$$

$$= \frac{626 \times 10^3}{8134 \times 520} = 0.15\,N/mm^2$$

This ultimate shear stress is not excessive, (see table 8.2) therefore $h = 600$ mm will be a suitable estimate.

5. Bending reinforcement – see figure 10.7(a)

At the column face which is the critical section

$$M_{Ed} = (239 \times 2.8 \times 1.2) \times \frac{1.2}{2}$$

$$= 482\,kN\,m$$

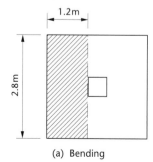

(a) Bending

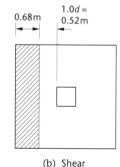

(b) Shear

2.8m

1.2m

0.68m

1.0d = 0.52m

Figure 10.7
Critical sections

For the concrete

$$M_{bal} = 0.167 f_{ck} b d^2$$
$$= 0.167 \times 30 \times 2800 \times 520^2 \times 10^{-6} = 3793 \, \text{kN m} \ (> 482)$$

$$A_s = \frac{M_{Ed}}{0.87 f_{yk} z}$$

From the lever-arm curve, figure 4.5, $l_a = 0.95$. Therefore:

$$A_s = \frac{482 \times 10^6}{0.87 \times 500 \times (0.95 \times 520)} = 2243 \, \text{mm}^2$$

Provide twelve H16 bars at 225 mm centres, $A_s = 2412 \, \text{mm}^2$. Therefore

$$\frac{100 A_s}{bd} = \frac{100 \times 2412}{2800 \times 520} = 0.165 \ (> 0.15 - \text{see table 6.8})$$

that is, the minimum steel area requirement is satisfied.

Maximum bar size

The steel stress should be calculated under the action of the quasi-permanent loading which can be estimated from equation 6.1 as follows:

$$f_s = \frac{f_{yk}(G_k + 0.3 Q_k)}{1.15(1.35 G_k + 1.5 Q_k)}$$
$$= \frac{500(1000 + 0.3 \times 350)}{1.15(1.35 \times 1000 + 1.5 \times 350)} = 256 \, \text{N/mm}^2$$

Therefore from table 6.9 the maximum allowable bar size is 16 mm. Hence, minimum area and bar size requirements as specified by the code for the purposes of crack control are met.

6. Final check of punching shear

The shear resistance of the concrete without shear reinforcement can be obtained from table 8.2 where

ρ_l can be taken as the average of the steel ratios in both directions

$$= \frac{A_s}{bd} = \frac{2412}{2800 \times 520} = 0.0017 \ (= 0.17\% < 2\%)$$

hence from table 8.2 $v_{Rd,c} = 0.4 \, \text{N/mm}^2$.

Therefore the shear resistance of the concrete, $V_{Rd,c}$ is given by:

$$V_{Rd,c} = v_{Rd,c} u d = 0.40 \times 8134 \times 520 \times 10^{-3} = 1691 \, \text{kN} \ (> V_{Ed} = 626 \, \text{kN})$$

7. Maximum Shear Force – see figure 10.7(b)

At the critical section for shear, $1.0d$ from the column face:

$$\text{Design shear } V_{Ed} = 239 \times 2.8 \times 0.68$$
$$= 455 \, \text{kN}$$

As before, $v_{Rd,c} = 0.40 \, \text{N/mm}^2$

$$\therefore V_{Rd,c} = v_{Rd,c} b d$$
$$= 0.40 \times 2800 \times 520 \times 10^{-3} = 582 \, \text{kN} \ (> V_{Ed} = 455 \, \text{kN})$$

Therefore no shear reinforcement is required.

Instead of assuming a footing weight of 150 kN at the start of this example it is possible to allow for the weight of the footing by using a net safe bearing pressure p_{net} where

$$p_{net} = 200 - h \times \text{unit weight of concrete}$$
$$= 200 - 0.6 \times 25 = 185.0 \, \text{kN/m}^2$$

Therefore

$$\text{Required base area} = \frac{1.0 \times \text{column load}}{p_{net}} = \frac{1000 + 350}{185.0} = 7.30 \, \text{m}^2$$

It should be noted that the self-weight of the footing or its effect must be included in the calculations at serviceability for determining the area of the base but at the ultimate limit state the self-weight should not be included.

Example 10.1 shows how to design a pad footing with a centrally located set of actions. If the actions are eccentric to the centroidal axis of the base then in the checking of punching shear the maximum shear stress, v_{Ed}, is multiplied by an enhancement factor β (> 1). This factor accounts for the non-linear distribution of stress around the critical perimeter due to the eccentricity of loading. Reference should be made to EC2 Clause 6.4.3 for the details of this design approach.

10.2 | Combined footings

Where two columns are close together it is sometimes necessary or convenient to combine their footings to form a continuous base. The dimensions of the footing should be chosen so that the resultant load passes through the centroid of the base area. This may be assumed to give a uniform bearing pressure under the footing and help to prevent differential settlement. For most structures the ratios of permanent and variable loads carried by each column are similar so that if the resultant passes through the centroid for the serviceability limit state then this will also be true – or very nearly – at the ultimate limit state, and hence in these cases a uniform pressure distribution may be considered for both limit states.

The shape of the footing may be rectangular or trapezoidal as shown in figure 10.8. The trapezoidal base has the disadvantage of detailing and cutting varying lengths of reinforcing bars; it is used where there is a large variation in the loads carried by the two columns and there are limitations on the length of the footing. Sometimes in order to strengthen the base and economise on concrete a beam is incorporated between the two columns so that the base is designed as an inverted T-section.

Centroid of base and
resultant load coincide

Figure 10.8
Combined bases

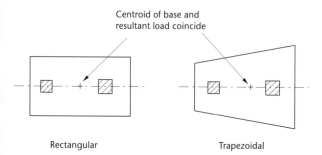

Rectangular

Trapezoidal

The proportions of the footing depend on many factors. If it is too long, there will be large longitudinal moments on the lengths projecting beyond the columns, whereas a short base will have a larger span moment between the columns and the greater width will cause large transverse moments. The thickness of the footing must be such that the shear stresses are not excessive.

EXAMPLE 10.2

Design of a combined footing

The footing supports two columns 300 mm square and 400 mm square with characteristic permanent and variable loads as shown in figure 10.9. The safe bearing pressure is 300 kN/m^2 and the characteristic material strengths are $f_{ck} = 30$ N/mm^2 and $f_{yk} = 500$ N/mm^2. Assume a base thickness $h = 850$ mm.

1. Base area (calculated at serviceability limit state, basing the design on the *Prescriptive Method*)

    Net safe bearing pressure $\rho_{net} = 300 - 25h = 300 - 25 \times 0.85$

    $$= 278.8 \text{ kN/m}^2$$

    Total load $= 1000 + 200 + 1400 + 300$

    $$= 2900 \text{ kN}$$

    Area of base required $= \dfrac{2900}{278.8}$

    $$= 10.4 \text{ m}^2$$

    Provide a rectangular base, 4.6 m × 2.3 m, area $= 10.58$ m^2.

Figure 10.9
Combined footing example

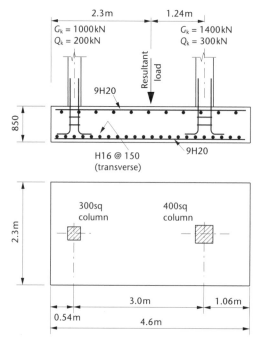

2. Resultant of column loads and centroid of base: taking moments about the centre line of the 400 mm square column

$$\bar{x} = \frac{1200 \times 3}{1200 + 1700} = 1.24\,\text{m}$$

The base is centred on this position of the resultant of the column loads as shown in figure 10.9.

3. Bearing pressure at the ultimate limit state (Load combination 1):

$$\text{Column loads} = 1.35 \times 1000 + 1.5 \times 200 + 1.35 \times 1400 + 1.5 \times 300$$
$$= 1650 + 2340 = 3990\,\text{kN}$$

therefore

$$\text{earth pressure} = \frac{3990}{4.6 \times 2.3} = 377\,\text{kN/m}^2$$

4. Assuming $d = 790$ mm for the longitudinal bars and with a mean $d = 780$ mm for punching shear calculations:

At the column face

$$\text{Maximum shear resistance, } V_{\text{Rd, max}} = 0.5ud\left[0.6\left(1 - \frac{f_{ck}}{250}\right)\right]\frac{f_{ck}}{1.5}$$

For 300 mm square column

$$V_{\text{Rd, max}} = 0.5ud\left[0.6\left(1 - \frac{f_{ck}}{250}\right)\right]\frac{f_{ck}}{1.5}$$
$$= 0.5 \times 1200 \times 780\left[0.6\left(1 - \frac{30}{250}\right)\right]\frac{30}{1.5} \times 10^{-3}$$
$$= 4942\,\text{kN} \quad (N_{\text{Ed}} = 1650\,\text{kN})$$

For 400 mm square column

$$V_{\text{Rd, max}} = 0.5ud\left[0.6\left(1 - \frac{f_{ck}}{250}\right)\right]\frac{f_{ck}}{1.5}$$
$$= 0.5 \times 1600 \times 780\left[0.6\left(1 - \frac{30}{250}\right)\right]\frac{30}{1.5} \times 10^{-3}$$
$$= 6589\,\text{kN} \quad (N_{\text{Ed}} = 2340\,\text{kN})$$

5. Longitudinal moments and shear forces: the shear-force and bending-moment diagrams at the ultimate limit state and for a net upward pressure of 377 kN/m^2 are shown in figure 10.10 overleaf.

6. Longitudinal bending

Maximum moment is at mid-span between the columns

$$A_s = \frac{M_{\text{Ed}}}{0.87f_{yk}z} = \frac{679 \times 10^6}{0.87 \times 500 \times 0.95 \times 790} = 2080\,\text{mm}^2$$

From table 6.8

$$A_{s,\,\text{min}} = \frac{0.15b_t d}{100} = 0.0015 \times 2300 \times 790 = 2726\,\text{mm}^2$$

Provide nine H20 at 270 mm centres, area = 2830 mm^2, top and bottom to meet the minimum area requirements.

Figure 10.10
Shear-force and bending-
moment diagrams

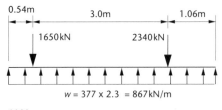

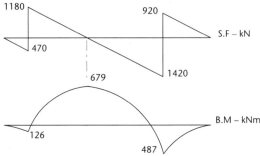

7. Transverse bending

$$M_{Ed} = 377 \times \frac{1.15^2}{2} = 249 \text{ kN m/m}$$

$$A_s = \frac{M_{Ed}}{0.87 f_{yk} z} = \frac{249 \times 10^6}{0.87 \times 500 \times 0.95 \times 770} = 783 \text{ mm}^2/\text{m}$$

But

$$\text{Minimum } A_s = \frac{0.15 bd}{100} = \frac{0.15 \times 1000 \times 770}{100} = 1155 \text{ mm}^2/\text{m}$$

Provide H16 bars at 150 mm centres, area = 1340 mm^2 per metre.

The transverse reinforcement should be placed at closer centres under the columns to allow for greater moments in those regions. For the purposes of crack control, the maximum bar size or maximum bar spacing should also be checked as in example 10.1.

8. Shear

Punching shear cannot be checked, since the critical perimeter 2.0d from the column face lies outside the base area. The critical section for shear is taken 1.0d from the column face. Therefore with $d = 780$ mm.

Design shear $V_{Ed} = 1420 - 377 \times 2.3(0.78 + 0.2) = 570$ kN

The shear resistance of the concrete without shear reinforcement can be obtained from table 8.2 where

ρ_1 can be taken as the average of the steel ratios in both directions

$$= 0.5 \sum \frac{A_s}{bd} = 0.5 \left[\frac{2830}{2300 \times 790} + \frac{1340}{1000 \times 770} \right] = 0.0016 \quad (= 0.16\% < 2\%)$$

hence from table 8.2 $v_{Rd,c} = 0.36 \text{ N/mm}^2$.

Therefore the shear resistance of the concrete, V_{Rd} is given by:

$$V_{Rd} = v_{Rd,c} bd = 0.36 \times 2300 \times 780 \times 10^{-3} = 645 \text{ kN} \quad (> V_{Ed} = 570 \text{ kN})$$

Therefore shear reinforcement is not required.

10.3 Strap footings

Strap footings, as shown in figure 10.11, are used where the base for an exterior column must not project beyond the property line. A strap beam is constructed between the exterior footing and the adjacent interior footing – the purpose of the strap is to restrain the overturning force due to the eccentric load on the exterior footing.

The base areas of the footings are proportioned so that the bearing pressures are uniform and equal under both bases. Thus it is necessary that the resultant of the loads on the two footings should pass through the centroid of the areas of the two bases. The strap beam between the footings should not bear against the soil, hence the ground directly under the beam should be loosened and left uncompacted. As well as the loadings indicated in figure 10.11 EC2 recommends that, where the action of compaction machinery could affect the tie beam, the beam should be designed for a minimum downward load of 10 kN/m.

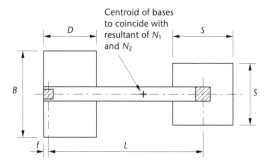

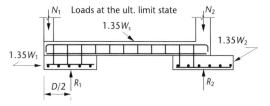

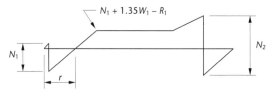

Shear Forces

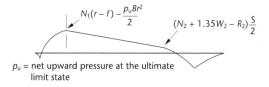

p_u = net upward pressure at the ultimate limit state

Bending Moments

Figure 10.11
Strap footing with shearing force and bending moments for the strap beam

To achieve suitable sizes for the footings several trial designs may be necessary. With reference to figure 10.11 the principal steps in the design are as follows.

1. Choose a trial width D for the rectangular outer footing and assume weights W_1 and W_2 for the footings and W_s for the strap beam.

2. Take moments about the centre line of the inner column in order to determine the reaction R_1 under the outer footing. The loadings should be those required for the serviceability limit state. Thus

$$(R_1 - W_1)\left(L + f - \frac{D}{2}\right) - N_1 L - W_s \frac{L}{2} = 0 \tag{10.5}$$

and solve for R_1. The width B of the outer footing is then given by

$$B = \frac{R_1}{pD}$$

where p is the safe bearing pressure.

3. Equate the vertical loads and reactions to determine the reaction R_2 under the inner footing. Thus

$$R_1 + R_2 - (N_1 + N_2 + W_1 + W_2 + W_s) = 0 \tag{10.6}$$

and solve for R_2. The size S of the square inner footing is then given by

$$S = \sqrt{\frac{R_2}{p}}$$

4. Check that the resultant of all the loads on the footings passes through the centroid of the areas of the two bases. If the resultant is too far away from the centroid then steps (1) to (4) must be repeated until there is adequate agreement.

5. Apply the loading associated with the ultimate limit state. Accordingly, revise equations 10.5 and 10.6 to determine the new values for R_1 and R_2. Hence calculate the bearing pressure p_u for this limit state. It may be assumed that the bearing pressures for this case are also equal and uniform, provided the ratios of dead load to imposed load are similar for both columns.

6. Design the inner footing as a square base with bending in both directions.

7. Design the outer footing as a base with bending in one direction and supported by the strap beam.

8. Design the strap beam. The maximum bending moment on the beam occurs at the point of zero shear as shown in figure 10.11. The shear on the beam is virtually constant, the slight decrease being caused by the beam's self-weight. The stirrups should be placed at a constant spacing but they should extend into the footings over the supports so as to give a monolithic foundation. The main tension steel is required at the top of the beam but reinforcement should also be provided in the bottom of the beam so as to cater for any differential settlement or downward loads on the beam.

10.4 | Strip footings

Strip footings are used under walls or under a line of closely spaced columns. Even where it is possible to have individual bases, it is often simpler and more economic to excavate and construct the formwork for a continuous base.

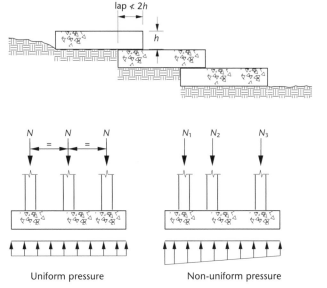

Figure 10.12
Stepped footing on a sloping site

Figure 10.13
Linear pressure distribution under a rigid strip footing

On a sloping site the foundations should be constructed on a horizontal bearing and stepped where necessary. At the steps the footings should be lapped as shown in figure 10.12.

The footings are analysed and designed as an inverted continuous beam subjected to the ground bearing pressures. With a thick rigid footing and a firm soil, a linear distribution of bearing pressure is considered. If the columns are equally spaced and equally loaded the pressure is uniformly distributed but if the loading is not symmetrical then the base is subjected to an eccentric load and the bearing pressure varies as shown in figure 10.13.

The bearing pressures will not be linear when the footing is not very rigid and the soil is soft and compressible. In these cases the bending-moment diagram would be quite unlike that for a continuous beam with firmly held supports and the moments could be quite large, particularly if the loading is unsymmetrical. For a large foundation it may be necessary to have a more detailed investigation of the soil pressures under the base in order to determine the bending moments and shearing forces.

Reinforcement is required in the bottom of the base to resist the transverse bending moments in addition to the reinforcement required for the longitudinal bending. Footings which support heavily loaded columns often require stirrups and bent-up bars to resist the shearing forces.

EXAMPLE 10.3

Design of a strip footing

Design a strip footing to carry 400 mm square columns equally spaced at 3.5 m centres. On each column the characteristic loads are 1000 kN permanent and 350 kN variable. The safe bearing pressure is 200 kN/m² and the characteristic material strengths are $f_{ck} = 30 \, \text{N/mm}^2$ and $f_{yk} = 500 \, \text{N/mm}^2$. Base the design on the *Prescriptive Method*.

1. Try a thickness of footing $= 800$ with $d = 740$ mm for the longitudinal reinforcement.

$$\text{Net bearing pressure, } \rho_{net} = 200 - 25h = 200 - 25 \times 0.8$$
$$= 180.0 \, \text{kN/m}^2$$
$$\text{Width of footing required} = \frac{1000 + 350}{180.0 \times 3.4} = 2.14 \, \text{m}$$

Provide a strip footing 2.2 m wide.

At the ultimate limit state

$$\text{column load, } N_{Ed} = 1.35 \times 1000 + 1.5 \times 350 = 1875 \, \text{kN}$$
$$\text{bearing pressure} = \frac{1875}{2.2 \times 3.5}$$
$$= 244 \, \text{kN/m}^2$$

2. *Punching shear* at the column face

Maximum shear resistance, $V_{Rd, max}$

$$= 0.5ud \left[0.6 \left(1 - \frac{f_{ck}}{250} \right) \right] \frac{f_{ck}}{1.5}$$
$$= 0.5(4 \times 400) \times 740 \times \left[0.6 \left(1 - \frac{30}{250} \right) \right] \frac{30}{1.5} \times 10^{-3}$$
$$= 6251 \, \text{kN}$$

By inspection, the normal shear on a section at the column face will be significantly less severe than this value.

3. *Longitudinal reinforcement*

Using the moment and shear coefficients for an equal-span continuous beam (figure 3.9), for an interior span

$$\text{moment at the columns } M_{Ed} = 244 \times 2.2 \times 3.5^2 \times 0.10$$
$$= 665 \, \text{kN m}$$

therefore

$$A_s = \frac{665 \times 10^6}{0.87 \times 500 \times 0.95 \times 740} = 2175 \, \text{mm}^2$$

From table 6.8

$$A_{s, min} = \frac{0.15b_t d}{100} = 0.0015 \times 2200 \times 740$$
$$= 2442 \, \text{mm}^2$$

Provide eight H20 bars at 300 mm centres, area $= 2510 \, \text{mm}^2$, bottom steel.

In the span

$$M_{Ed} = 244 \times 2.2 \times 3.5^2 \times 0.07$$
$$= 460 \, \text{kN m}$$

Therefore, as in the bottom face, provide eight H20 bars at 300 mm centres, area $= 2510 \, \text{mm}^2$, top steel (figure 10.14).

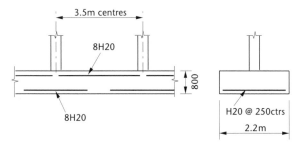

Figure 10.14
Strip footing with bending reinforcement

4. *Transverse reinforcement*

In the transverse direction the maximum moment can be calculated on the assumption that the 2.2 m wide footing is acting as a 1.1 m long cantilever for the purposes of calculating the design moment:

$$M_{Ed} = 244 \times \frac{1.1^2}{2} = 148 \text{ kNm/m}$$

$$A_s = \frac{148 \times 10^6}{0.87 \times 500 \times 0.95 \times 720} = 497 \text{ mm}^2/\text{m}$$

$$\text{Minimum } A_s = \frac{0.15 b_t d}{100} = 0.15 \times 1000 \times \frac{720}{100} = 1080 \text{ mm}^2/\text{m}$$

Provide H20 bars at 250 mm centres, area $= 1260 \text{ mm}^2/\text{m}$, bottom steel.

5. *Normal shear* will govern as the punching perimeter is outside the footing.

The critical section for shear is taken 1.0d from the column face. Therefore with $d = 740$ mm

$$\text{Design shear } V_{Ed} = 244 \times 2.2(3.5 \times 0.55 - 0.74 - 0.2)$$
$$= 529 \text{ kN}$$

(The coefficient of 0.55 is from figure 3.9.)

The shear resistance of the concrete without shear reinforcement can be obtained from table 8.2 where

ρ_1 can be taken as the average of the steel ratios in both directions

$$= 0.5 \sum \frac{A_s}{bd} = 0.5 \left[\frac{2510}{2200 \times 740} + \frac{1260}{1000 \times 720} \right] = 0.00165 \quad (= 0.165\% < 2\%)$$

hence from table 8.2 $v_{Rd,c} = 0.36 \text{ N/mm}^2$.

Therefore the shear resistance of the concrete, $V_{Rd,c}$ is given by:

$$V_{Rd,c} = v_{Rd,c} bd = 0.36 \times 2200 \times 740 \times 10^{-3} = 586 \text{ kN} \quad (> V_{Ed} = 529 \text{ kN})$$

Therefore shear reinforcement is not required.

10.5 | Raft foundations

A raft foundation transmits the loads to the ground by means of a reinforced concrete slab that is continuous over the base of the structure. The raft is able to span any areas of weaker soil and it spreads the loads over a wide area. Heavily loaded structures are often provided with one continuous base in preference to many closely-spaced, separate footings. Also where settlement is a problem, because of mining subsidence, it is

Figure 10.15
Raft foundations

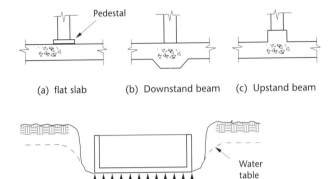

(a) flat slab   (b) Downstand beam   (c) Upstand beam

Figure 10.16
Raft foundation subject to
uplift

common practice to use a raft foundation in conjunction with a more flexible superstructure.

The simplest type of raft is a flat slab of uniform thickness supporting the columns. Where punching shears are large the columns may be provided with a pedestal at the base as shown in figure 10.15. The pedestal serves a similar function to the drop panel in a flat slab floor. Other, more heavily loaded rafts require the foundation to be strengthened by beams to form a ribbed construction. The beams may be downstanding, projecting below the slab or they may be upstanding as shown in figure 10.15. Downstanding beams have the disadvantage of disturbing the ground below the slab and the excavated trenches are often a nuisance during construction, while upstanding beams interrupt the clear floor area above the slab. To overcome this, a second slab is sometimes cast on top of the beams, so forming a cellular raft.

Rafts having a uniform slab, and without strengthening beams, are generally analysed and designed as an inverted flat slab floor subjected to earth bearing pressures. With regular column spacing and equal column loading, the coefficients tabulated in section 8.6 for flat slab floors are used to calculate the bending moments in the raft. The slab must be checked for punching shear around the columns and around pedestals, if they are used.

A raft with strengthening beams is designed as an inverted beam and slab floor. The slab is designed to span in two directions where there are supporting beams on all four sides. The beams are often subjected to high shearing forces which need to be resisted by a combination of stirrups and bent-up bars.

Raft foundations which are below the level of the water table, as in figure 10.16, should be checked to ensure that they are able to resist the uplift forces due to the hydrostatic pressure. This may be critical during construction before the weight of the superstructure is in place, and it may be necessary to provide extra weight to the raft and lower the water table by pumping. An alternative method is to anchor the slab down with short tension piles.

10.6 Piled foundations

Piles are used where the soil conditions are poor and it is uneconomical, or not possible, to provide adequate spread foundations. The piles must extend down to firm soil so that the load is carried by either (1) end bearing, (2) friction, or (3) a combination of both end bearing and friction. Concrete piles may be precast and driven into the ground, or they may be the cast-*in-situ* type which are bored or excavated.

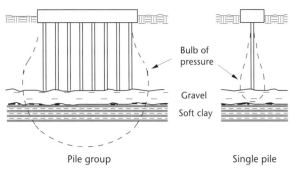

Figure 10.17
Bulbs of pressure

Bulb of
pressure

Gravel

Soft clay

Pile group   Single pile

A soils survey of a proposed site should be carried out to determine the depth to firm soil and the properties of the soil. This information will provide a guide to the lengths of pile required and the probable safe load capacity of the piles. On a large contract the safe loads are often determined from full-scale load tests on typical piles or groups of piles. With driven piles the safe load can be calculated from equations which relate the resistance of the pile to the measured set per blow and the driving force.

The load-carrying capacity of a group of piles is not necessarily a multiple of that for a single pile – it is often considerably less. For a large group of closely spaced friction piles the reduction can be of the order of one-third. In contrast, the load capacity of a group of end bearing piles on a thick stratum of rock or compact sand gravel is substantially the sum total of the resistance of each individual pile. Figure 10.17 shows the bulbs of pressure under piles and illustrates why the settlement of a group of piles is dependent on the soil properties at a greater depth.

The minimum spacing of piles, centre to centre, should not be less than (1) the pile perimeter for friction piles, or (2) twice the least width of the pile for end bearing piles. Bored piles are sometimes enlarged at the base so that they have a larger bearing area or a greater resistance to uplift.

A pile is designed as a short column unless it is slender and the surrounding soil is too weak to provide restraint. Precast piles must also be designed to resist the bending moments caused by lifting and stacking, and the head of the pile must be reinforced to withstand the impact of the driving hammer.

It is very difficult, if not impossible, to determine the true distribution of load of a pile group. Therefore, in general, it is more realistic to use methods that are simple but logical. A vertical load on a group of vertical piles with an axis of symmetry is considered to be distributed according to the following equation, which is similar in form to that for an eccentric load on a pad foundation:

$$P_n = \frac{N}{n} \pm \frac{Ne_{xx}}{I_{xx}} y_n \pm \frac{Ne_{yy}}{I_{yy}} x_n$$

where

P_n is the axial load on an individual pile

N is the vertical load on the pile group

n is the number of piles

e_{xx} and e_{yy} are the eccentricities of the load N about the centroidal axes XX and YY of the pile group

I_{xx} and I_{yy} are the second moments of area of the pile group about axes XX and YY

x_n and y_n are the distances of the individual pile from axes YY and XX, respectively.

EXAMPLE 10.4

Loads in a pile group

Determine the distribution between the individual piles of a 1000 kN vertical load acting at the position shown of the group of vertical piles shown in figure 10.18. To determine the centroid of the pile group take moments about line T–T.

$$\bar{y} = \frac{\sum y}{n} = \frac{2.0 + 2.0 + 3.0 + 3.0}{6} = 1.67 \, \text{m}$$

where n is the number of piles. Therefore the eccentricities of the load about the XX and YY centroidal axis are

$$e_{xx} = 2.0 - 1.67 = 0.33 \, \text{m}$$

and

$$e_{yy} = 0.2 \, \text{m}$$

$$I_{xx} = \sum y_n^2 \quad \text{with respect to the centroidal axis XX}$$
$$= 2 \times 1.67^2 + 2 \times 0.33^2 + 2 \times 1.33^2$$
$$= 9.33$$

Similarly

$$I_{yy} = \sum x_n^2 = 3 \times 1.0^2 + 3 \times 1.0^2$$
$$= 6.0$$

therefore

$$P_n = \frac{N}{n} \pm \frac{Ne_{xx}}{I_{xx}} y_n \pm \frac{Ne_{yy}}{I_{yy}} x_n$$
$$= \frac{1000}{6} \pm \frac{1000 \times 0.33}{9.33} y_n \pm \frac{1000 \times 0.2}{6.0} x_n$$
$$= 166.7 \pm 35.4 y_n \pm 33.3 x_n$$

Figure 10.18
Pile loading example

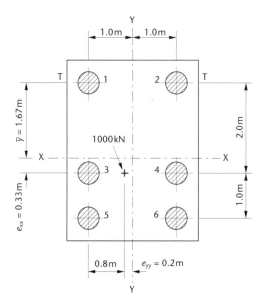

Therefore, substituting for y_n and x_n

$$P_1 = 166.7 - 35.4 \times 1.67 + 33.3 \times 1.0 = 140.9\,\text{kN}$$
$$P_2 = 166.7 - 35.4 \times 1.67 - 33.3 \times 1.0 = 74.3\,\text{kN}$$
$$P_3 = 166.7 + 35.4 \times 0.33 + 33.3 \times 1.0 = 211.7\,\text{kN}$$
$$P_4 = 166.7 + 35.4 \times 0.33 - 33.3 \times 1.0 = 145.1\,\text{kN}$$
$$P_5 = 166.7 + 35.4 \times 1.33 + 33.3 \times 1.0 = 247.1\,\text{kN}$$
$$P_6 = 166.7 + 35.4 \times 1.33 - 33.3 \times 1.0 = 180.5\,\text{kN}$$
$$\text{Total} = 999.6 \approx 1000\,\text{kN}$$

When a pile group is unsymmetrical about both co-ordinate axes it is necessary to consider the theory of bending about the principal axes which is dealt with in most textbooks on strength of materials. In this case the formulae for the pile loads are

$$P_n = \frac{N}{n} \pm Ay_n \pm Bx_n$$

where

$$A = \frac{N\left(e_{xx}\sum x_n^2 - e_{yy}\sum x_n y_n\right)}{\sum x_n^2 \sum y_n^2 - \left(\sum x_n y_n\right)^2}$$

and

$$B = \frac{N\left(e_{yy}\sum y_n^2 - e_{xx}\sum x_n y_n\right)}{\sum x_n^2 \sum y_n^2 - \left(\sum x_n y_n\right)^2}$$

Note that e_{xx} is the eccentricity about the XX axis, while e_{yy} is the eccentricity about the YY axis, as in figure 10.18.

Piled foundations are sometimes required to resist horizontal forces in addition to the vertical loads. If the horizontal forces are small they can often be resisted by the passive pressure of the soil against vertical piles, otherwise if the forces are not small then raking piles must be provided as shown in figure 10.19(a).

To determine the load in each pile either a static method or an elastic method is available. The static method is simply a graphical analysis using Bow's notation as illustrated in figure 10.19(b). This method assumes that the piles are pinned at their ends so that the induced loads are axial. The elastic method takes into account the displacements and rotations of the piles which may be considered pinned or fixed at their ends. The pile foundation is analysed in a similar manner to a plane frame or space frame and available computer programs are commonly used.

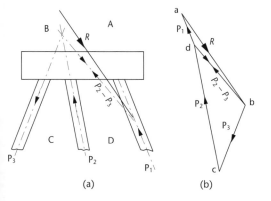

(a)   (b)

Figure 10.19
Forces in raking piles

10.7 | Design of pile caps

The pile cap must be rigid and capable of transferring the column loads to the piles. It should have sufficient thickness for anchorage of the column dowels and the pile reinforcement, and it must be checked for punching shear, diagonal shear, bending and bond. Piles are rarely positioned at the exact locations shown on the drawings, therefore this must be allowed for when designing and detailing the pile cap.

Two methods of design are common: design using beam theory or design using a truss analogy approach. In the former case the pile cap is treated as an inverted beam and is designed for the usual conditions of bending and shear. The truss analogy method is used to determine the reinforcement requirements where the span-to-depth ratio is less than 2 such that beam theory is not appropriate.

10.7.1 The truss analogy method

In the truss analogy the force from the supported column is assumed to be transmitted by a triangular truss action with concrete providing the compressive members of the truss and steel reinforcement providing the tensile tie force as shown in the two-pile cap in figure 10.20(a). The upper node of the truss is located at the centre of the loaded area and the lower nodes at the intersection of the tensile reinforcement with the centrelines of the piles. Where the piles are spaced at a distance greater than three times the pile diameter only the reinforcement within a distance of 1.5 times the pile diameter from the centre of the pile should be considered as effective in providing the tensile resistance within the truss.

Figure 10.20
Truss model for a two-pile cap

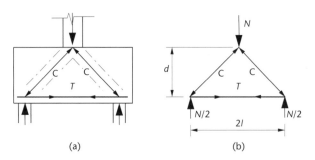

(a)   (b)

From the geometry of the force diagram in figure 10.20b:

$$\frac{T}{N/2} = \frac{l}{d}$$

therefore

$$T = \frac{Nl}{2d}$$

Hence

$$\text{required area of reinforcement} = \frac{T}{0.87f_{yk}} = \frac{N \times l}{2d \times 0.87f_{yk}} \tag{10.7}$$

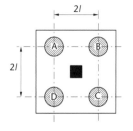

Figure 10.21
Four-pile cap

Where the pile cap is supported on a four-pile group, as shown in figure 10.21, the load can be considered to be transmitted equally by parallel pairs of trusses, such as

Table 10.4

Number of piles	Group arrangement	Tensile force
2	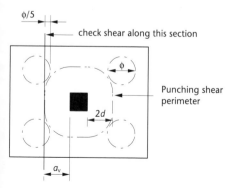	$T_{AB} = \dfrac{Nl}{2d}$
3		$T_{AB} = T_{BC} = T_{AC} = \dfrac{2Nl}{9d}$
4		$T_{AB} = T_{BC} = T_{CD} = T_{AD} = \dfrac{Nl}{4d}$

AB and CD, and equation 10.7 can be modified to give:

$$\text{required area of reinforcement in each truss} = \frac{T/2}{0.87f_{yk}} = \frac{N \times l}{4d \times 0.87f_{yk}} \qquad (10.8)$$

and this reinforcement should be provided in both directions in the bottom face of the pile-cap.

The truss theory may be extended to give the tensile force in pile caps with other configurations of pile groups. Table 10.4 gives the force for some common cases.

10.7.2 Design for shear

The shear capacity of a pile cap should be checked at the critical section taken to be 20 per cent of the pile diameter inside the face of the pile, as shown in figure 10.22. In determining the shear resistance, shear enhancement may be considered such that the shear force, V_{Ed}, may be decreased by $a_v/2d$ where a_v is the distance from the face of the column to the critical section. Where the spacing of the piles is less than or equal to

Figure 10.22
Critical sections for shear checks

three times the pile diameter, this enhancement may be applied across the whole of the critical section; otherwise it may only be applied to strips of width three times the pile diameter located central to each pile.

10.7.3   Design for punching shear

Where the spacing of the piles exceeds three times the pile diameter then the pile cap should be checked for punching shear using the method outlined in section 8.1.2. for slabs. The critical perimeter for punching shear is as shown in figure 10.22. The shear force at the column face should be checked to ensure that it is less than $0.5v_1 f_{cd} ud = 0.5v_1 (f_{ck}/1.5)ud$ where u is the perimeter of the column and the strength reduction factor, $v_1 = 0.6(1 - f_{ck}/250)$.

10.7.4   Reinforcement detailing

As for all members, normal detailing requirements must be checked. These include maximum and minimum steel areas, bar spacings, cover to reinforcement and anchorage lengths of the tension steel. The main tension reinforcement should continue past each pile and should be bent up vertically to provide a full anchorage length beyond the centreline of each pile. In orthogonal directions in the top and bottom faces of the pile cap a minimum steel area of $0.26(f_{ctm}/f_{yk})bd$ ($> 0.0013bd$) should be provided. It is normal to provide fully lapped horizontal links of size not less than 12 mm and at spacings of no greater than 250 mm, as shown in figure 10.23(b). The piles should be cut off so that they do not extend into the pile cap beyond the lower mat of reinforcing bars otherwise the punching shear strength may be reduced.

10.7.5   Sizing of the pile cap

In determining a suitable depth of pile cap table 10.5 may be used as a guide when there are up to six piles in the pile group.

Table 10.5   Depth of pile cap

Pile size (mm)	300	350	400	450	500	550	600	750
Cap depth (mm)	700	800	900	1000	1100	1200	1400	1800

EXAMPLE 10.5

Design of a pile cap

A group of four piles supports a 500 mm square column which transmits an ultimate axial load of 5000 kN. The piles are 450 mm diameter and are spaced at 1350 mm centres as shown. Design the pile cap for $f_{ck} = 30$ N/mm² and $f_{yk} = 500$ N/mm².

(a) Dimensions of pile cap

Try an overall depth of 1000 mm and an average effective depth of 875 mm. Allow the pile cap to extend 375 mm either side to give a 2100 mm square cap.

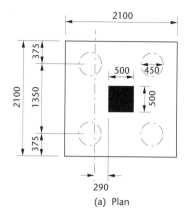

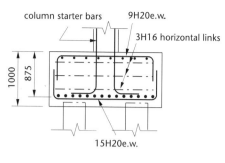

Figure 10.23
Pile-cap design example

(a) Plan   (b) Reinforcement details

(b) Design of main tension reinforcement

From equation 10.8, the required area of reinforcement in each truss is

$$A_s = \frac{T/2}{0.87f_{yk}} = \frac{N \times l}{4d \times 0.87f_{yk}}$$

$$= \frac{5000 \times 10^3 \times (1350/2)}{4 \times 875 \times 0.87 \times 500}$$

$$= 2216 \, \text{mm}^2$$

The total area of reinforcement required in each direction $2 \times A_s = 2 \times 2216 = 4432 \, \text{mm}^2$. As the piles are spaced at three times the pile diameter this reinforcement may be distributed uniformly across the section. Hence provide fifteen H20 bars, area $= 4710 \, \text{mm}^2$, at 140 mm centres in both directions:

$$\frac{100A_s}{bd} = \frac{100 \times 4710}{2100 \times 875} = 0.26 \quad \left(> 0.26\frac{f_{ctm}}{f_{yk}} = 0.15 \right)$$

(c) Check for shear

Shear force, V_{Ed}, along critical section $= 5000/2 = 2500 \, \text{kN}$ and to allow for shear enhancement this may be reduced to:

$$V_{Ed} = 2500 \times \frac{a_v}{2d} = 2500 \times \frac{290}{2 \times 875} = 414 \, \text{kN}$$

$$v_{Rd,c} = 0.12k(100\rho f_{ck})^{1/3} \quad \left(\geq 0.035k^{1.5}f_{ck}^{0.5} \right)$$

where: $k = 1 + \sqrt{200/d} = 1 + \sqrt{\dfrac{200}{850}} = 1.49 \quad (< 2)$ and $\rho = 0.0026$

$$\therefore \qquad v_{Rd,c} = 0.12k(100\rho f_{ck})^{1/3}$$

$$= 0.12 \times 1.49 \times (100 \times 0.0026 \times 30)^{1/3} = 0.35 \, \text{N/mm}^2$$

and $v_{Rd,c(min)} = 0.035k^{1.5}f_{ck}^{0.5} = 0.035 \times 1.49^{1.5} \times 30^{0.5} = 0.35 \, \text{N/mm}^2$

therefore the shear resistance of the concrete, $V_{Rd,c}$ is given by:

$$V_{Rd,c} = v_{Rd,c}bd$$

$$= 0.35 \times 2100 \times 875 \times 10^{-3} = 643 \, \text{kN} \quad (> V_{Ed} = 414 \, \text{kN})$$

(d) Check for punching shear

As the pile spacing is at three times the pile diameter no punching shear check is necessary. The shear at the column face should be checked:

Maximum shear resistance, $V_{Rd,max}$

$$= 0.5ud\left[0.6\left(1 - \frac{f_{ck}}{250}\right)\right]\frac{f_{ck}}{1.5}$$

$$= 0.5(4 \times 500) \times 875 \times \left[0.6\left(1 - \frac{30}{250}\right)\right]\frac{30}{1.5} \times 10^{-3}$$

$$= 9240\,kN \quad (> N_{Ed} = 5000\,kN)$$

10.8   Retaining walls

Such walls are usually required to resist a combination of earth and hydrostatic loadings. The fundamental requirement is that the wall is capable of holding the retained material in place without undue movement arising from deflection, overturning or sliding.

10.8.1   Types of retaining wall

Concrete retaining walls may be considered in terms of three basic categories: (1) gravity, (2) counterfort, and (3) cantilever. Within these groups many common variations exist, for example cantilever walls may have additional supporting ties into the retained material.

The structural action of each type is fundamentally different, but the techniques used in analysis, design and detailing are those normally used for concrete structures.

(i) Gravity walls

These are usually constructed of mass concrete, with reinforcement included in the faces to restrict thermal and shrinkage cracking. As illustrated in figure 10.24, reliance is placed on self-weight to satisfy stability requirements, both in respect of overturning and sliding.

It is generally taken as a requirement that under working conditions the resultant of the self-weight and overturning forces must lie within the middle third at the interface of the base and soil. This ensures that uplift is avoided at this interface, as described in section 10.1. Friction effects which resist sliding are thus maintained across the entire base.

Figure 10.24
Gravity wall

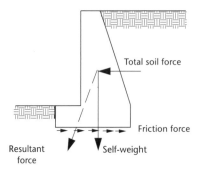

Total soil force

Friction force

Resultant force

Self-weight

Bending, shear, and deflections of such walls are usually insignificant in view of the large effective depth of the section. Distribution steel to control thermal cracking is necessary, however, and great care must be taken to reduce hydration temperatures by mix design, construction procedures and curing techniques.

(ii) Counterfort walls

This type of construction will probably be used where the overall height of the wall is too large to be constructed economically either in mass concrete or as a cantilever.

The basis of design of counterfort walls is that the earth pressures act on a thin wall which spans horizontally between the massive counterforts (figure 10.25). These must be sufficiently large to provide the necessary permanent load for stability requirements, possibly with the aid of the weight of backfill on an enlarged base. The counterforts must be designed with reinforcement to act as cantilevers to resist the considerable bending moments that are concentrated at these points.

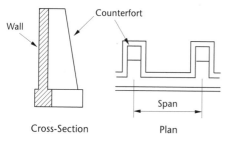

Figure 10.25
Counterfort wall

The spacing of counterforts will be governed by the above factors, coupled with the need to maintain a satisfactory span–depth ratio on the wall slab, which must be designed for bending as a continuous slab. The advantage of this form of construction is that the volume of concrete involved is considerably reduced, thereby removing many of the problems of large pours, and reducing the quantities of excavation. Balanced against this must be considered the generally increased shuttering complication and the probable need for increased reinforcement.

(iii) Cantilever walls

These are designed as vertical cantilevers spanning from a large rigid base which often relies on the weight of backfill on the base to provide stability. Two forms of this

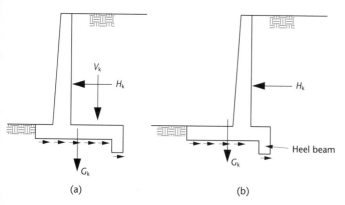

Figure 10.26
Cantilever walls

construction are illustrated in figure 10.26. In both cases, stability calculations follow similar procedures to those for gravity walls to ensure that the resultant force lies within the middle third of the base and that overturning and sliding requirements are met.

10.8.2　Analysis and design

The design of retaining walls may be split into three fundamental stages: (1) Stability analysis – ultimate limit state (EQU and GEO), (2) Bearing pressure analysis – ultimate limit state (GEO), and (3) Member design and detailing – ultimate limit state (STR) and serviceability limit states.

(i) Stability analysis

Under the action of the loads corresponding to the ultimate limit state (EQU), a retaining wall must be stable in terms of resistance to *overturning*. This is demonstrated by the simple case of a gravity wall as shown in figure 10.27.

The critical conditions for overturning are when a maximum horizontal force acts with a minimum vertical load. To guard against failure by overturning, it is usual to apply conservative factors of safety to the forces and loads. Table 10.1(c) gives the factors that are relevant to these calculations.

A partial factor of safety of $\gamma_G = 0.9$ is applied to the permanent load G_k if its effect is 'favourable', and the 'unfavourable' effects of the permanent earth pressure loading at the rear face of the wall are multiplied by a partial factor of safety of $\gamma_f = 1.1$. The 'unfavourable' effects of the variable surcharge loading, if any, are multiplied by a partial factor of safety of $\gamma_f = 1.5$.

For resistance to overturning, moments would normally be taken about the toe of the base, point A on figure 10.27. Thus the requirement is that

$$0.9G_k x \geq \gamma_f H_k y$$

Resistance to *sliding* is provided by friction between the underside of the base and the ground, and thus is also related to total self-weight G_k. Resistance provided by the passive earth pressure on the front face of the base may make some contribution, but since this material is often backfilled against the face, this resistance cannot be guaranteed and is usually ignored.

Failure by sliding is considered under the action of the loads corresponding to the ultimate limit state of GEO. Table 10.1 gives the factors that are relevant to these calculations.

Figure 10.27
Forces and pressures on a gravity wall

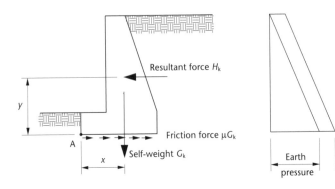

Resultant force H_k

y

Friction force μG_k

A

x

Self-weight G_k

Earth pressure

Surcharge pressure

A partial factor of safety of $\gamma_G = 1.0$ is applied to the permanent load G_k if its effect is 'favourable' (i.e. contributes to the sliding resistance) and the 'unfavourable' effects of the permanent earth pressure loading at the rear face of the wall are multiplied by a partial factor of safety of $\gamma_f = 1.35$. The 'unfavourable' effects of the variable surcharge loading are multiplied by a partial factor of safety of $\gamma_f = 1.5$.

Thus, if the coefficient of friction between base and soil is μ, the total friction force will be given by μG_k for the length of the wall of weight G_k; and the requirement is that

$$1.0 \mu G_k \geq \gamma_f H_k$$

where H_k is the horizontal force on this length of wall.

If this criterion is not met, a heel beam may be used, and the force due to the passive earth pressure over the face area of the heel may be included in resisting the sliding force. The partial load factor γ_f on the heel beam force should be taken as 1.0 to give the worst condition. To ensure the proper action of a heel beam, the front face must be cast directly against sound, undisturbed material, and it is important that this is not overlooked during construction.

In considering cantilever walls, a considerable amount of backfill is often placed on top of the base, and this is taken into account in the stability analysis. The forces acting in this case are shown in figure 10.28. In addition to G_k and H_k there is an additional vertical load V_k due to the material above the base acting a distance q from the toe. The worst condition for stability will be when this is at a minimum; therefore a partial load factor $\gamma_f = 0.9$ is used for consideration of overturning and 1.0 for consideration of sliding. The stability requirements then become

$$0.9 G_k x + 0.9 V_k q \geq \gamma_f H_k y \qquad \text{for overturning} \tag{10.9}$$

$$\mu(1.0 G_k + 1.0 V_k) \geq \gamma_f H_k \qquad \text{for sliding} \tag{10.10}$$

When a heel beam is provided the additional passive resistance of the earth must be included in equation 10.10.

Stability analysis, as described here, will normally suffice. However, if there is doubt about the foundation material in the region of the wall or the reliability of loading values, it may be necessary to perform a full slip-circle analysis, using techniques common to soil mechanics, or to use increased factors of safety.

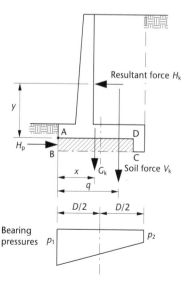

Figure 10.28
Forces on a cantilever wall

(ii) Bearing pressure analysis

The bearing pressures underneath retaining walls are assessed on the basis of the ultimate limit state (GEO) when determining the size of base that is required. The analysis will be similar to that discussed in section 10.1 with the foundation being subject to the combined effects of an eccentric vertical load, coupled with an overturning moment.

Considering a unit length of the cantilever wall (figure 10.28) the resultant moment about the centroidal axis of the base is

$$M = \gamma_{f1} H_k y + \gamma_{f2} G_k (D/2 - x) + \gamma_{f3} V_k (D/2 - q) \tag{10.11}$$

and the vertical load is

$$N = \gamma_{f2} G_k + \gamma_{f3} V_k \tag{10.12}$$

where in this case of the STR and GEO ultimate limit states the partial factors of safety are given in Table 10.1:

For load combination 1: $\gamma_{f1} = 1.35$ and $\gamma_{f2} = \gamma_{f3} = 1.0$

For load combination 2: $\gamma_{f1} = \gamma_{f2} = \gamma_{f3} = 1.0$

assuming that, for load combination 1, the effect of the moment due to the horizontal load on the maximum bearing pressure at the toe of the wall at A is 'unfavourable' whilst the moments of the self-weight of the wall and the earth acting on the heel of the wall act in the opposite sense and are thus 'favourable'. This assumption may need checking in individual cases and the appropriate partial factors applied depending on whether the effect of the load can be considered to be favourable or unfavourable.

The distribution of bearing pressures will be as shown in figure 10.28, provided the effective eccentricity lies within the 'middle third' of the base, that is

$$\frac{M}{N} \leq \frac{D}{6}$$

The maximum bearing pressure is then given by

$$p_1 = \frac{N}{D} + \frac{M}{I} \times \frac{D}{2}$$

where

$$I = D^3/12$$

Therefore

$$p_1 = \frac{N}{D} + \frac{6M}{D^2} \tag{10.13}$$

and

$$p_2 = \frac{N}{D} - \frac{6M}{D^2} \tag{10.14}$$

(iii) Member design and detailing

As with foundations, the design of bending and shear reinforcement is based on an analysis of the loads for the ultimate limit state (STR) , with the corresponding bearing pressures. Gravity walls will seldom require bending or shear steel, while the walls in counterfort and cantilever construction will be designed as slabs. The design of counterforts will generally be similar to that of a cantilever beam unless they are massive.

With a cantilever-type retaining wall the stem is designed to resist the moment caused by the force $\gamma_f H_f$, with γ_f values taken for load combination 1 if this load combination is deemed to be critical. For preliminary sizing, the thickness of the wall may be taken as 80 mm per metre depth of backfill.

The thickness of the base is usually of the same order as that of the stem. The heel and toe must be designed to resist the moments due to the upward earth bearing pressures and the downward weight of soil and base. The soil bearing pressures are calculated from equations 10.11 to 10.14, provided the resultant of the horizontal and vertical forces lies within the 'middle third'. Should the resultant lie outside the 'middle third', then the bearing pressures should be calculated using equation 10.4. The partial factors of safety γ_{f1}, γ_{f2} and γ_{f3} should be taken to provide a combination which gives the critical design condition (the worst of combinations 1 and 2).

Reinforcement detailing must follow the general rules for slabs and beams as appropriate. Particular care must be given to the detailing of reinforcement to limit shrinkage and thermal cracking. Gravity walls are particularly vulnerable because of the large concrete pours that are generally involved.

Restraints to thermal and shrinkage movement should be reduced to a minimum. However, this is counteracted in the construction of bases by the need for good friction between the base and soil; thus a sliding layer is not possible. Reinforcement in the bases must therefore be adequate to control the cracking caused by a high degree of restraint. Long walls restrained by rigid bases are particularly susceptible to cracking during thermal movement due to loss of hydration heat, and detailing must attempt to distribute these cracks to ensure acceptable widths. Complete vertical movement joints must be provided. These joints will often incorporate a shear key to prevent differential movement of adjacent sections of wall, and waterbars and sealers should be used.

The back faces of retaining walls will usually be subject to hydrostatic forces from groundwater. These may be reduced by the provision of a drainage path at the face of the wall. It is usual practice to provide such a drain by a layer of rubble or porous blocks as shown in figure 10.29, with pipes to remove the water, often through the front of the wall. In addition to reducing the hydrostatic pressure on the wall, the likelihood of leakage through the wall is reduced, and water is also less likely to reach and damage the soil beneath the foundations of the wall.

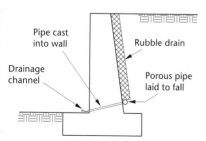

Figure 10.29
Drainage layer

EXAMPLE 10.6

Design of a retaining wall

The cantilever retaining wall shown in figure 10.30 supports a granular material of saturated density 1700kg/m². It is required to:

1. check the stability of the wall

2. determine the bearing pressures at the ultimate limit state, and

3. design the bending reinforcement using high-yield steel, $f_{yk} = 500\,\text{kN/mm}^2$ and concrete class C30/37.

(1) Stability

Horizontal force

It is assumed that the coefficient of active pressure $K_a = 0.33$, which is a typical value for a granular material. So the earth pressure is given by

$$p_a = K_a \rho g h$$

where ρ is the density of the backfill and h is the depth considered. Thus, at the base

$$p_a = 0.33 \times 1700 \times 10^{-3} \times 9.81 \times 4.9$$
$$= 27.0\,\text{kN/m}^2$$

Allowing for the minimum required surcharge of $10\,\text{kN/m}^2$ an additional horizontal pressure of

$$p_s = K_a \times 10 = 3.3\,\text{kN/m}^2$$

acts uniformly over the whole depth h.

Figure 10.30
Retaining wall design example

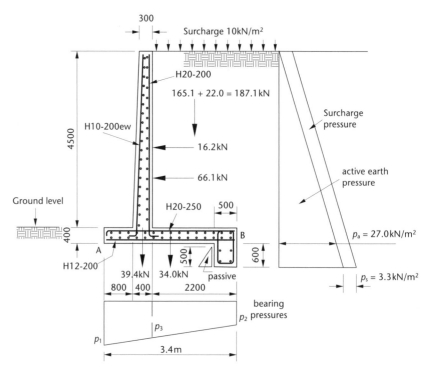

Therefore the horizontal force on 1 m length of wall is given by:

$$H_{k(earth)} = 0.5p_a h = 0.5 \times 27.0 \times 4.9 = 66.1 \text{ kN from the active earth pressure}$$

and

$$H_{k(sur)} = p_s h = 3.3 \times 4.9 = 16.2 \text{ kN from the surcharge pressure}$$

Vertical loads

(a) permanent loads

wall	$= \dfrac{1}{2}(0.4 + 0.3) \times 4.5 \times 25$	$= 39.4$
base	$= 0.4 \times 3.4 \times 25$	$= 34.0$
earth	$= 2.2 \times 4.5 \times 1700 \times 10^{-3} \times 9.81$	$= 165.1$
		Total $= 238.5$ kN

(b) variable loads

$$\text{surcharge} = 2.2 \times 10 \qquad\qquad = 22.0 \text{ kN}$$

The partial factors of safety as given in table 10.1 will be used.

(i) *Overturning:* taking moments about point A at the edge of the toe, at the ultimate limit state (EQU).

For the overturning (unfavourable) moment a factor of 1.1 is applied to the earth pressure and a factor of 1.5 to the surcharge pressure

$$\begin{aligned}
\text{overturning moment} &= \gamma_f H_{k(earth)} h/3 + \gamma_f H_{k(sur)} h/2 \\
&= (1.1 \times 66.1 \times 4.9/3) + (1.5 \times 16.2 \times 4.9/2) \\
&= 178 \text{ kN m}
\end{aligned}$$

For the restraining (favourable) moment a factor of 0.9 is applied to the permanent loads and 0 to the variable surcharge load

$$\begin{aligned}
\text{restraining moment} &= \gamma_f(39.4 \times 1.0 + 34.0 \times 1.7 + 165.1 \times 2.3) \\
&= 0.9 \times 476.9 \\
&= 429 \text{ kN m}
\end{aligned}$$

Thus the criterion for overturning is satisfied.

(ii) *Sliding:* from equation 10.10 it is necessary that

$$\mu(1.0G_k + 1.0V_k) \geq \gamma_f H_k \quad \text{for no heel beam}$$

For the sliding (unfavourable) effect a factor of 1.35 is applied to the earth pressure and a factor of 1.5 to the surcharge pressure

$$\begin{aligned}
\text{sliding force} &= 1.35 \times 66.1 + 1.50 \times 16.2 \\
&= 113.5 \text{ kN}
\end{aligned}$$

For the restraining (favourable) effect a factor of 1.0 is applied to the permanent loads and 0 to the variable surcharge load. Assuming a value of coefficient of friction $\mu = 0.45$

$$\begin{aligned}
\text{frictional resisting force} &= 0.45 \times 1.0 \times 238.5 \\
&= 107.3 \text{ kN}
\end{aligned}$$

Since the sliding force exceeds the frictional force, resistance must also be provided by the passive earth pressure acting against the heel beam and this force is given by

$$H_p = \gamma_f \times 0.5 K_p \rho g a^2$$

where K_p is the coefficient of passive pressure, assumed to be 3.5 for this granular material and a is the depth of the heel below the 0.5 m 'trench' allowance in front of the base. Therefore

$$H_p = 1.0 \times 0.5 \times 3.5 \times 1700 \times 10^{-3} \times 9.81 \times 0.5^2 = 7.3\,\text{kN}$$

Therefore total resisting force is

$$107.3 + 7.3 = 114.6\,\text{kN}$$

which marginally exceeds the sliding force.

(2) Bearing pressures at ultimate limit state (STR & GEO)

Consider load combination 1 as the critical combination that will give the maximum bearing pressure at the toe of the wall (see table 10.1), although in practice load combination 2 may have to be checked to determine if it gives a worse effect. Note that the weight of the earth and the surcharge loading exerts a moment about the base centreline that will *reduce* the maximum pressure at the toe of the wall. Hence the effect of the weight of the earth is taken as a *favourable* effect ($\gamma_f = 1$) and the weight of the surcharge load is also taken as a *favourable* effect ($\gamma_f = 0$) within the calculations below. The *unfavourable* effects of the lateral earth pressure and the lateral surcharge pressure are multiplied by factors of $\gamma_f = 1.35$ and $\gamma_f = 1.50$, respectively

From equations 10.13 and 10.14 the bearing pressures are given by

$$p = \frac{N}{D} \pm \frac{6M}{D^2}$$

where M is the moment about the base centreline. Therefore

$$M = \gamma_f(66.1 \times 4.9/3) + \gamma_f(16.2 \times 4.9/2) + \gamma_f \times 39.4(1.7 - 1.0)$$
$$- \gamma_f \times 165.1 \times (2.3 - 1.7)$$
$$= 1.35 \times 107.9 + 1.5 \times 39.7 + 1.35 \times 27.6 - 1.0 \times 99.1$$
$$= 145.7 + 59.6 + 37.3 - 99.1 = 143.5\,\text{kN m}$$

Therefore, bearing pressure at toe and heel of wall

$$p_1 = \frac{(1.35 \times (39.4 + 34.0) + 1.0 \times 165.1)}{3.4} \pm \frac{6 \times 143.5}{3.4^2}$$
$$= 77.7 \pm 74.5$$
$$= 152.2,\ 3.2\,\text{kN/m}^2 \quad (\text{as shown in figure 10.31})$$

Figure 10.31
Pressures under the base

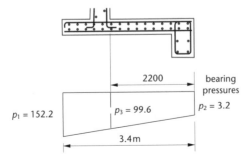

2200

bearing pressures

$p_1 = 152.2$

$p_3 = 99.6$

$p_2 = 3.2$

3.4m

(3) Bending reinforcement

(i) Wall

Horizontal force

$$= \gamma_f 0.5 K_a \rho g h^2 + \gamma_f p_s h$$
$$= 1.35 \times 0.5 \times 0.33 \times 1700 \times 10^{-3} \times 9.81 \times 4.5^2 + 1.50 \times 3.3 \times 4.5$$
$$= 75.2 + 22.3 = 97.5 \, \text{kN}$$

considering the effective span, the maximum moment is

$$M_{Ed} = 75.2 \times (0.2 + 4.5/3) + 22.3 \times (0.2 + 4.5/2) = 182.5 \, \text{kN m}$$
$$\frac{M_{Ed}}{bd^2 f_{ck}} = \frac{182.5 \times 10^6}{1000 \times 330^2 \times 30} = 0.056$$

for which $l_a = 0.95$ (figure 4.5). Therefore

$$A_s = \frac{182.5 \times 10^6}{0.95 \times 330 \times 0.87 \times 500}$$
$$= 1338 \, \text{mm}^2/\text{m}$$

Provide H20 bars at 200 mm centres ($A_s = 1570 \, \text{mm}^2$).

(ii) Base

The bearing pressures at the ultimate limit state are obtained from part (2) of these calculations. Using the figures from part (2):

$$\text{pressure } p_1 = 152.2 \, \text{kN/m}^2$$
$$p_2 = 3.2 \, \text{kN/m}^2$$

and in figure 10.31:

$$p_3 = 3.2 + (152.2 - 3.2)2.2/3.4 = 99.6 \, \text{kN/m}^2$$

Heel: taking moments about the stem centreline for the vertical loads and the bearing pressures

$$M_{Ed} = \gamma_f \times 34.0 \times \left(\frac{3.4}{2} - 1.0\right) + \gamma_f \times 165.1 \times 1.3 - 3.2 \times 2.2 \times 1.3$$
$$- (99.6 - 3.2) \times \frac{2.2}{2} \times \left(\frac{2.2}{3} + 0.2\right)$$
$$= 1.35 \times 23.8 + 1.0 \times 214.6 - 9.2 - 99.0$$
$$= 139 \, \text{kN m}$$

therefore

$$\frac{M_{Ed}}{bd^2 f_{ck}} = \frac{139 \times 10^6}{1000 \times 330^2 \times 30} = 0.043$$

for which $l_a = 0.95$ (figure 4.5). Therefore

$$A_s = \frac{139 \times 10^6}{0.95 \times 330 \times 0.87 \times 500}$$
$$= 1019 \, \text{mm}^2/\text{m}$$

Provide H20 bars at 250 mm centres ($A_s = 1260 \, \text{mm}^2/\text{m}$) , top steel.

Toe: taking moments about the stem centreline

$$M_{Ed} \approx \gamma_f \times 34.0 \times 0.6 \times \frac{0.8}{3.4} - 152.2 \times 0.8 \times 0.6$$

$$\approx 1.35 \times 4.8 - 73.1$$

$$\approx -67\,\text{kN m}$$

therefore

$$\frac{M_{Ed}}{bd^2f_{ck}} = \frac{67 \times 10^6}{1000 \times 330^2 \times 30} = 0.021$$

for which $l_a = 0.95$ (figure 4.5). Therefore:

$$A_s = \frac{67 \times 10^6}{0.95 \times 330 \times 0.87 \times 500} = 491\,\text{mm}^2/\text{m}$$

The minimum area for this, and for longitudinal distribution steel which is also required in the wall and the base, is given from table 6.8:

$$A_{s,\,\text{min}} = \frac{0.15b_t d}{100} = 0.0015 \times 1000 \times 330 = 495\,\text{mm}^2$$

Thus, provide H12 bars at 200 mm centres $(A_s = 566\,\text{mm}^2/\text{m})$, bottom and distribution steel.

Also steel should be provided in the compression face of the wall in order to prevent cracking – say, H10 bars at 200 mm centres each way.

Bending reinforcement is required in the heel beam to resist the moment due to the passive earth pressure. This reinforcement would probably be in the form of closed links.

CHAPTER 11

Prestressed concrete

CHAPTER INTRODUCTION

The analysis and design of prestressed concrete is a specialised field which cannot possibly be covered comprehensively in one chapter. This chapter concentrates therefore on the basic principles of prestressing, and the analysis and design of statically determinate members in bending for the serviceability and ultimate limit states.

A fundamental aim of prestressed concrete is to limit tensile stresses, and hence flexural cracking, in the concrete under working conditions. Design is therefore based initially on the requirements of the serviceability limit state. Subsequently considered are ultimate limit state criteria for bending and shear. In addition to the concrete stresses under working loads, deflections must be checked, and attention must also be paid to the construction stage when the prestress force is first applied to the immature concrete. This stage is known as the transfer condition.

The stages in the design of prestressed concrete may therefore be summarised as:

1. design for serviceability – cracking
2. check stresses at transfer
3. check deflections
4. check ultimate limit state – bending
5. design shear reinforcement for ultimate limit state.

They are illustrated by the flow chart in figure 11.1. $\longrightarrow$

When considering the basic design of a concrete section subject to prestress, the stress distribution due to the prestress must be combined with the stresses from the loading conditions to ensure that permissible stress limits are satisfied. Many analytical approaches have been developed to deal with this problem; however, it is considered that the method presented offers many advantages of simplicity and ease of manipulation in design.

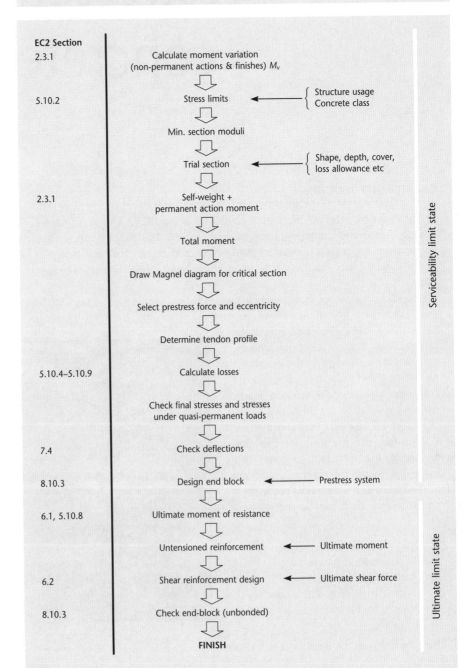

Figure 11.1
Prestressed concrete design flow chart

11.1 Principles of prestressing

In the design of a reinforced concrete beam subjected to bending it is accepted that the concrete in the tensile zone is cracked, and that all the tensile resistance is provided by the reinforcement. The stress that may be permitted in the reinforcement is limited by the need to keep the cracks in the concrete to acceptable widths under working conditions, thus there is no advantage to be gained from the use of the very high strength steels which are available. The design is therefore uneconomic in two respects: (1) dead weight includes 'useless' concrete in the tensile zone, and (2) economic use of steel resources is not possible.

'Prestressing' means the artificial creation of stresses in a structure before loading, so that the stresses which then exist under load are more favourable than would otherwise be the case. Since concrete is strong in compression the material in a beam will be used most efficiently if it can be maintained in a state of compression throughout. Provision of a longitudinal compressive force acting on a concrete beam may therefore overcome both of the disadvantages of reinforced concrete cited above. Not only is the concrete fully utilised, but also the need for conventional tension reinforcement is removed. The compressive force is usually provided by tensioned steel wires or strands which are anchored against the concrete and, since the stress in this steel is not an important factor in the behaviour of the beam but merely a means of applying the appropriate force, full advantage may be taken of very high strength steels.

The way in which the stresses due to bending and an applied compressive force may be combined is demonstrated in figure 11.2 for the case of an axially applied force acting over the length of a beam. The stress distribution at any section will equal the sum of the compression and bending stresses if it is assumed that the concrete behaves elastically. Thus it is possible to determine the applied force so that the combined stresses are always compressive.

By applying the compressive force eccentrically on the concrete cross-section, a further stress distribution, due to the bending effects of the couple thus created, is added to those shown in figure 11.2. This effect is illustrated in figure 11.3 and offers further advantages when attempting to produce working stresses within required limits.

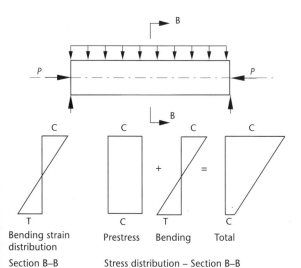

Bending strain
distribution

Section B–B

Prestress   Bending   Total

Stress distribution – Section B–B

Figure 11.2
Effects of axial prestress

Figure 11.3
Effects of eccentric prestress

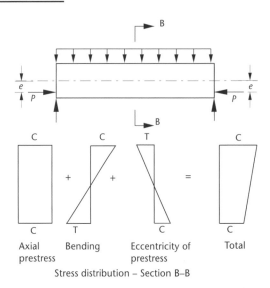

Stress distribution – Section B–B

Early attempts to achieve this effect were hampered both by the limited steel strengths available and by shrinkage and creep of the concrete under sustained compression, coupled with relaxation of the steel. This meant that the steel lost a large part of its initial pretension and as a result residual stresses were so small as to be useless. It is now possible, however, to produce stronger concretes which have good creep properties, and very high strength steels which can be stressed up to a high percentage of their 0.2 per cent proof stress are also available. For example, hard-drawn wires may carry stresses up to about three times those possible in grade 500 reinforcing steel. This not only results in savings of steel quantity, but also the effects of shrinkage and creep become relatively smaller and may typically amount to the loss of only about 25 per cent of the initial applied force. Thus, modern materials mean that the prestressing of concrete is a practical proposition, with the forces being provided by steel passing through the beam and anchored at each end while under high tensile load.

11.2 Methods of prestressing

Two basic techniques are commonly employed in the construction of prestressed concrete, their chief difference being whether the steel tensioning process is performed before or after the hardening of the concrete. The choice of method will be governed largely by the type and size of member coupled with the need for precast or *in situ* construction.

11.2.1 Pretensioning

In this method the steel wires or strands are stretched to the required tension and anchored to the ends of the moulds for the concrete. The concrete is cast around the tensioned steel, and when it has reached sufficient strength, the anchors are released and the force in the steel is transferred to the concrete by bond. In addition to long-term losses due to creep, shrinkage and relaxation, an immediate drop in prestress force occurs due to elastic shortening of the concrete. These features are illustrated in figure 11.4.

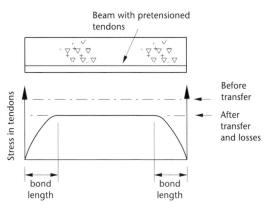

Figure 11.4
Tendon stresses –
pretensioning

Because of the dependence on bond, the tendons for this form of construction generally consist of small diameter wires or small strands which have good bond characteristics. Anchorage near the ends of these wires is often enhanced by the provision of small indentations in the surface of the wire.

The method is ideally suited for factory production where large numbers of identical units can be economically made under controlled conditions, a development of this being the 'long line' system where several units can be cast at once – end to end – and the tendons merely cut between each unit after release of the anchorages. An advantage of factory production of prestressed units is that specialised curing techniques such as steam curing can be employed to increase the rate of hardening of the concrete and to enable earlier 'transfer' of the stress to the concrete. This is particularly important where re-use of moulds is required, but it is essential that under no circumstances must calcium chloride be used as an accelerator because of its severe corrosive action on small diameter steel wires.

One major limitation of this approach is that tendons must be straight, which may cause difficulties when attempting to produce acceptable final stress levels throughout the length of a member. It may therefore be necessary to reduce either the prestress or eccentricity of force near the ends of a member, in which case tendons must either be 'debonded' or 'deflected'.

1. Debonding consists of applying a wrapping or coating to the steel to prevent bond developing with the surrounding concrete. Treating some of the wires in this way over part of their length allows the magnitude of effective prestress force to be varied along the length of a member.

2. Deflecting tendons is a more complex operation and is usually restricted to large members, such as bridge beams, where the individual members may be required to form part of a continuous structure in conjunction with *in situ* concrete slabs and sill beams. A typical arrangement for deflecting tendons is shown in figure 11.5, but it must be appreciated that substantial ancillary equipment is required to provide the necessary reactions.

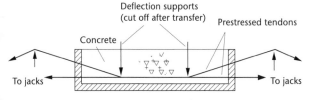

Figure 11.5
Tendon deflection

11.2.2 Post-tensioning

This method, which is the most suitable for *in situ* construction, involves the stressing against the hardened concrete of tendons or steel bars which are not bonded to the concrete. The tendons are passed through a flexible sheathing, which is cast into the concrete in the correct position. They are tensioned by jacking against the concrete, and anchored mechanically by means of steel thrust plates or anchorage blocks at each end of the member. Alternatively, steel bars threaded at their ends may be tensioned against bearing plates by means of tightening nuts. It is of course usually necessary to wait a considerable time between casting and stressing to permit the concrete to gain sufficient strength under *in situ* conditions.

The use of tendons consisting of a number of strands passing through flexible sheathing offers considerable advantages in that curved tendon profiles may be obtained. A post-tensioned structural member may be constructed from an assembly of separate pre-cast units which are constrained to act together by means of tensioned cables which are often curved as illustrated in figure 11.6. Alternatively, the member may be cast as one unit in the normal way but a light cage of untensioned reinforcing steel is necessary to hold the ducts in their correct position during concreting.

After stressing, the remaining space in the ducts may be left empty ('unbonded' construction), or more usually will be filled with grout under high pressure ('bonded' construction). Although this grout assists in transmitting forces between the steel and concrete under live loads, and improves the ultimate strength of the member, the principal use is to protect the highly stressed strands from corrosion. The quality of workmanship of grouting is thus critical to avoid air pockets which may permit corrosion. The bonding of the highly stressed steel with the surrounding concrete beam also greatly assists demolition, since the beam may then safely be 'chopped-up' into small lengths without releasing the energy stored in the steel.

Figure 11.6
Post-tensioned segmental construction

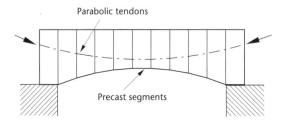

11.3 Analysis of concrete section under working loads

Since the object of prestressing is to maintain favourable stress conditions in a concrete member under load, the 'working load' for the member must be considered in terms of both maximum and minimum values. Thus at any section, the stresses produced by the prestress force must be considered in conjunction with the stresses caused by maximum and minimum values of applied moment.

Unlike reinforced concrete, the primary analysis of prestressed concrete is based on service conditions, and on the assumption that stresses in the concrete are limited to values which will correspond to elastic behaviour. In this section, the following assumptions are made in analysis.

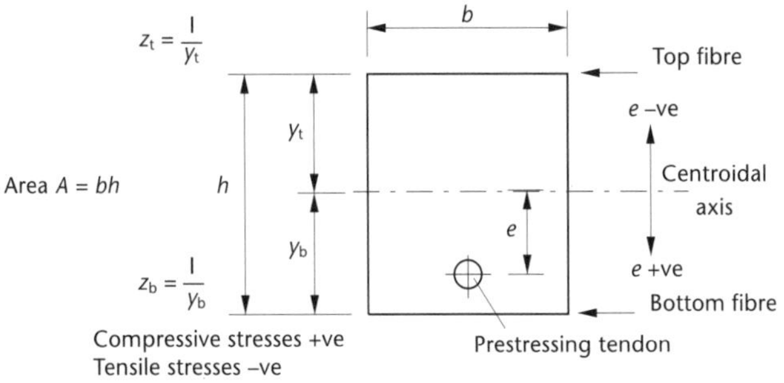

Figure 11.7
Sign convention and notation

1. Plane sections remain plane.

2. Stress–strain relationships are linear.

3. Bending occurs about a principal axis.

4. The prestressing force is the value remaining after all losses have occurred.

5. Changes in tendon stress due to applied loads on the member have negligible effect on the behaviour of the member.

6. Section properties are generally based on the gross concrete cross-section.

The stress in the steel is unimportant in the analysis of the concrete section under working conditions, it being the force provided by the steel that is considered in the analysis.

The sign conventions and notations used for the analysis are indicated in figure 11.7.

11.3.1 Member subjected to axial prestress force

If section BB of the member shown in figure 11.8 is subjected to moments ranging between M_{max} and M_{min}, the net stresses at the outer fibres of the beam are given by

under M_{max}
$$\left\{ \begin{aligned} f_t &= \frac{P}{A} + \frac{M_{max}}{z_t} \quad \text{at the top} & (11.1) \\ f_b &= \frac{P}{A} - \frac{M_{max}}{z_b} \quad \text{at the bottom} & (11.2) \end{aligned} \right.$$

under M_{min}
$$\left\{ \begin{aligned} f_t &= \frac{P}{A} + \frac{M_{min}}{z_t} \quad \text{at the top} & (11.3) \\ f_b &= \frac{P}{A} - \frac{M_{min}}{z_b} \quad \text{at the bottom} & (11.4) \end{aligned} \right.$$

where z_b and z_t are the elastic section moduli and P is the final prestress force.

The critical condition for tension in the beam is given by equation 11.2 which for no tension, that is $f_b = 0$, becomes

$$\frac{P}{A} = \frac{M_{max}}{z_b}$$

or

$$P = \frac{M_{max}A}{z_b} = \text{minimum prestress force required}$$

Figure 11.8
Stresses in member with axial
prestress force

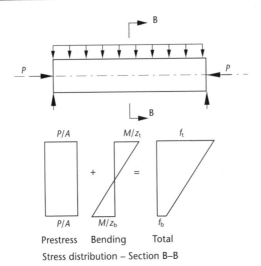

Stress distribution – Section B–B

For this value of prestress force, substitution in the other equations will yield the stresses in the beam under maximum load and also under minimum load. Similarly the stresses immediately after prestressing, before losses have occurred, may be calculated if the value of losses is known.

For example, the maximum stress in the top of the member is given by equation 11.1

$$f_t = \frac{P}{A} + \frac{M_{max}}{z_t}$$

where

$$P = \frac{M_{max}A}{z_b}$$

therefore

$$f_t = \frac{P}{A} + \frac{P}{A}\frac{z_b}{z_t}$$
$$= \frac{P}{A}\left(\frac{z_b + z_t}{z_t}\right)$$

It can be seen from the stress distributions in figure 11.8 that the top fibre is generally in considerable compression, while the bottom fibre is generally at lower stresses. Much better use of the concrete could be made if the stresses at both top and bottom can be caused to vary over the full range of permissible stresses for the two extreme loading conditions. This may be achieved by providing the force at an eccentricity e from the centroid.

11.3.2 Member subjected to eccentric prestress force

The stress distributions will be similar to those in section 11.3.1 but with the addition of the term $\pm Pe/z$ due to the eccentricity e of the prestressing force. For the position shown in figure 11.9, e will have a positive value. So that

under M_{max}
$$\begin{cases} f_t = \dfrac{P}{A} + \dfrac{M_{max}}{z_t} - \dfrac{Pe}{z_t} & \text{at the top} \end{cases} \tag{11.5}$$
$$\begin{cases} f_b = \dfrac{P}{A} - \dfrac{M_{max}}{z_b} + \dfrac{Pe}{z_b} & \text{at the bottom} \end{cases} \tag{11.6}$$

under M_{min}
$$\begin{cases} f_t = \dfrac{P}{A} + \dfrac{M_{min}}{z_t} - \dfrac{Pe}{z_t} & \text{at the top} \end{cases} \tag{11.7}$$
$$\begin{cases} f_b = \dfrac{P}{A} - \dfrac{M_{min}}{z_b} + \dfrac{Pe}{z_b} & \text{at the bottom} \end{cases} \tag{11.8}$$

Note that, as the prestressing force lies below the neutral axis, it has the effect of causing hogging moments in the section.

The critical condition for no tension in the bottom of the beam is again given by equation 11.6, which becomes

$$\frac{P}{A} - \frac{M_{max}}{z_b} + \frac{Pe}{z_b} = 0$$

or

$$P = \frac{M_{max}}{\left(\dfrac{z_b}{A} + e \right)} = \text{minimum prestress force required for no tension in bottom fibre}$$

Thus for a given value of prestress force P, the beam may carry a maximum moment of

$$M_{max} = P\left(\frac{z_b}{A} + e \right)$$

When compared with $M_{max} = P z_b / A$ for an axial prestress force it indicates an increase in moment carrying capacity of Pe.

The maximum stress in the top of the beam is given by equation 11.5 as

$$f_t = \frac{P}{A} + \frac{M_{max}}{z_t} - \frac{Pe}{z_t}$$

where

$$M_{max} = \frac{P z_b}{A} + Pe$$

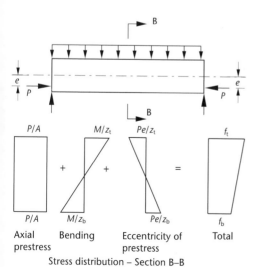

Figure 11.9
Stresses in member with eccentric prestress force

P/A M/z_t Pe/z_t f_t

$+$ $+$ $=$

P/A M/z_b Pe/z_b f_b

Axial prestress · Bending · Eccentricity of prestress · Total

Stress distribution – Section B–B

thus

$$f_t = \frac{P}{A} + \frac{Pz_b}{Az_t} + \frac{Pe}{z_t} - \frac{Pe}{z_t}$$

$$= \frac{P}{A}\left(\frac{z_b + z_t}{z_t}\right)$$

which is the same as that obtained in section 11.3.1 for an axially prestressed member. Thus the advantages of an eccentric prestress force with respect to the maximum moment-carrying capacity of a beam are apparent.

If the stress distributions of figure 11.9 are further examined, it can be seen that the differences in the net stress diagrams for the extreme loading cases are solely due to the differences between the applied moment terms M_{max} and M_{min}. It follows that by increasing the range of the stresses by the use of an eccentric prestress force the range of applied moments that the beam can carry is also increased. The minimum moment M_{min} that can be resisted is generally governed by the need to avoid tension in the top of the beam, as indicated in equation 11.7.

In the design of prestressed beams it is important that the minimum moment condition is not overlooked, especially when straight tendons are employed, as stresses near the ends of beams where moments are small may often exceed those at sections nearer mid-span. This feature is illustrated by the results obtained in example 11.1.

EXAMPLE 11.1

Calculation of prestress force and stresses

A rectangular beam 300×150 mm is simply supported over a 4 m span, and supports a live load of 10 kN/m. If a straight tendon is provided at an eccentricity of 65 mm below the centroid of the section, find the minimum prestress force necessary for no tension under live load at mid-span. Calculate the corresponding stresses under self-weight only at mid-span and at the ends of the member.

(a) Beam properties

$$\text{Self-weight} = 0.15 \times 0.3 \times 25 = 1.12 \text{ kN/m}$$

$$\text{Area} = 45 \times 10^3 \text{ mm}^2$$

$$\text{Section moduli } z_t = z_b = z = \frac{bh^2}{6} = \frac{150 \times 300^2}{6} = 2.25 \times 10^6 \text{ mm}^3$$

(b) Loadings (mid-span)

$$M_{max} = \frac{(10 + 1.12) \times 4^2}{8} = 22.2 \text{ kN m}$$

$$M_{min} = \frac{1.12 \times 4^2}{8} = 2.2 \text{ kN m}$$

(c) Calculate minimum prestress force

For no tension at the bottom under M_{max}

$$\frac{P}{A} - \frac{M_{max}}{z} + \frac{Pe}{z} = 0$$

where

$$e = 65 \text{ mm}$$

hence

$$P = \frac{M_{max}}{\left(\frac{z}{A} + e\right)} = \frac{22.2 \times 10^6 \times 10^{-3}}{\frac{2.25 \times 10^6}{45 \times 10^3} + 65}$$

$$= 193 \, kN$$

(d) Calculate stresses at mid-span under M_{min}

Stress at top $f_t = \dfrac{P}{A} + \dfrac{M_{min}}{z} - \dfrac{Pe}{z}$

where

$$\frac{P}{A} = \frac{193 \times 10^3}{45 \times 10^3} = 4.3 \, N/mm^2$$

$$\frac{M_{min}}{z} = \frac{2.2 \times 10^6}{2.25 \times 10^6} = 1.0 \, N/mm^2$$

$$\frac{Pe}{z} = \frac{193 \times 10^3 \times 65}{2.25 \times 10^6} = 5.6 \, N/mm^2$$

Hence

Stress at top $f_t = 4.3 + 1.0 - 5.6 = -0.3 \, N/mm^2$ (tension)

and

stress at bottom $f_b = \dfrac{P}{A} - \dfrac{M_{min}}{z} + \dfrac{Pe}{z}$

$$= 4.30 - 1.0 + 5.6 = +8.9 \, N/mm^2$$

The calculation shows that with minimum load it is possible for the beam to hog with tensile stresses in the top fibres. This is particularly likely at the initial transfer of the prestress force to the unloaded beam.

(e) Calculate stresses at ends

In this situation $M = 0$. Hence

$$f_t = \frac{P}{A} - \frac{Pe}{z} = 4.3 - 5.6 = -1.3 \, N/mm^2$$

and

$$f_b = \frac{P}{A} + \frac{Pe}{z} = 4.3 + 5.6 = 9.9 \, N/mm^2$$

11.4 Design for the serviceability limit state

The design of a prestressed concrete member is based on maintaining the concrete stresses within specified limits at all stages of the life of the member. Hence the primary design is based on the serviceability limit state, with the concrete stress limits based on the acceptable degree of flexural cracking, the necessity to prevent excessive creep and the need to ensure that excessive compression does not result in longitudinal and micro cracking.

Guidance regarding the allowable concrete compressive stress in bending is given in EC2 as limited to:

(i) $0.6f_{ck}$ under the action of characteristic loads

and (ii) $0.45f_{ck}$ under the action of the *quasi-permanent* loads.

The *quasi-permanent* loads are the permanent and prestressing load, $G_k + P_{m,t}$, plus a proportion of the characteristic variable imposed load. This proportion is taken as 0.3 for dwellings, offices and stores, 0.6 for parking areas and 0.0 for snow and wind loading.

If the tensile stress in the concrete is limited to the values of f_{ctm} given in table 6.11 then all stresses can be calculated on the assumption that the section is uncracked and the gross concrete section is resisting bending. If this is not the case then calculations may have to be based on a cracked section. Limited cracking is permissible depending on whether the beam is pre- or post-tensioned and the appropriate exposure class. Generally for prestressed members with bonded tendons crack widths should be limited to 0.2 mm under the action of the *frequent* loading combination taken as the permanent characteristic and prestressing load, $G_k + P_{m,t}$, plus a proportion of the characteristic variable imposed load as given by equation 2.3 and table 2.4. In some, more aggressive exposure conditions, the possibility of decompression under the *quasi permanent* load conditions may need to be considered.

At initial transfer of prestress to the concrete, the prestress force will be considerably higher than the 'long-term' value as a result of subsequent losses which are due to a number of causes including elastic shortening, creep and shrinkage of the concrete member. Estimation of losses is described in section 11.4.7. Since these losses commence immediately, the condition at transfer represents a transitory stage in the life of a member and further consideration should be given to limiting both compressive and tensile stresses at this stage. In addition, the concrete, at this stage, is usually relatively immature and not at full strength and hence transfer is a critical stage which should be considered carefully. The compressive stress at transfer should be limited to $0.6f_{ck}$ where f_{ck} is based on the strength on the concrete at transfer. The tensile stress should be limited to 1 N/mm^2 for sections designed not to be in tension in service. Where limited flexural stress under service loads is permitted, some limited tensile stress is permitted at transfer.

The choice of whether to permit cracking to take place or not will depend on a number of factors which include conditions of exposure and the nature of loading. If a member consists of precast segments with mortar joints, or if it is essential that cracking should not occur, then it will be designed to be in compression under all load conditions. However a more efficient use of materials can be made if the tensile strength of the concrete, f_{ctm}, given in table 6.11 is utilised. Provided these stresses are not exceeded then the section can be designed, based on the gross uncracked section.

Unless the section is designed to be fully in compression under the characteristic loads, a minimum amount of bonded reinforcement should be provided to control cracking. This is calculated in an identical manner to the minimum requirement for reinforced concrete (see section 6.1.5) with the allowance that a percentage of the prestressing tendons can be counted towards this minimum area.

The design of prestressing requirements is based on the manipulation of the four basic expressions given in section 11.3.2 describing the stress distribution across the concrete section. These are used in conjunction with permissible stresses appropriate to the type of member and covering the following conditions:

1. Initial transfer of prestress force with the associated loading (often just the beam's self-weight);
2. At service, after prestress losses, with minimum and maximum characteristic loading;
3. At service with the quasi-permanent loading.

The loadings must encompass the full range that the member will encounter during its life, and the minimum values will thus be governed by the construction techniques used. The partial factors of safety applied to these loads will be those for serviceability limit state, that is 1.0 for both permanent and variable loads. The quasi-permanent loading situation is considered with only a proportion of the characteristic variable load acting.

For a beam with a cantilever span or a continuous beam it is necessary to consider the loading patterns of the live loads at service in order to determine the minimum and maximum moments. For a single-span, simply supported beam it is usually the minimum moment at transfer and the maximum moment at service that will govern, as shown in figure 11.10. From figure 11.10 the governing equations for a single-span beam are:

At transfer

$$\frac{P_0}{A} - \frac{P_0 e}{z_t} + \frac{M_{min}}{z_t} = f_t' \geq f_{min}' \tag{11.9}*$$

$$\frac{P_0}{A} + \frac{P_0 e}{z_b} - \frac{M_{min}}{z_b} = f_b' \leq f_{max}' \tag{11.10}*$$

At service

$$\frac{KP_0}{A} - \frac{KP_0 e}{z_t} + \frac{M_{max}}{z_t} = f_t \leq f_{max} \tag{11.11}*$$

$$\frac{KP_0}{A} + \frac{KP_0 e}{z_b} - \frac{M_{max}}{z_b} = f_b \geq f_{min} \tag{11.12}*$$

where f_{max}', f_{min}', f_{max} and f_{min} are the appropriate permissible stresses at transfer and serviceability conditions. P_0 is the prestressing force at transfer and K is a loss factor that accounts for the prestress losses – for example, $K = 0.8$ for 20 per cent loss.

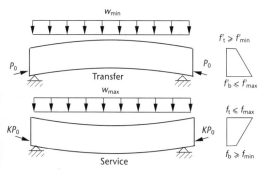

Figure 11.10
Prestressed beam at transfer and service

11.4.1 Determination of minimum section properties

The two pairs of expressions can be combined as follows:

11.9 and 11.11

$$(M_{max} - KM_{min}) \le (f_{max} - Kf'_{min})z_t \tag{11.13}$$

11.10 and 11.12

$$(M_{max} - KM_{min}) \le (Kf_{max} - f_{min})z_b \tag{11.14}$$

Hence, if $(M_{max} - KM_{min})$ is written as M_v, the moment variation

$$z_t \ge \frac{M_v}{(f_{max} - Kf'_{min})} \tag{11.15}$$

and

$$z_b \ge \frac{M_v}{(Kf'_{max} - f_{min})} \tag{11.16}$$

In equations 11.15 and 11.16, for z_t and z_b it can be assumed with sufficient accuracy, for preliminary sizing that M_{max} will depend on both the imposed and dead (self-weight) load and M_{min} will depend on the dead (self-weight) load only, so that in effect the calculations for M_v become independent of the self-weight of the beam.

These minimum values of section moduli must be satisfied by the chosen section in order that a prestress force and eccentricity exist which will permit the stress limits to be met; but to ensure that practical considerations are met the chosen section must have a margin above the minimum values calculated above. The equations for minimum moduli depend on the difference between maximum and minimum values of moment. The maximum moment on the section has not directly been included in these figures, thus it is possible that the resulting prestress force may not be economic or practicable. However, it is found in the majority of cases that if a section is chosen which satisfies these minimum requirements, coupled with any other specified requirements regarding the shape of the section, then a satisfactory design is usually possible. The ratio of acceptable span to depth for a prestressed beam cannot be categorised on the basis of deflections as easily as for reinforced concrete. In the absence of any other criteria, the following formulae may be used as a guide and will generally produce reasonably conservative designs for post-tensioned members.

$$\text{span} \le 36\,\text{m} \qquad h = \frac{\text{span}}{25} + 0.1\,\text{m}$$

$$\text{span} > 36\,\text{m} \qquad h = \frac{\text{span}}{20}\,\text{m}$$

In the case of short-span members it may be possible to use very much greater span–depth ratios quite satisfactorily, although the resulting prestress forces may become very high.

Other factors which must be considered at this stage include the slenderness ratio of beams, where the same criteria apply as for reinforced concrete, and the possibility of web and flange splitting in flanged members.

EXAMPLE 11.2

Selection of cross-section

Select a rectangular section for a post-tensioned beam to carry, in addition to its own self-weight, a uniformly distributed load of 3 kN/m over a simply supported span of 10 m. The member is to be designed with a concrete strength class C40/50 and is restrained against torsion at the ends and at mid-span. Assume 20 per cent loss of prestress ($K = 0.8$).

Design concrete stresses

At service:

$$f_{max} = 0.6f_{ck} = 0.6 \times 40 = 24 \, \text{N/mm}^2;$$
$$f_{min} = 0.0 \, \text{N/mm}^2$$

At transfer:

$$f'_{max} = 16 \, \text{N/mm}^2 \approx 0.6 \text{ strength at transfer};$$
$$f'_{min} = -1.0 \, \text{N/mm}^2$$
$$M_v = 3.0 \times 10^2/8 = 37.5 \, \text{kN m}$$

From equations 11.15 and 11.16:

$$z_t \geq \frac{M_v}{(f_{max} - Kf'_{min})} = \frac{37.5 \times 10^6}{(24 - 0.8\{-1\})} = 1.50 \times 10^6 \, \text{mm}^3$$

$$z_b \geq \frac{M_v}{(Kf'_{max} - f_{min})} = \frac{37.5 \times 10^6}{(0.8 \times 16 - 0.0)} = 2.93 \times 10^6 \, \text{mm}^3$$

Take $b = 200$ mm. Hence

$$z = 200h^2/6 \geq 2.93 \times 10^6$$

Therefore

$$h \geq \sqrt{(2.93 \times 10^6 \times 6/200)} = 297 \, \text{mm}$$

The minimum depth of beam is therefore 297 mm and to allow a margin in subsequent detailed design a depth of 350 mm would be appropriate as a first attempt.

To prevent lateral buckling EC2 specifies a maximum span/breadth ratio requirement:

$$\frac{l_{ot}}{b} \leq \frac{50}{(h/b)^{1/3}} \quad \text{with } h/b \leq 2.5$$

where l_{ot} = the distance between torsional restraints = 5.0 m in this example.

$$\text{Actual } \frac{l_{ot}}{b} = \frac{5000}{200} = 25$$

$$\text{maximum } \frac{l_{ot}}{b} = \frac{50}{(350/200)^{1/3}} = 41.5$$

hence the chosen dimensions are satisfactory as an initial estimate of the required beam size.

11.4.2 Design of prestress force

The inequalities of equations 11.9 to 11.12 may be rearranged to give expressions for the minimum required prestress force for a given eccentricity:

$$P_0 \leq \frac{(z_t f_{max} - M_{max})}{K(z_t/A - e)} \tag{11.17}$$

$$P_0 \geq \frac{(z_t f'_{min} - M_{min})}{(z_t/A - e)} \tag{11.18}$$

$$P_0 \geq \frac{(z_b f_{min} + M_{max})}{K(z_b/A + e)} \tag{11.19}$$

$$P_0 \leq \frac{(z_b f'_{max} + M_{min})}{(z_b/A + e)} \tag{11.20}$$

Note that in equations 11.17 and 11.18 it is possible that the denominator term, $(z_t/A - e)$, might be negative if $e > z_t/A$. In this case, the sense of the inequality would have to change as the effect of dividing an inequality by a negative number is to change its sense.

These equations give a range within which the prestress force must lie to ensure that the allowable stress conditions are met at all stages in the life of the member. In the case of a simply supported beam, the design prestress force will generally be based on the minimum value which satisfies these equations at the critical section for bending in the member.

Although a range of values of permissible prestress force can be found, this makes no allowance for the fact that the corresponding eccentricity must lie within the beam. It is therefore necessary to consider the effect of limiting the eccentricity to a maximum practical value for the section under consideration. Such limits will include consideration of the required minimum cover to the prestressing tendons which will depend on the exposure and structural class assumed for the design. The effect of this limitation will be most severe when considering the maximum moments acting on the section, that is, the inequalities of equations 11.11 and 11.12.

If the limiting value for maximum eccentricity e_{max}, depends on cover requirements, equation 11.11 becomes

$$M_{max} \leq f_{max} z_t - K P_0 \left(\frac{z_t}{A} - e_{max} \right) \tag{11.21}$$

and equation 11.12 becomes

$$M_{max} \leq K P_0 \left(\frac{z_b}{A} + e_{max} \right) - f_{min} z_b \tag{11.22}$$

These represent linear relationships between M_{max} and P_0. For the case of a beam subject to sagging moments e_{max} will generally be positive in value, thus equation 11.22 is of positive slope and represents a lower limit to P_0. It can also be shown that for most practical cases $[(z_t/A) - e_{max}] < 0$, thus equation 11.21 is similarly a lower limit of positive, though smaller slope.

Figure 11.11 represents the general form of these expressions, and it can be seen clearly that providing a prestress force in excess of Y' produces only small benefits of additional moment capacity. The value of Y' is given by the intersection of these two expressions, when

$$K P_0 \left(\frac{z_b}{A} + e_{max} \right) - f_{min} z_b = f_{max} z_t - K P_0 \left(\frac{z_t}{A} - e_{max} \right)$$

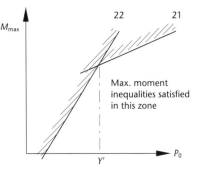

Figure 11.11
Maximum moment and
prestress force relationship

thus

$$P_0 = \frac{f_{max}z_t + f_{min}z_b}{K\left(\frac{z_b + z_t}{A}\right)} \tag{11.23}$$

Thus the value of prestress force $P_0 = Y'$ may be conveniently considered as a maximum economic value beyond which any increase in prestress force would be matched by a diminishing rate of increase in moment-carrying capacity. If a force larger than this limit is required for a given section it may be more economic to increase the size of this section.

EXAMPLE 11.3

Calculation of prestress force

The 10 metre span beam in example 11.2 was determined to have a breadth of 200 mm and a depth of 350 mm ($z_b = z_t = 4.08 \times 10^6$ mm^3). Determine the minimum initial prestress force required for an assumed maximum eccentricity of 75 mm.

From example 11.2:

$f'_{max} = 16 \, \text{N/mm}^2 \qquad f_{max} = 24 \, \text{N/mm}^2$

$f'_{min} = -1.0 \, \text{N/mm}^2 \quad f_{min} = 0.0 \, \text{N/mm}^2$

Self-weight of beam $= 0.2 \times 0.35 \times 25 = 1.75 \, \text{kN/m}$

$M_{min} = 1.75 \times 10^2/8 = 21.9 \, \text{kN m}$

$M_{max} = 3.0 \times 10^2/8 + 21.9 = 59.4 \, \text{kN m}$

(a) From equation 11.17:

$$P_0 \leq \frac{(z_t f_{max} - M_{max})}{K(z_t/A - e_{max})}$$

$$\leq \frac{(4.08 \times 10^6 \times 24 - 59.4 \times 10^6)}{0.8(4.08 \times 10^6/70\,000 - 75)} \times 10^{-3}$$

and allowing for the division by the negative denominator

$P_0 \geq -2881 \, \text{kN}$

Similarly from equations 11.18 to 11.20:

$$P_0 \leq +1555 \, \text{kN}$$
$$P_0 \geq +557 \, \text{kN}$$
$$P_0 \leq +654 \, \text{kN}$$

The minimum value of prestress force is therefore 557 kN with an upper limit of 654 kN.

(b) Check the upper economic limit to prestress force

From equation 11.23:

$$P_0 \leq \frac{f_{max} z_t + f_{min} z_b}{K \left(\dfrac{z_b + z_t}{A} \right)} = \frac{24zA}{2Kz}$$

$$\leq 12A/K$$
$$\leq 12 \times (350 \times 200) \times 10^{-3}/0.8$$
$$\leq 1050 \, \text{kN}$$

Since this is greater than the upper limit already established from equation 11.20 a design with an initial prestressing force between 557 kN and 654 kN will be acceptable.

11.4.3 Stresses under the quasi-permanent loading

The calculation in example 11.3 is based on the characteristic loads. Once a value of prestress force lying between the minimum and upper limit value is chosen, the compressive stress at the top of the section under the quasi-permanent loads should also be calculated and compared with the lesser allowable value of $0.45f_{ck}$. If this proves to be critical then the section may have to be redesigned taking the quasi-permanent load condition as more critical than the characteristic load condition.

EXAMPLE 11.4

Stress under quasi-permanent loads

For the previous example, using minimum prestress force of 557 kN, check the stress condition under the quasi-permanent loading condition. Assume that the 3 kN/m imposed load consists of a permanent load of 2 kN/m as finishes and 1.0 kN/m variable load. Take 30 per cent of the variable load contributing to the quasi-permanent load.

From the previous example:

Moment due to self-weight	$= 21.9 \, \text{kN m}$
Moment due to finishes	$= 2 \times 10^2/8$
	$= 25.0 \, \text{kN m}$
Moment due to variable load	$= 1 \times 10^2/8$
	$= 12.5 \, \text{kN m}$
Quasi-permanent moment	$= 21.9 + 25.0 + (0.3 \times 12.5)$
	$= 50.65 \, \text{kN m}$

Stress at the top of section is given by:

$$f_t = \frac{KP_0}{A} - \frac{KP_0 e}{z_t} + \frac{M}{z_t}$$

$$= \frac{0.8 \times 557 \times 10^3}{70\,000} - \frac{0.8 \times 557 \times 10^3 \times 75}{4.08 \times 10^6} + \frac{50.65 \times 10^6}{4.08 \times 10^6}$$

$$= 6.37 - 8.19 + 12.41$$

$$= 10.59 \, \text{N/mm}^2$$

Allowable compressive stress $= 0.45 f_{ck} = 0.45 \times 40 = 18 \, \text{N/mm}^2$.

Hence the maximum compressive stress is less than the allowable figure.

11.4.4 Magnel diagram construction

Equations 11.17 to 11.20 can be used to determine a range of possible values of prestress force for a given or assumed eccentricity. For different assumed values of eccentricity further limits on the prestress force can be determined in an identical manner although the calculations would be tedious and repetitive. In addition, it is possible to assume values of eccentricity for which there is no solution for the prestress force as the upper and lower limits could overlap.

A much more useful approach to design can be developed if the equations are treated graphically as follows. Equations 11.9 to 11.12 can be rearranged into the following form:

$$\frac{1}{P_0} \geq \frac{K(1/A - e/z_t)}{(f_{max} - M_{max}/z_t)} \qquad \{\text{equation 11.11}\} \qquad (11.24)$$

$$\frac{1}{P_0} \geq \frac{(1/A - e/z_t)}{(f'_{min} - M_{min}/z_t)} \qquad \{\text{equation 11.9}\} \qquad (11.25)$$

$$\frac{1}{P_0} \leq \frac{K(1/A + e/z_b)}{(f_{min} + M_{max}/z_b)} \qquad \{\text{equation 11.12}\} \qquad (11.26)$$

$$\frac{1}{P_0} \geq \frac{(1/A + e/z_b)}{(f'_{max} + M_{min}/Z_b)} \qquad \{\text{equation 11.10}\} \qquad (11.27)$$

These equations now express linear relationships between $1/P_0$ and e. Note that in equation 11.25 the sense of the inequality has been reversed to account for the fact that the denominator is negative (f'_{min} is negative according to the chosen sign convention). The relationships can be plotted as shown in figure 11.12(a) and (b) and the area of the graph to one side of each line, as defined by the inequality, can be eliminated, resulting in an area of graph within which any combination of force and eccentricity will simultaneously satisfy all four inequalities and hence will provide a satisfactory design. The lines marked 1 to 4 correspond to equations 11.24 to 11.27 respectively. This form of construction is known as a *Magnel Diagram*.

The additional line (5) shown on the diagram corresponds to a possible physical limitation of the maximum eccentricity allowing for the overall depth of section, cover to the prestressing tendons, provision of shear links and so on. Two separate figures are shown as it is possible for line 1, derived from equation 11.24, to have either a positive or a negative slope depending on whether f_{max} is greater or less than M_{max}/z_t.

Figure 11.12
Magnel diagram construction

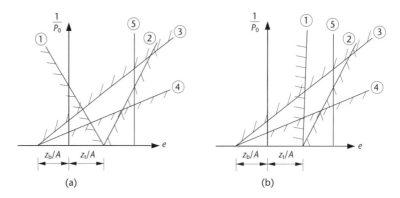

(a)   (b)

The Magnel diagram is a powerful design tool as it covers all possible solutions of the inequality equations and enables a range of prestress force and eccentricity values to be investigated. Values of minimum and maximum prestress force can be readily read from the diagram as can intermediate values where the range of possible eccentricities for a chosen force can be easily determined. The diagram also shows that the minimum prestress force (largest value of $1/P_0$) corresponds to the maximum eccentricity, and as the eccentricity is reduced the prestress force must be increased to compensate.

EXAMPLE 11.5

Construction of Magnel diagram

Construct the Magnel diagram for the beam given in example 11.2 and determine the minimum and maximum possible values of prestress force. Assume a maximum possible eccentricity of 125 mm allowing for cover etc. to the tendons.

From the previous examples:

$$f'_{max} = 16\,\text{N/mm}^2 \qquad f_{max} = 24\,\text{N/mm}^2$$

$$f'_{min} = -1.0\,\text{N/mm}^2 \quad f_{min} = 0.0\,\text{N/mm}^2$$

$$M_{min} = 21.9\,\text{kN m} \qquad M_{max} = 59.4\,\text{kN m}$$

$$K = 0.8 \qquad\qquad z_b = z_t = 4.08 \times 10^6\,\text{mm}^3$$

$$A = 70\,000\,\text{mm}^2$$

From equation 11.24:

$$\frac{1}{P_0} \geq \frac{K(1/A - e/z_t)}{(f_{max} - M_{max}/z_t)}$$

$$\geq 0.8\left(\frac{1}{70\,000} - \frac{e}{4.08 \times 10^6}\right) \times 10^3 \bigg/ \left(24 - \frac{59.4 \times 10^6}{4.08 \times 10^6}\right)$$

which can be re-arranged to give:

$$\frac{10^6}{P_0} \geq 1210 - 20.77e$$

and similarly from the other three inequalities, equations 11.25 to 11.27:

$$\frac{10^6}{P_0} \geq -2243 + 38.50e$$

$$\frac{10^6}{P_0} \leq 785 + 13.5e$$

$$\frac{10^6}{P_0} \geq 669 + 11.5e$$

These inequalities are plotted on the Magnel diagram in figure 11.13 and the zone bounded by the four lines defines an area in which all possible design solutions lie. The line of maximum possible eccentricity is also plotted but, as it lies outside the zone bounded by the four inequalities, does not place any restriction on the possible solutions.

From figure 11.13 it can be seen that the maximum and minimum values of prestress force are given by:

Maximum $10^6/P_0 = 2415$; hence minimum $P_0 = 414\,\text{kN}$ ($e = 121\,\text{mm}$)

Minimum $10^6/P_0 = 862$; hence maximum $P_0 = 1160\,\text{kN}$ ($e = 17\,\text{mm}$)

The intersection of the two lines at position A on the diagram corresponds to a value of $P_0 = 1050\,\text{kN}$, established in example 11.3 as the maximum economical value of prestress force for this section (see equation 11.23). Hence the intersection of these two lines should be taken as the maximum prestress force and, as can be seen, this information can be readily determined from the diagram without the need for further calculation.

The Magnel diagram can now be used to investigate other possible solutions for the design prestressing force and eccentricity. For a fixed value of prestress force (and hence fixed value of $1/P_0$) the corresponding range of permissible eccentricity can be read directly from the diagram. Alternatively, if the eccentricity is fixed, the diagram can be used to investigate the range of possible prestress force for the given eccentricity.

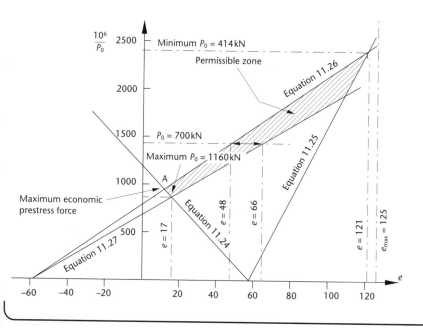

Figure 11.13
Magnel diagram for example 11.5

11.4.5 Design of tendon profiles

Having obtained a value of prestress force which will permit all stress conditions to be satisfied at the critical section, it is necessary to determine the eccentricity at which this force must be provided, not only at the critical section but also throughout the length of the member.

At any section along the member, e is the only unknown term in the four equations 11.9 to 11.12 and these will yield two upper and two lower limits which must all be simultaneously satisfied. This requirement must be met at all sections throughout the member and will reflect both variations of moment, prestress force and section properties along the member.

The design expressions can be rewritten as:

At transfer

$$e \leq \left[\frac{z_t}{A} - \frac{f'_{min} z_t}{P_0} \right] + \frac{M_{min}}{P_0} \tag{11.28}$$

$$e \leq \left[-\frac{z_b}{A} + \frac{f'_{max} z_b}{P_0} \right] + \frac{M_{min}}{P_0} \tag{11.29}$$

At service

$$e \geq \left[\frac{z_t}{A} - \frac{f_{max} z_t}{K P_0} \right] + \frac{M_{max}}{K P_0} \tag{11.30}$$

$$e \geq \left[-\frac{z_b}{A} + \frac{f_{min} z_b}{K P_0} \right] + \frac{M_{max}}{K P_0} \tag{11.31}$$

Equations 11.28–11.31 can be evaluated at any section to determine the range of eccentricities within which the resultant force P_0 must lie. The moments M_{max} and M_{min} are those relating to the section being considered.

For a member of constant cross-section, if minor changes in prestress force along the length are neglected, the terms in brackets in the above expressions are constant. Therefore the zone within which the centroid must lie is governed by the shape of the bending moment envelopes, as shown in figure 11.14.

In the case of uniform loading the bending moment envelopes are parabolic, hence the usual practice is to provide parabolic tendon profiles if a straight profile will not fit within the zone. At the critical section, the zone is generally narrow and reduces to zero if the value of the prestress force is taken as the minimum value from the Magnel diagram. At sections away from the critical section, the zone becomes increasingly greater than the minimum required.

Figure 11.14
Cable zone limits

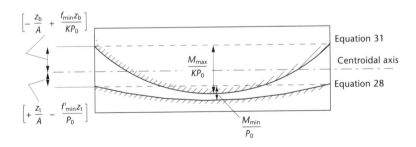

EXAMPLE 11.6

Calculation of cable zone

Determine the cable zone limits at mid-span and ends of the member designed in examples 11.2 to 11.5 for a constant initial prestress force of 700 kN. Data for this question are given in the previous examples.

(a) Ends of beam

Limits to cable eccentricity are given by equation 11.29, which at the end section can be readily shown, for this example, to be more critical than equation 11.28:

$$e \leq \left[-\frac{z_b}{A} + \frac{f'_{max} z_b}{P_0} \right] + \frac{M_{min}}{P_0}$$

and equation 11.31:

$$e \geq \left[-\frac{z_b}{A} + \frac{f_{min} z_b}{K P_0} \right] + \frac{M_{max}}{K P_0}$$

As there are no moments due to external loading at the end of a simply supported beam equation 11.29 becomes

$$e \leq \left[-\frac{4.08 \times 10^6}{(350 \times 200)} + \frac{16 \times 4.08 \times 10^6}{700 \times 10^3} \right] + 0$$
$$\leq -58.28 + 93.25$$
$$\leq 35 \, mm$$

Similarly equation 11.31 becomes

$$e \geq \left[-\frac{4.08 \times 10^6}{(350 \times 200)} + 0 \right] + 0$$
$$\geq -58.29 \, mm$$

At the ends of the beam where the moments are zero, and for $z_t = z_b$, the inequality expressions can apply with the tendon eccentricities above or below the neutral axis (e positive or negative). So that e must lie within the range $\pm 35 \, mm$.

(b) Mid-span

Equation 11.28 becomes:

$$e \leq \left[\frac{4.08 \times 10^6}{(350 \times 200)} - \frac{(-1)4.08 \times 10^6}{700 \times 10^3} \right] + \frac{21.9 \times 10^6}{700 \times 10^3}$$
$$\leq 64.1 + 31$$
$$\leq 95.1 \, mm$$

Equation 11.29 might be more critical than equation 11.28 and should be also checked. From equation 11.29:

$$e \leq \left[-\frac{4.08 \times 10^6}{(350 \times 200)} + \frac{16 \times 4.08 \times 10^6}{700 \times 10^3} \right] + \frac{21.9 \times 10^6}{700 \times 10^3}$$
$$\leq -58.3 + 93.3 + 31$$
$$\leq 66 \, mm$$

Hence equation 11.29 is critical and the eccentricity must be less than 66 mm.

Equation 11.31 gives

$$e \geq \left[-\frac{4.08 \times 10^6}{(350 \times 200)} + 0 \right] + \frac{59.4 \times 10^6}{0.8 \times 700 \times 10^3}$$

$$\geq -58.3 + 106.1$$

$$\geq 47.8 \, \text{mm}$$

Hence at mid-span the resultant of the tendon force must lie at an eccentricity in the range of 47.8 to 66 mm.

Provided that the tendons can be arranged so that their resultant force lies within the calculated limits then the design will be acceptable.

If a Magnel diagram for the stress condition at mid-span had been drawn, as in example 11.5, then the eccentricity range could have been determined directly from the diagram without further calculation. For tendons with a combined prestress force at transfer of $P_0 = 700 \, \text{kN}$ $(10^6/P_0 = 1428)$, plotting this value on the diagram of figure 11.13 will give the range of possible eccentricity between 48 mm and 66 mm.

11.4.6 Width of cable zone

From the Magnel diagram of figure 11.13 it can be seen that for any chosen value of prestress force there is an eccentricity range within which the resultant tendon force must lie. As the force approaches a value corresponding to the top and bottom limits of the diagram the width of the available cable zone diminishes until at the very extremities the upper and lower limits of eccentricity coincide, giving zero width of cable zone.

Practically, therefore, a prestress force will be chosen which has a value in between the upper and lower limits of permissible prestress force whilst, at the same time, ensuring that, for the chosen force, a reasonable width of cable zone exists. The prestressing cables must also satisfy requirements of cover, minimum spacing between tendons, available size of tendons and so on. A number of alternative tendon combinations and configurations are likely to be tried so that all requirements are simultaneously met. The advantage of the Magnel diagram is that a range of alternatives can be quickly considered without the necessity for any further calculation, as illustrated at the end of example 11.6.

11.4.7 Prestress losses

From the time that the prestressing force is first applied to the concrete member, losses of this force will take place because of the following causes:

1. Elastic shortening of the concrete.
2. Creep of the concrete under sustained compression.
3. Relaxation of the prestressing steel under sustained tension.
4. Shrinkage of the concrete.

These losses will occur whichever form of construction is used, although the effects of elastic shortening will generally be much reduced when post-tensioning is used. This is because stressing is a sequential procedure, and not instantaneous as with pre-tensioning. Creep and shrinkage losses depend to a large extent on the properties of the concrete with particular reference to the maturity at the time of stressing. In pre-tensioning, where the concrete is usually relatively immature at transfer, these losses may therefore be expected to be higher than in post-tensioning.

In addition to losses from these causes, which will generally total between 20 and 30 per cent of the initial prestress force at transfer, further losses occur in post-tensioned concrete during the stressing procedure. These are due to friction between the strands and the duct, especially where curved profiles are used, and to mechanical anchorage slip during the stressing operation. Both these factors depend on the actual system of ducts, anchorages and stressing equipment that are used.

Thus although the basic losses are generally highest in pre-tensioned members, in some instances overall losses in post-tensioned members may be of similar magnitude.

Elastic shortening

The concrete will immediately shorten elastically when subjected to compression, and the steel will generally shorten by a similar amount (as in pre-tensioning) with a corresponding loss of prestress force. To calculate this it is necessary to obtain the compressive strain at the level of the steel.

If the transfer force is P_0 and the force after elastic losses is P' then

$$P' = P_0 - \text{loss in force}$$

and the corresponding stress in the concrete at the level of the tendon

$$\sigma_{cp} = \frac{P'}{A} + \frac{(P'e) \times e}{I} + \sigma_{cg}$$

where σ_{cg} is the stress due to self-weight which will be relatively small when averaged over the length of the member and may thus be neglected. Hence

$$\sigma_{cp} = \frac{P'}{A}\left(1 + \frac{e^2 A}{I}\right)$$

and concrete strain $= \sigma_{cp}/E_{cm}$, thus reduction in steel strain $= \sigma_{cp}/E_{cm}$ and

$$\text{reduction in steel stress} = \left(\frac{\sigma_{cp}}{E_{cm}}\right)E_s = \alpha_e \sigma_{cp}$$

thus with $A_p =$ area of tendons

$$\begin{aligned}
\text{loss in prestress force} &= \alpha_e \sigma_{cp} A_p \\
&= \alpha_e \frac{A_p}{A} P'\left(1 + \frac{e^2 A}{I}\right)
\end{aligned}$$

hence

$$P' = P_0 - \alpha_e \frac{A_p}{A} P'\left(1 + \frac{e^2 A}{I}\right)$$

so that

$$\text{remaining prestress force } P' = \frac{P_0}{1 + \alpha_e \dfrac{A_p}{A}\left(1 + \dfrac{e^2 A}{I}\right)}$$

In pre-tensioned construction this full loss will be present; however when post-tensioning the effect will only apply to previously tensioned cables and although a detailed calculation could be undertaken it is normally adequate to assume 50 per cent of the above losses. In this case the remaining prestress force is

$$P' = \frac{P_0}{1 + 0.5\alpha_e \frac{A_p}{A}\left(1 + \frac{e^2 A}{I}\right)}$$

and it is this value which applies to subsequent loss calculations. In calculating α_e, E_{cm} may be taken from table 6.11 where f_{ck} should be taken as the transfer strength of the concrete.

Creep of concrete

The sustained compressive stress on the concrete will also cause a long-term shortening due to creep, which will similarly reduce the prestress force. As above, it is the stress in the concrete at the level of the steel which is important, that is

$$\sigma_{cp} = \frac{P'}{A}\left(1 + \frac{e^2 A}{I}\right)$$

and

loss of steel stress $= E_s \sigma_{cp} \times$ specific creep strain

then

$$\text{loss of prestress force} = E_s \frac{A_p}{A} P'\left(1 + \frac{e^2 A}{I}\right) \times \text{specific creep strain}$$

The value of specific creep used in this calculation will be influenced by the factors discussed in section 6.3.2, and may be obtained from the values of the final creep coefficient $\phi(\infty, t_0)$ given in table 6.12 in chapter 6 using the relationship

$$\text{Specific creep strain} = \frac{\phi(\infty, t_0)}{1.05 E_{cm}} \Bigg/ \text{N/mm}^2$$

Table 6.12 may be used where the concrete stress does not exceed $0.45 f_{ck}$ at transfer, where f_{ck} relates to the concrete strength at transfer.

Relaxation of steel

Despite developments in prestressing steel manufacture, relaxation of the wire or strand under sustained tension may still be expected to be a significant factor. The precise value will depend upon whether pre-tensioning or post-tensioning is used and the characteristics of the steel type. Equations allowing for method of construction are given in EC2 section 3.3.2(7) which should be applied to 1000-hour relaxation values provided by the manufacturer. The amount of relaxation will also depend upon the initial tendon load relative to its breaking load. In most practical situations the transfer steel stress is about 70 per cent of the characteristic strength and relaxation losses are likely to be approximately 4–10 per cent of the tendon load remaining after transfer.

Shrinkage of concrete

This is based on empirical figures for shrinkage/unit length of concrete (ε_{cs}) for particular curing conditions and transfer maturity as discussed in chapter 6. Typical values range from 230×10^{-6} for UK outdoor exposure (80% relative humidity) to

550×10^{-6} for indoor exposure, (50% relative humidity), depending on the notional size of the member. See table 6.13.

The loss in steel stress is thus given by $\varepsilon_{cs} E_s$, hence

loss in prestress force $= \varepsilon_{cs} E_s A_p$

Friction in ducts (post-tensioning only)

When a post-tensioned cable is stressed, it will move relative to the duct and other cables within the duct. Friction will tend to resist this movement hence reducing the effective prestress force at positions remote from the jacking point. This effect may be divided into unintentional profile variations, and those due to designed curvature of ducts.

(a) 'Wobble' effects in straight ducts will usually be present. If $P_0 =$ jack force, and $P_x =$ cable force at distance x from jack then it is generally estimated that

$$P_x = P_0 e^{-\mu k x}$$

where $e =$ base of napierian logs (2.718) and $k =$ unintentional angular displacement per unit length generally in the range of 0.005 to 0.01 radians/metre.

(b) Duct curvature will generally cause greater prestress force losses, and is given by

$$P_x = P_0 e^{-\mu \theta}$$

where $\mu =$ coefficient of friction (typically 0.17 for cold drawn wire and 0.19 for strand) and θ is the sum of the angular displacements over the distance x. If the duct curvature is not constant, the profile must be subdivided into sections, each assumed to have constant curvature, in which case P_0 is taken as the force at the jacking end of the section, x as the length of the segment and P_x the force at the end remote from the jack, which then becomes P_0 for the next section and so on.

The above effects may be combined to produce an effective prestress force diagram for a member. If friction losses are high, it may be worthwhile to jack simultaneously from both ends, in which case the two diagrams may be superimposed, maintaining symmetry of prestress force relative to the length of the member.

Losses at anchorages (post-tensioned only)

When post-tensioned tendons are 'locked off' at the anchorages there is invariably some loss of prestress due to slippage of the anchorage. Advice should be sought from the manufacturers of the anchorage systems or from European technical approval documents.

Code formula for time-dependent losses

Time-dependent losses due to creep, shrinkage and relaxation can be calculated separately, as indicated above, or the combined formula, as given in EC2, can be used to determine the variation of stress, $\delta\sigma_{p,c+s+r}$ at location x and at time t, where

$$\delta\sigma_{p,c+s+r} = \frac{\varepsilon_{cs} E_s + 0.8\delta\sigma_{pr} + \alpha_e \phi(t, t_0)\sigma_{c, Qp}}{1 + \alpha_e \dfrac{A_p}{A}\left[\left(1 + \dfrac{A}{I}e^2\right)(1 + 0.8\phi(t, t_0))\right]} \tag{11.32}$$

where $\sigma_{c,Qp}$ is the initial stress in the concrete adjacent to the tendons due to prestress, self-weight and other quasi-permanent actions, $\delta\sigma_{pr}$ is the variation of stress in the tendons due to relaxation and the other terms in the formula are as previously defined.

EXAMPLE 11.7

Estimation of prestress losses at mid-span

A post-tensioned beam shown in figure 11.15 is stressed by two tendons with a parabolic profile and having a total cross-sectional area $A_p = 7500\,\text{mm}^2$. The total initial prestress force is $P_0 = 10\,500\,\text{kN}$ and the total characteristic strength is $P_{pk} = 14000\,\text{kN}$.

Figure 11.15
Post-tensioned beam

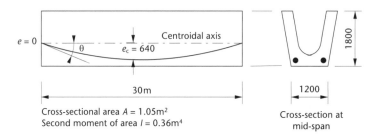

30m

Cross-sectional area $A = 1.05\text{m}^2$
Second moment of area $I = 0.36\text{m}^4$

1200

Cross-section at
mid-span

Assume the following data for estimating losses:

Coefficient of friction $\mu = 0.19$

wobble factor $k = 0.01/\text{metre}$

Elastic modulus E_{cm} (transfer) $= 32\,\text{kN/mm}^2$

$E_s = 205\,\text{kN/mm}^2$

Creep coefficient $\phi(\infty, t_0) = 1.6$

Shrinkage strain $\varepsilon_{cs} = 330 \times 10^{-6}$

The tendon supplier specifies class 2 strands with a 1000 hour relaxation loss of 2.5 per cent at 70 per cent of the characteristic strength.

(1) Friction

The equation of the parabola is $y = Cx^2$ and with the origin at mid-span when $x = 15\,000$, $y = 640$, so that $C = 640/15000^2 = 2.844 \times 10^{-6}$.

The gradient θ at the ends is given by

$$\theta = dy/dx = 2Cx = 2 \times 2.844 \times 10^{-6} \times 15\,000$$
$$= 0.0853\,\text{radians}$$

At mid-span

$$\text{loss } \Delta P(x) = P_0\left(1 - e^{-\mu(\theta+kx)}\right)$$
$$= P_0\left(1 - e^{-0.19(0.0853+0.01\times15)}\right)$$
$$= 0.044P_0 = 460\,\text{kN} = 4.4 \text{ per cent}$$

(2) Elastic shortening for post-tensioned construction

$$P' = \frac{P}{1 + 0.5\alpha_e \dfrac{A_p}{A}\left(1 + e^2\dfrac{A}{I}\right)}$$

Take the average eccentricity for the parabolic tendon as $5/8e_c = 5/8 \times 640 = 400\,\text{mm}$ and $\alpha_e = E_s/E_{cm} = 205/32 = 6.41$.

$$P' = \cfrac{P_0}{1 + 0.5 \times 6.41 \times \cfrac{7.5 \times 10^3}{1.05 \times 10^6}\left(1 + 400^2 \cfrac{1.05 \times 10^6}{0.36 \times 10^{12}}\right)}$$

$$= 0.968 P_0 = 10\,160\,\text{kN}$$

Loss $\Delta P = 10\,500 - 10\,160 = 340\,\text{kN} = 3.2$ per cent

Total short-term losses $= 460 + 340 = 800\,\text{kN}$

$P' = P_0 -$ short-term losses

$\quad = 10\,500 - 800 = 9700\,\text{kN}$

(3) Creep

Loss $\Delta P = \phi \dfrac{E_s A_p}{(1.05 E_{cm})A}\left(1 + e^2\dfrac{A}{I}\right)P'$

$$= 1.6\dfrac{205 \times 10^3}{(1.05 \times 32) \times 10^3} \times \dfrac{7.5 \times 10^3}{1.05 \times 10^6}\left(1 + 400^2\dfrac{1.05 \times 10^6}{0.36 \times 10^{12}}\right)9700$$

$$= 992\,\text{kN}\ (= 9.4\ \text{per cent of}\ P_0)$$

(4) Shrinkage

Loss $\Delta P = \varepsilon_{cs} E_s A_p$

$$= 330 \times 10^{-6} \times 205 \times 7.5 \times 10^3$$

$$= 507\,\text{kN}\ (= 4.8\ \text{per cent of}\ P_0)$$

(5) Relaxation

Long-term relaxation loss factor $= 2.5$ for class 2 strand estimated from equation 3.29 of EC2

$$\text{loss}\ \Delta P = (2.5 \times 2.5/100)P' = 0.0625 \times 9700$$

$$= 606\,\text{kN}\ (= 5.8\ \text{per cent of}\ P_0)$$

Total estimated losses $= 800 + 992 + 507 + 606 = 2905\,\text{kN}$

$$= 28\ \text{per cent of}\ P_0$$

11.4.8 Calculation of deflections

The anticipated deflection of a prestressed member must always be checked since span–effective depth ratios are not specified in the code for prestressed concrete members. The deflection due to the eccentric prestress force must be evaluated and added to that from the normal permanent and variable load on the member. In the majority of cases, particularly where the member is designed to be uncracked under full load, a simple linear elastic analysis based on the gross concrete section will be sufficient to give a reasonable and realistic estimate of deflections.

Where the member is designed such that, under the characteristic loads, the tensile strength exceeds the cracking strength of the concrete, f_{ctm}, it may be necessary to base the calculation of deflection on the cracked concrete section and reference should be made to the Code for the method of dealing with this situation.

The basic requirements which should generally be satisfied in respect of deflections are similar to those of a reinforced concrete beam (section 6.3) which are:

1. Deflection under the action of the *quasi-permanent load* $\leq$ span/250 measured below the level of the supports;

2. Span/500 maximum movement after other elements, which are susceptible to damage by movement, are applied.

The evaluation of deflections due to prestress loading can be obtained by double integration of the expression

$$M_x = Pe_x = EI\frac{d^2y}{dx^2}$$

over the length of the member, although this calculation can prove tedious for complex tendon profiles.

The simple case of straight tendons in a uniform member however, yields $M = -Pe = $ a constant, which is the situation evaluated in section 6.3.3 to yield a maximum mid-span deflection of $ML^2/8EI = -PeL^2/8EI$. If the cables lie below the centroidal axis, e is positive, and the deflection due to prestress is then negative, that is upwards.

Another common case of a symmetrical parabolic tendon profile in a beam of constant section can also be evaluated quite simply by considering the bending-moment distribution in terms of an equivalent uniformly distributed load.

For the beam in figure 11.16 the moment due to prestress loading at any section is $M_x = -Pe_x$ but since e_x is parabolic, the prestress loading may be likened to a uniformly distributed load w_e on a simply supported beam; then mid-span moment

$$M = \frac{w_e L^2}{8} = -Pe_c$$

thus

$$w_e = \frac{-8Pe_c}{L^2}$$

But since the mid-span deflection due to a uniformly distributed load w over a span L is given by

$$y = \frac{5}{384}\frac{wL^4}{EI}$$

the deflection due to w_e is

$$y = -\frac{5}{48}\frac{(Pe_c)L^2}{EI}$$

Figure 11.16
Parabolic tendon profile

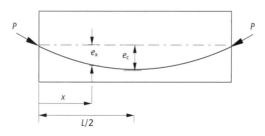

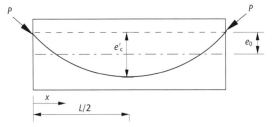

Figure 11.17
Parabolic tendon profile
eccentric at ends of beam

If the prestress force does not lie at the centroid of the section at the ends of the beam, but at an eccentricity e_0 as shown in figure 11.17, the expression for deflection must be modified. It can be shown that the deflection is the same as that caused by a force P acting at a constant eccentricity e_0 throughout the length of the member, plus a force P following a parabolic profile with mid-span eccentricity e'_c as shown in figure 11.17.

The mid-span deflection thus becomes

$$y = \frac{(Pe_0)L^2}{8EI} - \frac{5}{48}\frac{(Pe'_c)L^2}{EI}$$

Deflections due to more complex tendon profiles are most conveniently estimated on the basis of coefficients which can be evaluated for commonly occurring arrangements. These are on the basis $y = (KL^2)/EI$ where K incorporates the variations of curvature due to prestress along the member length.

There are three principal stages in the life of a prestressed member at which deflections may be critical and may need to be assessed.

1. At transfer – a check of actual deflection at transfer for comparison with estimated values is a useful guide that a prestressed beam has been correctly constructed.

2. Under dead load, before application of finishes – deflections must be evaluated to permit subsequent movement and possible damage to be estimated.

3. Long-term under full quasi-permanent actions – deflections are required, both to determine the subsequent movement and also to assess the appearance of the final structure.

Short-term deflections will be based on materials properties associated with characteristic strengths ($\gamma_m = 1$) and with actual loading ($\gamma_f = 1$). Long-term assessment however must not only take into account loss in prestress force, but also the effects of creep both on the applied loading and the prestress loading components of the deflection. Creep is allowed for by using an effective modulus of elasticity for the concrete, as discussed in section 6.3.2.

Thus if $E_{c(t_0)}$ is the instantaneous value, the effective value after creep is given by

$$E_{c,\,eff} = \frac{E_{c(t_0)}}{1 + \phi(\infty,\,t_0)}$$

where the value of $\phi(\infty, t_0)$, the creep coefficient can be obtained from table 6.12

It can be shown in some instances that when net upward deflections occur, these often increase because of creep, thus the most critical downward deflection may well be before creep losses occur, while the most critical upward deflection may be long-term. This further complicates a procedure which already has many uncertainties as discussed in chapter 6; thus deflections must always be regarded as estimates only.

EXAMPLE 11.8

Calculation of deflection

Estimate transfer and long-term deflections for a $200 \times 350\,\text{mm}$ beam of 10 m span. The prestressing tendon has a parabolic profile with mid-span eccentricity $= 75\,\text{mm}$ and the end eccentricity $= 0$ at both ends. The initial prestress force at transfer, P_0, is 560 kN and there are 20 per cent losses. The imposed load consists of 2.0 kN/m finishes and 1.0 kN/m variable load. $E_{cm} = 35\,\text{kN/mm}^2$ and the creep factor $\phi(\infty, t_0) = 2.0$.

$$\text{Self-weight} = 0.2 \times 0.35 \times 25 = 1.75\,\text{kN/m}$$

$$I = \frac{bh^3}{12} = \frac{200 \times 350^3}{12} = 715 \times 10^6\,\text{mm}^4$$

(a) At transfer

$$\begin{aligned}
\text{Deflection } y_a &= \frac{5}{384} \frac{w_{\min} L^4}{E_{cm} I} - \frac{5}{48} \frac{(P_0 e_c) L^2}{E_{cm} I} \\
&= \frac{5}{384} \frac{1.75 \times 10^4 \times 10^{12}}{35 \times 10^3 \times 715 \times 10^6} - \frac{5}{48} \frac{560 \times 10^3 \times 75 \times 10^2 \times 10^6}{35 \times 10^3 \times 715 \times 10^6} \\
&= 9.1 - 17.5 \\
&= -8\,\text{mm (upwards)}
\end{aligned}$$

(b) At application of finishes

Assume that only a small proportion of prestress losses have occurred:

$$\text{Weight of finishes} = 2.0\,\text{kN/m}$$

therefore

$$\begin{aligned}
y_b &= y_a - \frac{5 \times 2.0 \times 10^4 \times 10^{12}}{384 \times 35 \times 10^3 \times 715 \times 10^6} \\
&= -8 + 10\,\text{mm} = 2\,\text{mm (downwards)}
\end{aligned}$$

(c) In the long term due to the quasi-permanent action plus prestress force after losses

Assuming 30 per cent of the variable load contributes to the quasi-permanent action:

$$\text{Quasi-permanent action} = \text{self-weight} + \text{finishes} + 0.3 \times \text{variable load}$$

$$= 1.75 + 2.0 + 0.3 \times 1.0 = 4.05\,\text{kN/m}$$

$$\text{Prestress forces after losses} = 0.8P_0 = 0.8 \times 560 = 448\,\text{kN}$$

$$E_{c,\text{eff}} = \frac{E_{cm}}{(1 + \phi(\infty, t_0))} = \frac{35}{(1 + 2.0)} = 11.7\,\text{kN/mm}^2$$

$$\begin{aligned}
y_c &= \frac{5}{384} \frac{4.05 \times 10^4 \times 10^{12}}{11.7 \times 10^3 \times 715 \times 10^6} - \frac{5}{48} \frac{448 \times 10^3 \times 75 \times 10^2 \times 10^6}{11.3 \times 10^3 \times 715 \times 10^6} \\
&= 63.0 - 43.3 = 20\,\text{mm (downwards)} \le \text{span}/250 = 40\,\text{mm}
\end{aligned}$$

Therefore satisfactory.

(d) Movement after application of finishes

$$y_d = y_c - y_b = 20 - 2 = 18\,\text{mm} \le \text{span}/500 = 20\,\text{mm (satisfactory)}.$$

11.4.9   End blocks

In pre-tensioned members, the prestress force is transferred to the concrete by bond over a definite length at each end of the member. The transfer of stress to the concrete is thus gradual. In post-tensioned members however, the force is concentrated over a small area at the end faces of the member, and this leads to high-tensile forces at right angles to the direction of the compression force. This effect will extend some distance from the end of the member until the compression has distributed itself across the full concrete cross-section. This region is known as the 'end block' and must be heavily reinforced by steel to resist the bursting tension forces. End block reinforcement will generally consist of closed links which surround the anchorages, and the quantities provided are usually obtained from empirical methods.

Typical 'flow lines' of compressive stress are shown in figure 11.18, from which it can be seen that whatever type of anchorage is used, the required distribution can be expected to have been attained at a distance from the loaded face equal to the lateral dimension of the member. This is relatively independent of the anchorage type.

In designing the end block it is necessary to check that the bearing stress behind the anchorage plate due to the prestressing force does not exceed the limiting stress, f_{Rdu}, given by

$$f_{Rdu} = 0.67 f_{ck} (A_{c1}/A_{c0})^{0.5} \leq 2.0 f_{ck}$$

where

A_{c0} is the loaded area of the anchorage plate

A_{c1} is the maximum area, having the same shape as A_{c0} which can be inscribed in the total area A_c, as shown in figure 11.19(a)

The lateral tensile bursting forces can be established by the use of a statically determinate strut and tie model where it is assumed that the load is carried by a truss consisting of concrete struts and links of reinforcement acting as steel ties. In carrying out these calculations a partial factor of safety of $\gamma_p = 1.2$ is applied to the prestressing

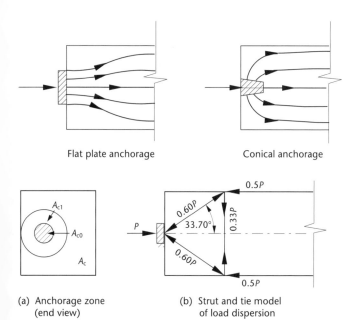

Flat plate anchorage   Conical anchorage

Figure 11.18
Stress distribution in end blocks

(a) Anchorage zone
     (end view)

(b) Strut and tie model
     of load dispersion

Figure 11.19
Bursting tensile force in end blocks

force. EC2 suggests that in determining the geometry of this truss the prestressing force can be assumed to disperse at an angle of 33.7° to the longitudinal axis of the beam as shown in figure 11.19(b). The compressive stresses in the assumed struts should not exceed $0.4\left(1 - \dfrac{f_{ck}}{250}\right)f_{ck}$ and the reinforcement is designed to act at a design strength of $0.87f_{yk}$. However if the stress in the reinforcement is limited to $300\,\text{N/mm}^2$ then no checks on crack widths are necessary. This reinforcement, in the form of closed links, is then distributed over a length of the end-block equal to the greater lateral dimension of the block, this length being the length over which it is assumed that the lateral tensile stresses are acting.

EXAMPLE 11.9

Design of end block reinforcement

The beam in figure 11.20 is stressed by four identical 100 mm diameter conical anchorages located as shown, with a jacking force of 250 kN applied to each. The area may be subdivided into four equal end zones of 200×150 mm each. Determine the reinforcement required around the anchorages: $f_{ck} = 40\,\text{N/mm}^2$, $f_{yk} = 500\,\text{N/mm}^2$.

Consider one anchor.

(a)   Check bearing stress under the anchor

$$\text{Actual bearing stress} = \frac{\gamma_p \times \text{Prestressing force}}{\text{Loaded area}}$$

$$= \frac{1.2 \times 250 \times 10^3}{\pi \times 100^2/4}$$

$$= 38.2\,\text{N/mm}^2$$

Allowable bearing stress $f_{Rdu} = 0.67 f_{ck}(A_{c1}/A_{c0})^{0.5}$

$$= 0.67 \times 40 \left(\frac{\pi \times 150^2/4}{\pi \times 100^2/4}\right)^{0.5}$$

$$= 40.2\,\text{N/mm}^2 \quad (> 38.2)$$

Figure 11.20
End block reinforcement example

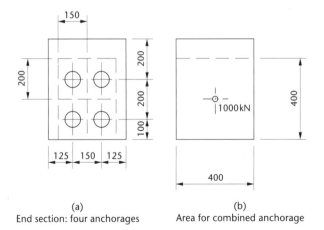

(a)
End section: four anchorages

(b)
Area for combined anchorage

(b) Reinforcement

From figure 11.19b, the tensile force in the tie of the equivalent truss is given by

$$T = 0.33 \times 1.2 \times 250 = 100 \, \text{kN}$$

Area of tensile steel required (assuming stress in the steel is limited to 300 N/mm^2)

$$A_s = \frac{100 \times 10^3}{300}$$
$$= 330 \, \text{mm}^2$$

This can be provided by three 10 mm closed links (471 mm^2) at, say, 50, 125 and 200 mm from the end face; that is, distributed over a length equal to the largest dimension of the anchorage block (200 mm). Note that in each direction there are two legs of each link acting to resist the tensile force.

(c) Check compressive stress in the struts

Allowable compressive stress $= 0.4(1 - f_{ck}/250)f_{ck}$

$$= 0.4(1 - 40/250)40 = 13.44 \, \text{N/mm}^2$$

$$\text{Actual stress in strut} = \frac{\text{Force in strut}}{\text{Cross-sectional area}}$$

$$= \frac{0.60 \times 1.2 \times 250 \times 10^3}{(200 \times 150 \times \cos 33.7°)}$$

$$= 7.21 \, \text{N/mm}^2$$

The effect of the combined anchorage can be considered by considering the total prestress force of 1000 kN acting on an effective end block of 400 × 400 mm.

The tensile force in the tie of the equivalent truss is given by

$$T = 0.33 \times 1.2 \times 1000 = 400 \, \text{kN}$$

Area of tensile steel required

$$A_s = \frac{400 \times 10^3}{300}$$
$$= 1333 \, \text{mm}^2$$

This can be provided by six 12 mm closed links (1358 mm^2) distributed over a length equal to the largest dimension of the anchorage block, that is, 400 mm.

11.5 | Analysis and design at the ultimate limit state

After a prestressed member has been designed to satisfy serviceability requirements, a check must be carried out to ensure that the ultimate moment of resistance and shear resistance are adequate to satisfy the requirements of the ultimate limit state. The partial factors of safety on loads and materials for this analysis are the normal values for the ultimate limit state which are given in chapter 2. However, in consideration of the effect of the prestress force this force should be multiplied by a partial factor of safety, γ_p, of 0.9 (UK National Annex) when the prestress force is considered to be, as is usual, a 'favourable effect'.

11.5.1 Analysis of the section

As the loads on a prestressed member increase above the working values, cracking occurs and the prestressing steel begins to behave as conventional reinforcement. The behaviour of the member at the ultimate limit state is exactly as that of an ordinary reinforced concrete member except that the initial strain in the steel must be taken into account in the calculations. The section may easily be analysed by the use of the equivalent rectangular stress block described in chapter 4.

Although illustrated by a simple example this method may be applied to a cross-section of any shape which may have an arrangement of prestressing wires or tendons. Use is made of the stress–strain curve for the prestressing steel shown in figure 11.21 to calculate tension forces in each layer of steel. The total steel strain is that due to bending added to the initial strain in the steel resulting from prestress. For a series of assumed neutral axis positions, the total tension capacity is compared with the compressive force developed by a uniform stress of $0.567f_{ck}$, and when reasonable agreement is obtained, the moment of resistance can be evaluated.

Figure 11.21
Stress-strain curve for prestressing steel

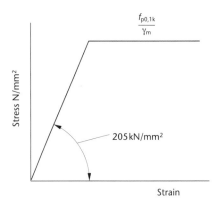

EXAMPLE 11.10

Calculation of ultimate moment of resistance

The section of a pretensioned beam shown in figure 11.22 is stressed by ten 5 mm wires of 0.1% proof stress $f_{p0,1k} = 1600\,\text{N/mm}^2$. If these wires are initially stressed to $1120\,\text{N/mm}^2$ and 30 per cent losses are anticipated, estimate the ultimate moment of resistance of the section if class C35/45 concrete is used. The stress–strain curve for prestressing wire is shown in figure 11.23.

Area of 5 mm wire $= \pi \times 5^2/4 = 19.6\,\text{mm}^2$

Stress in steel after losses $= \gamma_p \times 1120 \times 0.7 = 0.9 \times 1120 \times 0.7 = 705\,\text{N/mm}^2$

therefore

Strain in steel after losses $= \dfrac{f_s}{E_s} = \dfrac{705}{205 \times 10^3} = 0.0034$

which is less than ε_y, the yield strain.

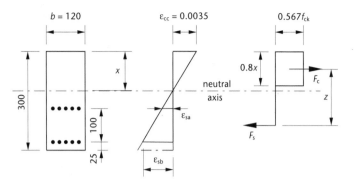

Figure 11.22
Ultimate moment of resistance example

Section   Bending Strains   Stress Block

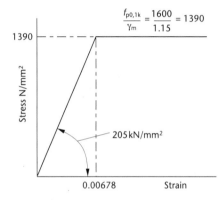

$$\frac{f_{p0,1k}}{\gamma_m} = \frac{1600}{1.15} = 1390$$

Figure 11.23
Stress–strain curve for prestressing wire

A depth x of neutral axis must be found for which the compressive force F_c in the concrete is balanced by the tensile force F_s in the steel. Then the ultimate moment of resistance is given by

$$M_u = F_c z = F_s z \tag{11.33}$$

where z is the lever arm between F_c and F_s.

As a first attempt try $x = 130$ mm, approximately equal to $0.5d$.

(a) Steel strains

Final steel strain, ε_s = prestress strain + bending strain, ε_s'

(In calculating ε_s' the initial concrete strain due to prestress can be ignored without undue error).

Top layer

$$\varepsilon_{sa} = 0.0034 + \varepsilon_{sa}'$$

therefore

$$\varepsilon_{sa} = 0.0034 + +\frac{(175 - x)}{x}\varepsilon_{cc} \tag{11.34}$$

$$= 0.0034 + \frac{(175 - 130)}{130}0.0035$$

$$= 0.0046$$

Bottom layer

$$\varepsilon_{sb} = 0.0034 + \varepsilon'_{sb}$$

$$= 0.0034 + \frac{(275 - x)}{x}\varepsilon_{cc} \tag{11.35}$$

$$= 0.0034 + \frac{(275 - 130)}{130}0.0035$$

$$= 0.0073$$

(b) Steel stresses

From the stress–strain curve the corresponding steel stresses are:

Top layer

$$f_{sa} = \varepsilon_{sa} \times E_s \tag{11.36}$$

$$= 0.0046 \times 205 \times 10^3$$

$$= 943 \text{ N/mm}^2$$

and

$$f_{sb} = 1390 \text{ N/mm}^2$$

as the strain in the bottom steel exceeds the yield strain ($\varepsilon_y = 0.00678$).

(c) Forces in steel and concrete

Steel tensile force $F_s = \sum f_s A_s = (f_{sa} + f_{sb})5 \times 19.6$ (11.37)

$$= (943 + 1390) \times 98$$

$$= 229 \times 10^3 \text{ N}$$

With a rectangular stress block

Concrete compressive force $F_c = 0.567 f_{ck} b \times 0.8x$ (11.38)

$$= 0.567 \times 35 \times 120 \times 0.8 \times 130$$

$$= 248 \times 10^3 \text{ N}$$

The force F_c in the concrete is larger than the force F_s in the steel, therefore a smaller depth of neutral axis must be tried.

Table 11.1 shows the results of calculations for further trial depths of neutral axis. For $x = 110$, F_c became smaller than F_s, therefore $x = 120$ and 123 were tried and it was then found that $F_s = F_c$.

Table 11.1

x (mm)	Strains		Stresses		Forces	
	ε_{sa}	ε_{sb}	f_{sa}	f_{sb}	F_s	F_c
	($\times 10^3$)		(N/mm^2)		(kN)	
130	4.6	7.3	943	1390	229	248
110	5.5	8.6	1121	1390	246	210
120	5.0	7.9	1026	1390	237	229
123	4.9	7.7	1000	1390	234	234

In terms of the tensile force in the steel, the ultimate moment of resistance of the section is given by

$$M_u = F_s z = \sum [f_s A_s(d - 0.4x)] \qquad (11.39)$$
$$= 5 \times 19.6[1000(175 - 0.4 \times 123) + 1390(275 - 0.4 \times 123)]$$
$$= 43.1 \times 10^6 \, \text{N mm}$$

If x had been incorrectly chosen as 130 mm then using equation 11.39 M_u would equal 42.0 kN m, or in terms of the concrete

$$M_u = 0.567 f_{ck} b \times 0.8xz$$
$$\approx 0.567 \times 35 \times 120 \times 0.8 \times 130(225 - 0.4 \times 130) \times 10^{-6}$$
$$\approx 43 \, \text{kN m}$$

Comparing the average of these two values of M_u (= 42.5 kN m) with the correct answer, it can be seen that a slight error in the position of the neutral axis does not have any significant effect on the calculated moment of resistance.

11.5.2 Design of additional reinforcement

If it is found that the ultimate limit state requirements are not met, additional untensioned or partially tensioned steel may be added to increase the ultimate moment of resistance.

EXAMPLE 11.11

Design of untensioned reinforcement

Design untensioned high yield reinforcement ($f_{yk} = 500 \, \text{N/mm}^2$) for the rectangular beam section shown in figure 11.24 which is stressed by five 5 mm wires, if the ultimate moment of resistance is to exceed 40 kN m for class 40/50 concrete. The characteristic strength of tensioned steel, $f_{po, 1k} = 1600 \, \text{N/mm}^2$.

(a) Check ultimate moment of resistance

Maximum tensile force if prestressing steel yielded

$$= 0.9 \times \left[5 \times 19.6 \times \frac{1600}{1.15} \right] \times 10^{-3} = 123 \, \text{kN}$$

Concrete compressive area to balance $= \dfrac{123 \times 10^3}{0.567 \times 40} = 0.8 \times 120x$

thus neutral-axis depth $x = 56$ mm.

Assuming prestrain as calculated in example 11.10

total steel strain = prestrain + bending strain

$$= 0.0034 + \frac{(d - x)}{x} \times 0.0035$$

$$= 0.0034 + \frac{219}{56} \times 0.0035 = 0.0171 \quad (> \text{yield})$$

Lever arm $= 275 - 0.40 \times 56 = 253$ mm

Figure 11.24
Ultimate moment of
resistance example

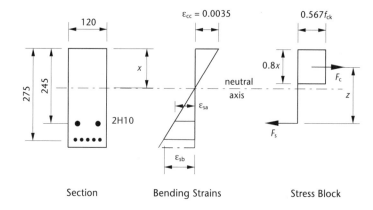

hence

ultimate moment of resistance $= 253 \times 123 \times 10^{-3} = 31.1 \, \text{kN m}$

Untensioned steel is therefore required to permit the beam to support an ultimate moment of 40 kNm.

Additional moment capacity to be provided $= 40 - 31.1 = 8.9 \, \text{kN m}$

Effective depth of additional steel $= 245 \, \text{mm}$

then

lever arm to additional steel $\approx 220 \, \text{mm}$

and

additional tension force required $= \dfrac{8900}{220} = 40.5 \, \text{kN}$

thus

estimated area of untensioned steel required at its yield stress

$$= \frac{40\,500}{0.87 \times 500} = 93 \, \text{mm}^2$$

Try two 10 mm diameter bars ($157 \, \text{mm}^2$).

(b) Check steel strain

If additional steel has yielded, force in two H10 bars $= 157 \times 500 \times 10^{-3}/1.15$ $= 68.3 \, \text{kN}$, therefore

total tensile force if all the steel has yielded $= 123 + 68.3$
$$= 191.3 \, \text{kN}$$

thus

depth of neutral axis at ultimate $= \dfrac{191.3 \times 10^3}{0.567 \times 40 \times 120 \times 0.8}$
$$= 88 \, \text{mm}$$

Therefore

prestressing steel strain $\varepsilon_{sb} = \dfrac{275 - 88}{88} \times 0.0035 + 0.0034$
$$= 0.0108 \quad (> \text{yield})$$

and

$$\text{untensioned steel strain } \varepsilon_{sa} = \frac{245 - 88}{88} \times 0.0035$$
$$= 0.0062$$

This value is greater than the yield strain of 0.00217 from section 4.1.2.

(c) Check ultimate moment of resistance

Taking moments about the centre of compression

$$M_u = 123(275 - 0.40x) + 68.3(245 - 0.40x)$$
$$= [123(275 - 0.40 \times 88) + 68.3(245 - 0.40 \times 88)]10^{-3}$$
$$= 43.8 \, \text{kN m}$$

If it had been found in (b) that either the prestressing steel or untensioned steel had not yielded, then a trial and error approach similar to example 11.10 would have been necessary.

11.5.3 Shear

Shear in prestressed concrete is considered at the ultimate limit state. Design for shear therefore involves the most severe loading conditions, with the usual partial factors of safety being applied to the actions for the ultimate limit state being considered.

The response of a member in resisting shear is similar to that for reinforced concrete, but with the additional effects of the compression due to the prestressing force. This will increase the shear resistance considerably and this is taken into account in EC2 by enhancing the equation for the shear capacity ($V_{Rd,c}$) of the section without shear reinforcement. With a few slight modifications, the Code gives an almost identical approach, based on the *Variable Strut Inclination Method* of shear design, in prestressed sections as is used in reinforced concrete sections as outlined in Chapter 5.

In calculating the design shear force, V_{Ed}, it is permissible to take into account the vertical component of force in any inclined tendons which will tend to act in a direction that resists shear, thus enhancing the shear capacity of the section. In such a case the prestressing force should be multiplied by the partial factor of safety, $\gamma_p = 0.9$.

Sections that do not require designed shear reinforcement

In regions of prestressed beams where shear forces are small and, taking into account any beneficial effect of forces attributable to inclined prestressing tendons, the concrete section on its own may have sufficient shear capacity ($V_{Rd,c}$) to resist the ultimate shear force (V_{Ed}). Notwithstanding this it is usual to provide a minimum amount of shear links unless the beam is a minor member such as a short-span, lightly loaded lintel.

The concrete shear strength ($V_{Rd,c}$) is given by the empirical expression:

$$V_{Rd,c} = \left[0.12k(100\rho_1 f_{ck})^{1/3} + 0.15\sigma_{cp}\right]b_w d \qquad (11.40)$$

with a minimum value of:

$$V_{Rd,c} = \left[0.035k^{3/2}f_{ck}^{1/2} + 0.15\sigma_{cp}\right]b_w d \qquad (11.41)$$

where:

$V_{Rd,c}$ = the design shear resistance of the section without shear reinforcement

$$k = \left(1 + \sqrt{\frac{200}{d}}\right) \leq 2.0 \text{ with } d \text{ expressed in mm}$$

$$\rho_1 = \frac{A_{sl}}{b_w d} \leq 0.02$$

A_{sl} = the area of tensile reinforcement that extends beyond the section being considered by at least a full anchorage length plus one effective depth (d)

b_w = the smallest width of the section in the tensile area (mm)

σ_{cp} = axial stress in section due to prestress ($\gamma_p K P_0 / A$) ($< 0.133 f_{ck}$)

It can be seen that equations 11.40 and 11.41 are practically identical to equations 5.1 and 5.2 for shear in reinforced concrete sections. The additional term of $0.15\sigma_{cp}$ indicates that the effect of the prestress is to enhance the shear capacity of the section by 15% of the longitudinal stress due to prestressing.

Shear strength without shear reinforcement – regions uncracked in bending (special case)

For the special case of a *single span beam*, in regions which are uncracked in bending (i.e where sagging moments are relatively small near to the supports), the shear strength of the concrete section could be governed by the development of excessive tensile stresses in the concrete. These regions are defined as where the flexural tensile stress in the uncracked section does not exceed f_{ctk}/γ_c, where f_{ctk} is the characteristic axial tensile strength of the concrete. The applicable equations in EC2 can be developed as follows.

At an uncracked section, a Mohr's circle analysis of a beam element shown in figure 11.25 which is subject to a longitudinal compressive stress, f_c and a shear stress v_{co} gives the principal tensile stress as:

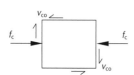

Figure 11.25
Stress in uncracked section

$$f_t = \sqrt{\left[\left(\frac{f_c}{2}\right)^2 + v_{co}^2\right]} - \left(\frac{f_c}{2}\right)$$

This can be re-arranged to give the shear stress

$$v_{co} = \sqrt{(f_t^2 + f_c f_t)}$$

The actual shear stress at any level of a beam subject to a shear force, V, can be shown to be:

$$v_{co} = \frac{V(A\bar{y})}{bI} \quad \text{or} \quad V = \frac{v_{co} b I}{(A\bar{y})}$$

where $A\bar{y}$ is the first moment of area of the part of the section above the level considered about the centroidal axis of the beam, as shown in figure 11.26, b is the breadth of the section at the level considered and I is the second moment of area of the whole section about its centroidal axis.

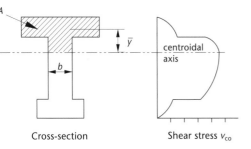

Figure 11.26
Shear stress distribution

Cross-section

Shear stress v_{co}

Hence if f_{ctd} is the limiting value of principal tensile stress, the ultimate shear resistance $V_{Rd,c}$ of the uncracked section becomes:

$$V_{Rd,c} = \frac{bI}{A\bar{y}}\sqrt{(f_{ctd}^2 + f_c f_{ctd})}$$

This equation forms the basis of the design equation given in EC2 which is expressed as:

$$V_{Rd,c} = \frac{b_w I}{A\bar{y}}\sqrt{(f_{ctd}^2 + \alpha_1 \sigma_{cp} f_{ctd})} \qquad (11.42)$$

where:

σ_{cp} = axial stress in section due to prestress $(\gamma_p K P_0/A)$

f_{ctd} = the design tensile strength of the concrete $(= f_{ctk}/\gamma_c)$

α_1 = 1 for post-tensioned tendons;

$\leq$ 1 for pretensioned tendons, and in this case the value of α_1 is given in EC2 according to the distance of the section being considered in relation to the transmission length of the tendon.

EC2 states that, for the special case of a simply supported beam, equation 11.42 should be used in those regions where the flexural tensile stress in the uncracked section does not exceed f_{ctk}/γ_c and where the beam is cracked in bending equation 11.40 should be used. Determining where the beam is uncracked at the ultimate limit state is not straight-forward and, in practice, both these equations should be applied at each section considered and the lowest of the two values calculated then taken as the shear capacity of the section.

The variable strut inclination method for sections that do require shear reinforcement

As previously noted the design for shear and the provision of shear reinforcement in prestressed concrete is practically identical to that for reinforced concrete and is summarised below.

(1) The diagonal compressive strut and the angle θ

The maximum design shear force that a section can carry $(V_{Rd,max})$ is governed by the requirement that excessive compressive stresses should not occur in the diagonal compressive struts of the assumed truss, leading possibly to compressive failure of the concrete. The maximum shear force is given by:

$$V_{Rd,max} = \frac{\alpha_{cw}b_w z v_1 f_{ck}}{[1.5(\cot\theta + \tan\theta)]}$$

where $z = 0.9d$ and $v_1 = 0.6(1 - f_{ck}/250)$. Hence:

$$\begin{aligned} V_{Rd,max} &\leq \frac{\alpha_{cw}b_w 0.9 d 0.6(1 - f_{ck}/250)f_{ck}}{[1.5(\cot\theta + \tan\theta)]} \\ &\leq \frac{\alpha_{cw}0.36 b_w d(1 - f_{ck}/250)f_{ck}}{[\cot\theta + \tan\theta]} \end{aligned} \tag{11.43}$$

This equation is practically identical to equation 5.4 in Chapter 5 except that it includes a coefficient α_{cw} given by:

$$\begin{aligned} \alpha_{cw} &= 1 + 1.5\sigma_{cp}/f_{ck} && \text{for } 0 < \sigma_{cp} \leq 0.167 f_{ck} \\ \alpha_{cw} &= 1.25 && \text{for } 0.167 f_{ck} < \sigma_{cp} \leq 0.333 f_{ck} \\ \alpha_{cw} &= 2.5(1 - 1.5\sigma_{cp}/f_{ck}) && \text{for } 0.333 f_{ck} < \sigma_{cp} < 0.667 f_{ck} \end{aligned}$$

where σ_{cp} = the mean compressive stress, taken as positive, in the concrete due to the prestress force.

For the two limiting values of $\cot\theta$ comparison with equations 5.6 and 5.7 gives:

with $\cot\theta = 2.5$: $\qquad V_{Rd,max(22)} = \alpha_{cw}0.124 b_w d(1 - f_{ck}/250)f_{ck}$ $\qquad$ (11.44)

and with $\cot\theta = 1.0$: $\qquad V_{Rd,max(45)} = \alpha_{cw}0.18 b_w d(1 - f_{ck}/250)f_{ck}$ $\qquad$ (11.45)

and for values of θ that lie between these two limiting values the required value of θ can be obtained by equating V_{Ed} to $V_{Rd,max}$. Thus the equation, analogous to equation 5.8, for the calculation of θ is as follows:

$$\theta = 0.5\sin^{-1}\left\{ \frac{V_{Ed}}{\alpha_{cw}0.18 b_w d(1 - f_{ck}/250)f_{ck}} \right\} \leq 45° \tag{11.46a*}$$

which alternatively can be expressed as:

$$\theta = 0.5\sin^{-1}\left\{ \frac{V_{Ef}}{V_{Rd,max(45)}} \right\} \leq 45° \tag{11.46b}$$

where V_{Ef} is the shear force at the section being considered and the calculated value of the angle θ can then be used to determine $\cot\theta$ and to calculate the shear reinforcement A_{sw}/s at that section from equation 11.47 below (when $22° < \theta < 45°$).

If the web of the section contains grouted ducts with diameter greater than one-eighth of the web thickness, in the calculation of $V_{Rd,max}$, the web thickness should be reduced by one-half of the sum of the duct diameters measured at the most unfavourable section of the web. For non-grouted ducts, grouted plastic ducts and unbonded tendons the web thickness should be reduced by 1.2 times the sum of the duct diameters. If the design shear force exceeds $V_{Rd,max}$ then it will be necessary to increase the size of the section.

(2) The vertical shear reinforcement

As in reinforced concrete, shear reinforcement must be provided to resist the shear force if it can not be sustained by the concrete section including the enhanced shear resistance due to the effect of prestress. If vertical shear reinforcement is provided then it is assumed that all the shear force is carried by this reinforcement and the contribution of the concrete is neglected. Equation 5.9, as derived in chapter 5 can be used to determine the amount and spacing of shear reinforcement and is reproduced below:

$$\frac{A_{sw}}{s} = \frac{V_{Ed}}{0.78 df_{yk}\cot\theta} \qquad \left(> \frac{A_{sw,min}}{s} = \frac{0.08f_{ck}^{1/2}b_w}{f_{yk}} \right) \tag{11.47}$$

For the limiting values of $\cot\theta$:

$$\text{with } \cot\theta = 2.5: \quad \frac{A_{sw}}{s} = \frac{V_{ed}}{0.78df_{yk}\cot\theta} = \frac{V_{Ed}}{0.78df_{yk}2.5} = \frac{V_{Ed}}{1.95df_{yk}} \tag{11.48}$$

$$\text{with } \cot\theta = 1.0: \quad \frac{A_{sw}}{s} = \frac{V_{ed}}{0.78df_{yk}\cot\theta} = \frac{V_{Ed}}{0.78df_{yk}1.0} = \frac{V_{Ed}}{0.78df_{yk}} \tag{11.49}$$

(3) Additional longitudinal force

When using this method of shear design it is necessary to provide for the tensile force generated in the tension chord of the assumed truss. For vertical links the tensile force (ΔF_{td}) to be provided in the tensile zone is given by:

$$\Delta F_{td} = 0.5V_{Ed}\cot\theta = 0.87f_{yk}A_{sl} \tag{11.50}*$$

To provide for this force untensioned longitudinal reinforcement of area A_{sl}, working at its full design strength, must be provided to ensure that this force is developed and/or the profile of the prestressing tendons may be adjusted to provide greater resistance to the design shear force.

Summary of the design procedure with vertical links

1. Calculate the ultimate design shear forces V_{Ed} along the beam's span. Then, at set intervals along the beam's span, commencing at the section of maximum shear at the supports, follow the steps below. The section of maximum shear can be taken to be at the face of the support where the maximum shear force is V_{Ef}.

2. Check using equations 11.40 and 11.41 and, if relevant, equation 11.42 that shear reinforcement is required. If appropriate, the vertical component of the force in any inclined tendons should be taken into account in determining the shear resistance. If it is required continue as follows otherwise go directly to step 5.

3. Check the crushing strength $V_{Rd,max}$ of the concrete diagonal strut. For most cases the angle of inclination of the strut is $\theta = 22°$, with $\cot\theta = 2.5$ and $\tan\theta = 0.4$ so that from equation 11.43

$$V_{Rd,max} = \frac{\alpha_{cw}0.36b_w d(1 - f_{ck}/250)f_{ck}}{(\cot\theta + \tan\theta)}$$

and if $V_{Rd,max} \geq V_{Ef}$ with $\theta = 22°$ and $\cot\theta = 2.5$ then go directly to step 4, but if $V_{Rd,max} < V_{Ef}$ then $\theta > 22°$ and therefore θ must be calculated from equation 11.46 as:

$$\theta = 0.5\sin^{-1}\left\{\frac{V_{Ef}}{\alpha_{cs}0.18b_w d(1 - f_{ck}/250)f_{ck}}\right\} \leq 45°$$

If this calculation gives a value of θ greater than 45° then the beam should be re-sized or a higher class of concrete could be used.

4. The shear links required can be calculated from equation 11.47

$$\frac{A_{sw}}{s} = \frac{V_{Ed}}{0.78 d f_{yk} \cot \theta}$$

where A_{sw} is the cross-sectional area of the legs of the links ($2 \times \pi \phi^2 / 4$ for single stirrups).

For a predominately uniformly distributed load the maximum shear V_{Ed} can be calculated at a distance d from the face of the support and the shear reinforcement should continue to the face of the support.

The shear resistance for the links actually specified is

$$V_{min} = \frac{A_{sw}}{s} \times 0.78 d f_{yk} \cot \theta$$

and this value will be used together with the shear force envelope to determine the curtailment position of each set of designed links.

5. Calculate the minimum links required by EC2 from

$$\frac{A_{sw,\,min}}{s} = \frac{0.08 f_{ck}^{0.5} b_w}{f_{yk}}$$

6. Calculate the additional longitudinal tensile force caused by the shear

$$\Delta F_{td} = 0.5 V_{Ed} \cot \theta$$

The above procedure should be repeated at different sections along the beam, as illustrated in the following example

EXAMPLE 11.12

Design of shear reinforcement

The beam cross-section shown in figure 11.27 is constant over a 30 m simply supported span with a parabolic tendon profile and an eccentricity varying between 300 mm at the ends and 750 mm at mid-span, measured below the neutral axis in both cases. The beam supports an ultimate uniformly distributed load of 40 kN/m and $f_{ck} = 35$ N/mm^2.

Figure 11.27
Shear reinforcement example

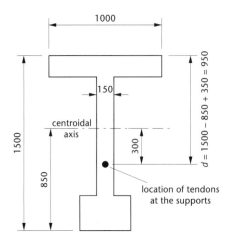

Given data:

Prestress force after losses $= 2590\,\text{kN}$

$I = 145\,106 \times 10^6\,\text{mm}^4$

$A = 500 \times 10^3\,\text{mm}^2$

$A_p = 3450\,\text{mm}^2$

$f_{yk} = 500\,\text{N/mm}^2$ for the shear links

$f_{ctk} = 2.2\,\text{N/mm}^2$

The calculations will be presented for a section at the support and then repeated and tabulated at 3 m intervals along the span.

(1) Calculate shear force at the section

Although the maximum shear force can be taken at the face of the support, in this example we will, for illustrative purposes, take the section at the middle of the support itself. Hence:

$$V_{Ed} = 40 \times 30/2 = 600\,\text{kN}$$

(2) Check if shear reinforcement is required

From equation 11.40 the concrete shear strength is given by:

$$V_{Rd,c} = \left[0.12k(100\rho_1 f_{ck})^{1/3} + 0.15\sigma_{cp}\right]b_w d$$

where:

$$d = 1.5 - 0.85 + e = 1.5 - 0.85 + 0.3 = 0.95\,\text{m at the support}$$

$$k = \left(1 + \sqrt{\frac{200}{d}}\right) = \left(1 + \sqrt{\frac{200}{950}}\right) = 1.46 \quad (\leq 2.0)$$

$$\rho_1 = \frac{A_{s1}}{b_w d} = \frac{3450}{150 \times 950} = 0.0242 \quad (> 0.02) \therefore \rho_1 = 0.02$$

$$\sigma_{cp} = \gamma_{cp} K P_0 / A = 0.9 \times 2590 \times 10^3 / (500 \times 10^3)$$

$$= 4.66\,\text{N/mm}^2 \quad \{\leq 0.133 f_{ck} = 0.133 \times 35 = 4.66\,\text{OK}\}$$

Hence:

$$V_{Rd,c} = \left[0.12k(100\rho_1 f_{ck})^{1/3} + 0.15\sigma_{cp}\right]b_w d$$

$$= \left[0.12 \times 1.46(100 \times 0.02 \times 35)^{1/3} + 0.15 \times 4.66\right]150 \times 950 \times 10^{-3}$$

$$= 202\,\text{kN}$$

Note: a check on equation 11.41 will show that the minimum value of $V_{Rd,c}$ as given by equation 11.41 is not critical in this case.

As this is a simply supported beam equation 11.42 should also be used to check the shear capacity of the concrete section. From equation 11.42:

$$V_{Rd,c} = \frac{b_w I}{A\bar{y}}\sqrt{(f_{ctd}^2 + \alpha_1 \sigma_{cp} f_{ctd})}$$

where:

σ_{cp} = axial stress in section due to prestress = 4.66 N/mm^2, as before

f_{ctd} = the design tensile strength of the concrete = 2.2/1.5 = 1.47 N/mm^2

α_1 = 1 for post-tensioned tendons.

Hence by reference to the dimensions shown in figure 11.25:

$$V_{Rd,c} = \frac{b_w I}{A\bar{y}} \sqrt{(f_{ctd}^2 + \alpha_1 \sigma_{cp} f_{ctd})}$$

$$= \frac{150 \times 145\,106 \times 10^6}{[(1000 \times 175 \times 562.5) + (150 \times 475 \times 237.5)]} \sqrt{(1.47^2 + 1 \times 4.66 \times 1.47)} \times 10^{-3}$$

$$= 566\,kN$$

This is considerably greater than the figure of 202 kN calculated from equation 11.40. We will take the lower value of 202 kN as representing the shear capacity of the concrete section

The effective resistance of the section is the sum of the shear resistance of the concrete, $V_{Rd,c}$, plus that of the vertical shear resistance of the inclined tendons.

Shear strength including the shear resistance of the inclined tendons

The vertical component of the prestress force is $P \sin \beta$ where β = tendon slope. The tendon profile is $y = Cx^2$ with the origin of the cable profile taken at mid-span; hence at $x = 15000$, $y = 750 - 300 = 450$ and

$$450 = C \times 15\,000^2$$

$$C = 2.0 \times 10^{-6}$$

Therefore the tendon profile is $y = 2.0 \times 10^{-6}x^2$ and tendon slope = $dy/dx = 2Cx$.

At end

$$dy/dx = 2 \times 2.0 \times 10^{-6} \times 15\,000 = 0.060 = \tan \beta$$

Hence,

$$\beta = 3.43° \quad \text{and} \quad \sin \beta \approx \tan \beta = 0.06$$

Therefore vertical component, V_t, of prestress force at the supports is:

$$V_t = 2590 \sin \beta = 2590 \times 0.06 = 155\,kN$$

and the total shear capacity is:

$$V_{Rd,c} + V_t = 202 + \gamma_p \times 155 = 202 + 0.9 \times 155$$

$$= 342\,kN \text{ at the supports.}$$

At the end of the beam the design shear force is $(40 \times 30/2) = 600\,kN$ and hence the shear capacity of the concrete section is inadequate and shear reinforcement must be provided.

(3) Check the crushing strength $V_{Rd,max}$ of the concrete diagonal strut

A check must be made to ensure that the shear force does not cause excessive compression to develop in the diagonal struts of the assumed truss.

From equation 11.44 ($\cot \theta = 2.5$):

$$V_{Rd,max(22)} = \alpha_{cw} 0.124 b_w d (1 - f_{ck}/250) f_{ck}$$

where the value of α_{cw} depends on the magnitude of σ_{cp} given by:

$$\sigma_{cp} = \gamma_p K P_0 / A = 0.9 \times 2590 \times 10^3 / (500 \times 10^3) = 4.66\,N/mm^2$$

Hence:

$$\frac{\sigma_{cp}}{f_{ck}} = \frac{4.66}{35} = 0.133 \ (< 0.167)$$

$$\therefore \alpha_{cw} = 1 + 1.5 \frac{\sigma_{cp}}{f_{ck}} = 1 + 1.5 \times 0.133 = 1.200$$

$$\therefore V_{Rd, max(22)} = \alpha_{cw} 0.124 b_w d (1 - f_{ck}/250) f_{ck}$$

$$= 1.200 \times 0.124 \times 150 \times 950 \left(1 - \frac{35}{250}\right) \times 35 \times 10^{-3} = 638 \, kN$$

As the shear force at the end of the beam is 600 kN then the upper limit to the shear force is not exceeded.

(4) Calculate the area and spacing of links

Where the shear force exceeds the capacity of the concrete section, allowing for the enhancement from the inclined tendon force, shear reinforcement must be provide to resist the net shear force taking into account the beneficial effect of the inclined tendons. From equation 11.48 this is given by:

$$\frac{A_{sw}}{s} = \frac{V_{Ed}}{1.95 df_{yk}} = \frac{(600 - 0.9 \times 155) \times 10^3}{1.95 \times 950 \times 500} = 0.497$$

(5) Calculate the minimum link requirement

$$\frac{A_{sw, min}}{s} = \frac{0.08 f_{ck}^{1/2} b_w}{f_{yk}} = \frac{0.08 \times 35^{1/2} \times 150}{500} = 0.14$$

Therefore provide 10 mm links at 300 mm centres ($A_{sw}/s = 0.523$) such that the shear resistance of the links actually specified is:

$$V_{min} = \frac{A_{sw}}{s} \times 0.78 df_{yk} \cot \theta$$

$$= 0.523 \times 0.78 \times 950 \times 500 \times 2.5 \times 10^{-3} = 484 \, kN$$

(6) Calculate the additional longitudinal force

The additional longitudinal tensile force is:

$$\Delta F_{td} = 0.5 V_{Ed} \cot \theta = 0.5 \times (600 - 0.9 \times 155) \times 2.5 = 575 \, kN$$

Hence:

$$A_{sl} = \frac{\Delta F_{td}}{0.87 f_{yk}} = \frac{575 \times 10^3}{0.87 \times 500} = 1322 \, mm^2$$

This additional longitudinal steel can be provided for by four untensioned H25 bars (1960 mm^2) located at the bottom of the beam's cross-section and fully anchored past the point required using hooks and bends as necessary. Untensioned longitudinal reinforcement must be provided at every cross-section to resist the longitudinal tensile force due to shear and the above calculation must be repeated at each section to determine the longitudinal steel requirement.

All of the above calculations can be repeated at other cross-sections and are tabulated in table 11.2 from which it can be seen that, from mid-span to a section approximately 9 m from mid-span, nominal shear reinforcement is required and in the outer 6 m of the span fully designed shear reinforcement is required. This can be provided as

Table 11.2 Shear calculations at 3 m intervals

	x (m)	D (mm)	$V_{Rd,c}$ (kN)[1]	Prestress $\gamma_p \times V_t$ (kN)	(1) V_{Ed} (kN)	(2) $V_{Rd,c} + \gamma_p V_t$ (kN)	(3) $V_{Rd,max}$ (kN)[2]		(4) A_{sw}/s	(5) $A_{sw}/s_{(min)}$	(6) ΔF_{td} (kN)
Mid-span	0	1400	281	0	0	281	941	Minimum		0.14	0
	3	1382	278	28	120	306	928	reinforcement		0.14	115
	6	1328	270	56	240	325	892	only		0.14	230
	9	1238	255	84	360	339	832	Reinforcement	0.229		345
	12	1112	234	112	480	346	747	carries all the	0.339		460
End-span	15	950	202	140	600	342	638	shear force	0.497		575

1 Equation 11.40.
2 Equation 11.44.

Figure 11.28
Shear resistance diagram

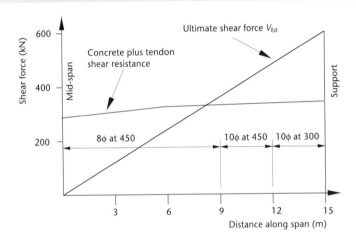

(figure 11.28) 10 mm links at 300 mm centres in the outer 3 metres ($A_{sw}/s = 0.523$) changing to 10 mm links at 450 mm centres ($A_{sw}/s = 0.349$) between 3 and 6 m from the end of the beam and then 8 mm at 450 centres ($A_{sw}/s = 0.223$) throughout the rest of the span.

CHAPTER 12

Composite construction

CHAPTER INTRODUCTION

Many buildings are constructed with a steel framework composed of steel beams and steel columns but mostly with a concrete floor slab. A much stiffer and stronger structure can be achieved by ensuring that the steel beams and concrete slabs act together as composite and so, effectively, monolithic units. This composite behaviour is obtained by providing shear connections at the interface between the steel beam and the concrete slab as shown in figure 12.1. These shear connections resist the horizontal shear at the interface and prevent slippage between the beam and the slab. The shear connectors are usually in the form of steel studs welded to the top flange of the beam and embedded in the concrete slab.

The steel beam will usually be a universal I-beam. Other alternatives are a castellated beam or a lattice girder as shown in figure 12.2. These alternative types of beam provide greater depth for the floor system and openings for the passage of service conduits such as for heating and air conditioning systems.

Two other types of composite floor system are shown in figure 12.3. The stub girder system consists of a main beam with transverse secondary beams supported on the top flange. Short lengths of stub members similar in section to the secondary beams are also connected to the top flange of the main beam. The stub beams and the secondary beams are connected to the slab with steel studs as shown.

→

Figure 12.1
Composite beam sections

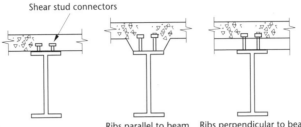

Shear stud connectors

Ribs parallel to beam   Ribs perpendicular to beam

Composite beam with ribbed slab

Figure 12.2
Composite floor beams

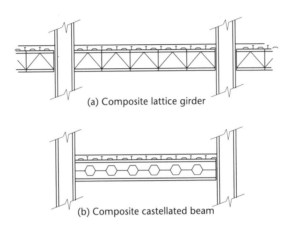

(a) Composite lattice girder

(b) Composite castellated beam

Figure 12.3
Composite floor systems

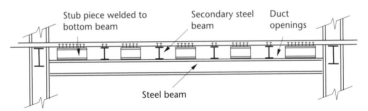

Stub piece welded to bottom beam   Secondary steel beam   Duct openings

Steel beam

(a) Typical sub-girder system

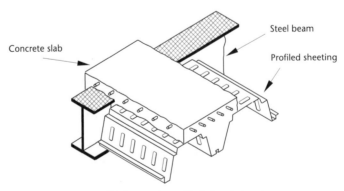

Concrete slab

Steel beam

Profiled sheeting

(b) Composite profiled slim deck system

$\longrightarrow$

The Slimdec system shown in Figure 12.3 is manufactured by Corus. The special steel beams have a patterned tread on the top flange that provides an enhanced bond with the concrete slab so that a composite action can be developed without the use of shear studs. Deep ribbed profiled sheeting is used to support the slab with the deep ribs resting on the bottom flange of the beam. With this arrangement the steel beam is partially encased by the concrete which provides it with better fire resistance. Openings for services can be cut in the web of the beam between the concrete ribs.

The concrete slab itself can also be constructed as a composite member using the profiled steel decking on the soffit of the slab as shown in figure 12.4. The steel decking acts as the tension reinforcement for the slab and also as permanent shuttering capable of supporting the weight of the wet concrete. It is fabricated with ribs and slots to form a key and bond with the concrete. Properties of the steel decking and safe load tables for the decking and the composite floors are obtainable from the manufacturing companies.

Many composite beams are designed as simply supported non-continuous beams. Beams that are continuous require moment resisting connections at the columns and additional reinforcing bars in the slab over the support.

The method of construction may be either:

- Propped
- Unpropped

With propped construction temporary props are placed under the steel beam during construction of the floor and the props carry all the construction loads. After the concrete has hardened the props are removed and then the loads are supported by the composite beam. The use of temporary props has the disadvantage of the lack of clear space under the floor during construction and the extra cost of longer construction times.

Unpropped construction requires that the steel beam itself must support the construction loads and the steel beam has to be designed for this condition, which may govern the size of beam required. The beam can only act as a composite section when the concrete in the slab has hardened. This also means that the deflection at service is greater than that of a propped beam as the final deflection is the sum of the deflection of the steel beam during construction plus the deflection of the composite section due to the additional loading that takes place after construction. The calculations for this are shown in example 12.4 which sets out the serviceability checks for an unpropped beam.

As there are differences in the design procedures for these two types of construction it is important that the construction method should be established at the outset.

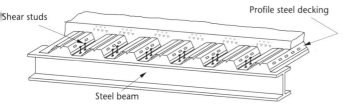

Shear studs

Profile steel decking

Steel beam

Figure 12.4
Composite slab with steel decking

12.1 The design procedure

The design procedure for composite beams follows the requirements of:

(a) EC2, (EN1992-1-1) for the design of concrete structures,

(b) EC3 (EN 1993-1-1) for the design of steel structures, and

(c) EC4 (EN1994-1-1) for the design of composite steel and concrete structures.

At the time of writing this chapter the UK National Annex for EC3 and EC4, and the Concise Eurocodes are not available. Parts of these codes are quite complex; for example the list of symbols for the three codes extends to 21 pages. It is intended in this chapter to try and simplify many of the complications and enable the reader to gain a grasp of the basic principles of the design of composite beams.

12.1.1 Effective width of the concrete flange (EC4, cl 5.4.1.2)

An early step in the design of the composite beam section is to determine the effective breadth b_{eff} of the concrete flange.

For building structures at mid-span or an internal support

$$b_{eff} = \sum b_{ei}$$

where b_{ei} is the effective width of the concrete flange on each side of the steel web and is taken as $L_e/8$, but not greater than half the distance to the centre of the adjacent beam. The length L_e is the approximate distance between points of zero bending moment which can be taken as $L/2$ for the mid-span of a continuous beam, or L for a one-span simply supported beam. The length L is the span of the beam being considered.

For example, for a continuous beam with a span of $L = 16$ m and the adjacent beams being at 5 m centre to centre the effective breadth, b_{eff}, of the concrete flange is

$$b_{eff} = 2 \times L_e/8 = 2 \times 0.5 \times 16/8 = 2.0 \, \text{m}$$

If the beam was a one-span simply supported beam the effective breadth, b_{eff}, would be 4.0 m.

12.1.2 The principal stages in the design

These stages are listed with brief descriptions as follows:

(1) Preliminary sizing

The depth of a universal steel beam may be taken as approximately the span/20 for a simply supported span and the span/24 for a continuous beam. The yield strength, f_y, and the section classification of the steel beam should be determined.

(2) During construction (for unpropped construction only)

The loading is taken as the self-weight of the steel beam with any shuttering or steel decking, the weight of the wet concrete and an imposed construction load of at least $0.75 \, \text{kN/m}^2$. The following design checks are required:

(a) At the ultimate limit state

Check the strength of the steel section in bending and shear.

(b) At the serviceability limit state

Check the deflection of the steel beam.

(3) Bending and shear of the composite section at the ultimate limit state

Check the ultimate moment of resistance of the composite section and compare it with the ultimate design moment. Check the shear strength of the steel beam.

(4) Design of the shear connectors and the transverse steel at the ultimate limit state

The shear connecters are required to resist the horizontal shear at the interface of the steel and the concrete so that the steel beam and the concrete flange act as a composite unit. The shear connectors can be either a full shear connection or a partial shear connection depending on the design and detailing requirements.

Transverse reinforcement is required to resist the longitudinal shear in the concrete flange and to prevent cracking of the concrete in the region of the shear connectors.

(5) Bending and deflection at the serviceability limit state for the composite beam

The deflection of the beam is checked to ensure it is not excessive and so causing cracking of the architectural finishes.

12.2   Design of the steel beam for conditions during construction (for unpropped beams only)

The steel beam must be designed to support a dead load of its estimated self-weight, the weight of wet concrete and the weight of the profiled steel decking or the formwork, plus a construction live load of at least 0.75 kN/m^2 covering the floor area.

A preliminary depth for the sizing of the steel beam can be taken as the span/20 for a one-span simply supported beam.

(a) At the ultimate limit state

(i) Bending

The plastic section modulus $W_{pl, y}$, for the steel beam may be calculated from

$$W_{pl, y} = \frac{M_{Ed}}{f_y}$$

(12.1)

where

M_{Ed} is the ultimate design moment

f_y is the design strength of the steel as obtained from EC3, table 3.1

This assumes that the compression flange of the steel beam is adequately restrained against buckling by the steel decking for the slab and the steel section used can be classified as a plastic or compact section as defined in EC3, sections 5.5 and 5.6.

(ii) Shear

The shear is considered to be carried by the steel beam alone at the construction stage and also for the final composite beam.

The ultimate shear strength of a rolled I-beam is based on the following shear area, A_v, of the section

$$A_v = A_a - 2bt_f + (t_w + 2r)t_f \quad \text{but not less than} \quad \eta h_w t_w \tag{12.2}$$

where A_a is the cross-sectional area of the steel beam and h_w is the overall depth of the web. η can be taken as 1.0.

The other dimensions of the cross-section are defined in figure 12.5.

Figure 12.5
Dimensions for an
I-section beam

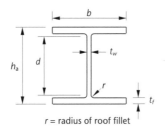

r = radius of roof fillet

For class 1 and class 2 I-beams with a predominately uniformly distributed load the design shear stresses are seldom excessive and the shear area, A_v may be conservatively taken as the web area so that

$$A_v = d \times t_w \tag{12.3}$$

where d is the depth of the straight portion of the web.

The design plastic shear resistance $V_{pl,Rd}$ of the section is given by:

$$V_{pl,Rd} = \frac{A_v f_y}{\gamma_{MO}\sqrt{3}} \tag{12.4}$$

where $\gamma_{MO} = 1.0$ is the material partial factor of safety for the steel.

(b) At the serviceability limit state

The deflection δ at mid-span for a uniformly distributed load on a steel beam is given by:

$$\delta = \frac{5wL^4}{384 E_a I_a} \tag{12.5}$$

where

    w is the serviceability load per metre at construction

    L is the beam's span

    E_a is the elastic modulus of the steel $= 210\,\text{kN/mm}^2$

    I_a is the second moment of area of the steel section

The deflections at the construction stage due to the permanent loads are locked into the beam as the concrete hardens.

EXAMPLE 12.1

Design of steel beam for construction loads

Figure 12.6 shows the section of an unpropped composite beam. Check the strength of the universal $457 \times 191 \times 74$ kg/m steel beam for the loading at construction. The steel is grade S355 with $f_y = 355$ N/mm^2 and the plastic modulus for the steel section is $W_{pl,y} = 1653$ cm^3. The one-span simply supported beam spans 9.0 metres and the width of loading on the concrete flange is 3.0 metres.

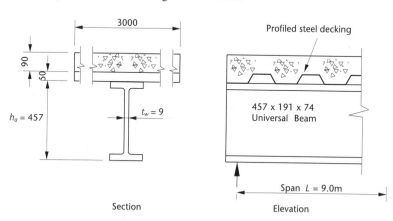

Figure 12.6
Construction design example

(a) Steel strength and classification of the steel beam (see EC3, tables 3,1 and 5.2)

The web thickness, $t_w = 9.0$ mm and the flange thickness, $t_f = 14.4$ mm, and both are less than 40 mm. Therefore from EC3, section 3.2, table 3.1 the yield strength, $f_y = 355$ N/mm^2.

From EC3, section 5.6, table 5.2

$$\varepsilon = \sqrt{\frac{235}{f_y}} = 0.81$$

$$\frac{c}{t} = \frac{d}{t_w} = \frac{407.6}{9.0} = 45.3 < 72 \times \varepsilon = 58.3$$

therefore the steel section is class 1.

(b) Loading at construction

Average depth of concrete slab and ribs $= 90 + 50/2 = 115$ mm

Weight of concrete	$= 0.115 \times 25 \times 3$	$= 8.62$ kN/m
Steel deck	$= 0.15 \times 3$	$= 0.45$
Steel beam	$= 74 \times 9.81 \times 10^{-3}$	$= 0.73$
Total dead load		$= 9.8$ kN/m

Imposed construction load $= 0.75 \times 3$ $\qquad = 2.25$ kN/m

Ultimate load $= 1.35 G_k + 1.5 Q_k = (1.35 \times 9.8 + 1.5 \times 2.25)$

$$= 16.6 \text{ kN/metre}$$

(c) Bending

Maximum bending moment $= wL^2/8 = (16.6 \times 9^2)/8 = 168\,\text{kN m}$

Moment of resistance of steel section $= W_{\text{pl, y}}f_y = 1653 \times 355 \times 10^{-3}$

$$= 587\,\text{kNm} > 168\,\text{kNm} \quad \text{OK}$$

(d) Shear

Maximum shear force $V = wL/2 = 16.6 \times 9/2 = 74.7\,\text{kN}$

Shear resistance of section $= V_{\text{pl, Rd}} = \dfrac{A_v f_y}{\gamma_{\text{MO}}\sqrt{3}}$

For the steel section the web depth, $d = 407.6\,\text{mm}$ and the web thickness $t = 9\,\text{mm}$. Using the conservative value of

$A_v = dt_w = 407.6 \times 9.0 = 3.67 \times 10^3\,\text{mm}^2$

Shear resistance of section $= V_{\text{pl, Rd}} = \dfrac{A_v f_y}{\gamma_{\text{MO}}\sqrt{3}} = \dfrac{3.67 \times 10^3 \times 355}{1.0 \times \sqrt{3}} \times 10^{-3}$

$$= 752\,\text{kN} > 74.7\,\text{kN}$$

From the calculations for bending and shear it can be seen the loading on the beam during construction is relatively low compared to the strength of the beam. Also, the steel decking with the corrugations at right angles to the span gives lateral and torsional restraint to the steel beam. For these reasons it is considered unnecessary to carry out the involved calculations for lateral and torsional stability which are described in EC3, Design of Steel Structures.

For the calculation of the deflection of the steel beam during construction at the serviceability limit state see Example 12.4.

12.3 The composite section at the ultimate limit state

At the ultimate limit state it is necessary to check the composite section for its moment capacity and its shear strength, and compare them against the maximum design ultimate moment and shear.

12.3.1 Moment capacity with full shear connection

The moment capacity M_c of the composite section is derived in terms of the tensile or compressive strengths of the various elements of the section as follows:

Resistance of the concrete flange	$R_{\text{cf}} = 0.567 f_{\text{ck}} b_{\text{eff}}(h - h_{\text{p}})$
Resistance of the steel section	$R_s = f_y A_a$
Resistance of the steel flange	$R_{\text{sf}} = f_y b t_f$
Resistance of overall web depth	$R_w = R_s - 2R_{\text{sf}}$
Resistance of clear web depth	$R_v = f_y d t_w$
Resistance of the concrete above the neutral axis	$R_{\text{cx}} = 0.567 f_{\text{ck}} b_{\text{eff}} x$
Resistance of the steel flange above the neutral axis	$R_{\text{sx}} = f_y b x_1$
Resistance of the web over distance x_2	$R_{\text{wx}} = f_y t_w x_2$

The dimensions used in these expressions are defined in figures 12.7 and 12.8.

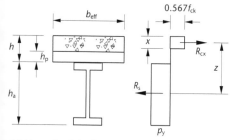

Figure 12.7
Composite section dimensions

Note: d is the distance between the fillets of the steel section

It is important to note in the figures that the stress block for the concrete extends to the depth of the neutral axis as specified in EC4 for composite design.

There are three possible locations of the neutral axis as shown in figure 12.8. These are:

(a) The neutral axis in the concrete flange;
(b) The neutral axis in the steel flange;
(c) The neutral axis in the steel web.

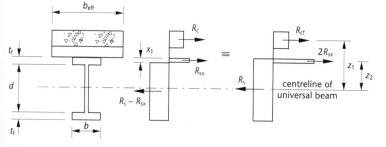

Figure 12.8
Stress blocks at the ultimate limit state

(a) Neutral axis in the concrete flange $x < h$: $R_{cf} > R_s$

(b) Neutral axis in the steel flange $h < x < h + t_f$ and $R_s > R_{cf} > R_w$

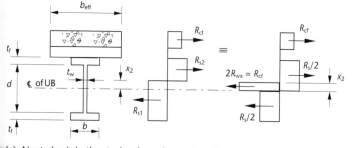

(c) Neutral axis in the steel web $x > h + t_f$: $R_{cf} < R_w$

The location of the neutral axis is determined from the equilibrium equation of the resistance forces R at the section.

i.e. $\sum R = 0$

The moment of resistance at the section is then obtained by taking moments about a convenient axis such as the centreline of the steel section, so that

$$M_c = \sum(Rz)$$

where z is the lever arm about a chosen axis for the resistance R.

For cases (b) and (c) the analysis is facilitated by considering an equivalent system of the resistance forces as shown in the relevant diagrams.

(a) Neutral axis in the concrete flange, x < h Figure 12.8(a)

This condition occurs when $R_{cf} > R_s$.

Then the depth x of the neutral axis is given by

$$R_{cx} = 0.567f_{ck}b_{eff}x = R_s$$

Therefore $x = \dfrac{R_s}{0.567f_{ck}b_{eff}} = \dfrac{R_s(h - h_p)}{R_{cf}}$

The moment of resistance is

$$M_c = R_s z$$

where the lever arm z is

$$z = (h_a/2 + h - x/2)$$

Therefore

$$M_c = R_s\left\{\frac{h_a}{2} + h - \frac{R_s}{R_{cf}}\frac{(h - h_p)}{2}\right\} \qquad (12.6)$$

(b) Neutral axis in the steel flange, h < x < h + t_f (Figure 12.8(b))

This condition occurs when $R_s > R_{cf} > R_w$.

For the equilibrium of the resistance forces

$$R_{cf} + 2R_{sx} = R_s$$

i.e. $2R_{sx} = 2f_y bx_1 = R_s - R_{cf}$

and $x_1 = \dfrac{(R_s - R_{cf})}{2f_y b} = \dfrac{(R_s - R_{cf})t_f}{2R_{sf}}$

where b is the breadth and t_f is the thickness of the steel flange.

The moment of resistance is given by

$$M_c = R_{cf}z_1 + 2R_{sx}z_2$$
$$= R_{cf}z_1 + (R_s - R_{cf})z_2$$

where z_1 and z_2 are the lever arms as shown in figure 12.8(b) and

$$z_1 = (h_a + h + h_p)/2$$
$$z_2 = (h_a - x_1)/2$$

Therefore substituting for z_1, z_2 and x_1 and rearranging

$$M_c = \frac{R_s h_a}{2} + \frac{R_{cf}(h + h_p)}{2} - \frac{(R_s - R_{cf})^2 t_f}{4R_{sf}} \qquad (12.7)$$

c) Neutral axis in the steel web, $x > h + t_f$ (Figure 12.8(c))

This condition occurs when $R_{cf} < R_w$ and is mostly associated with built-up beam sections with small top flanges and larger bottom flanges and also with stiffened webs to avoid web buckling.

For equilibrium of the equivalent arrangement of the resistance forces

$$2R_{wx} = 2f_y t_w x_2 = R_{cf}$$

Therefore

$$x_2 = \frac{R_{cf}}{2f_y t_w} = \frac{dR_{cf}}{2R_v}$$

where x_2 is the distance between the neutral axis and the centreline of the steel section.

The moment of resistance of the composite section is the moment of the two couples produced by R_s and R_{cf} with $2R_{wx}$ so that

$$M_c = M_s + \frac{R_{cf}(h_a + h + h_p - x_2)}{2}$$

or

$$M_c = M_s + \frac{R_{cf}(h_a + h + h_p)}{2} - \frac{R_{cf}^2 d}{4R_v} \qquad (12.8)$$

EXAMPLE 12.2

Moment of resistance of a composite section

Determine the moment of resistance of the composite section shown in figure 12.9 . The universal $457 \times 191 \times 74$ kg/m steel beam has a cross-sectional area of $A_a = 94.6\,\text{cm}^2$ and is grade S355 steel with $f_y = 355\,\text{N/mm}^2$.

Use concrete class C25/30 with characteristic cylinder strength $f_{ck} = 25\,\text{N/mm}^2$.

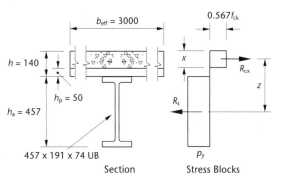

Figure 12.9
Moment of resistance example

(a) From first principles

Resistance of concrete flange $R_{cf} = 0.567 f_{ck} b_{eff} (h - h_p)$
$$= 0.567 \times 25 \times 3000 \times (140 - 50) \times 10^{-3}$$
$$= 3827 \, \text{kN}$$

Resistance of steel beam $\quad R_s = f_y A_a = 355 \times 9460 \times 10^{-3} = 3358 \, \text{kN}$

As $R_s < R_{cf}$ the neutral axis is within the concrete flange.

Determine the depth of neutral axis:
$$0.567 f_{ck} b_{eff} x = R_s = 3358 \, \text{kN}$$

Therefore $\quad x = \dfrac{3358 \times 10^3}{0.567 \times 25 \times 3000} = 79.0 \, \text{mm}$

Moment of resistance:

Lever arm z to the centre of the steel section is
$$z = (h_a/2 + h - x/2) = 457/2 + 140 - 79.0/2 = 329 \, \text{mm}$$
$$M_c = R_s z = 3358 \times 329 \times 10^{-3} = 1105 \, \text{kN m}$$

(b) Alternatively using the design equations derived

From part (a) the neutral axis is within the concrete flange therefore, from equation 12.6, the moment of resistance of the section is given by

$$M_c = R_s \left\{ \frac{h_a}{2} + h - \frac{R_s}{R_{cf}} \frac{(h - h_p)}{2} \right\}$$
$$= 3358 \left\{ \frac{457}{2} + 140 - \frac{3358}{3827} \frac{(140 - 50)}{2} \right\} 10^{-3} = 1105 \, \text{kNm}$$

12.3.2 The shear strength V_{Rd} of the composite section

For the composite section, as for the construction stage, section 12.2a(ii), the shear is resisted by the shear area A_v of the steel beam and the shear resistance V_{Rd} is given by

$$V_{Rd} = V_{pl, Rd} = \frac{A_v f_y}{\gamma_{MO} \sqrt{3}} \tag{12.9}$$

and the shear area A_v is given by

$$A_v = A_a - 2 b t_f + (t_w + 2r) t_f \tag{12.10}$$

where A_a is the cross-sectional area of the steel beam and the other dimensions of the cross section are defined in figure 12.5.

For class 1 and class 2 I-beams with a predominately uniformly distributed load the design shear stress is seldom excessive and the shear area, A_v, may be safely taken conservatively as the area of the web.

$$A_v = d \times t_w$$

For beams where high shear forces and moments occur at the same section such that $V_{Ed} > 0.5 V_{Rd}$ it is necessary to use a reduced moment capacity for the composite section by reducing the bending stress in the steel web as described in EC4 section 6.2.2.4.

12.4 | Design of shear connectors

The shear connectors are required to prevent slippage between the concrete flange and the steel beam thus enabling the concrete and steel to act as a composite unit. Stud shear connectors welded to the steel flange are the most common type used. The head on the stud acts to prevent the vertical lifting or prising of the concrete away from the steel beam.

Figure 12.10(a) shows the slippage that occurs without shear connectors. The slippage is a maximum at the supported end of the beam where the shear V and the rate of change of moment dM/dx are a maximum. The slippage reduces to zero at mid-span where the moment is a maximum and shear $V = 0$ for a uniformly distributed load.

The connectors restrain the slippage by resisting the horizontal shear at the interface of the concrete and the steel. The design is carried out for the conditions at the ultimate limit state.

The design shear resistance, P_{Rd}, of a headed stud automatically welded is given by EC4 as the lesser value of the following two equations:

$$P_{Rd} = \frac{0.8 f_u \pi d^2/4}{\gamma_v} \qquad (12.11a)$$

or:

$$P_{Rd} = \frac{0.29 \alpha d^2 \sqrt{f_{ck} E_{cm}}}{\gamma_v} \qquad (12.11b)$$

$$\alpha = 0.2 \left(\frac{h_{sc}}{d} + 1 \right) \quad \text{for } 3 \leq h_{sc}/d \leq 4$$

$$\alpha = 1 \qquad \text{for } h_{sc}/d > 4$$

where

γ_v is the partial safety factor $= 1.25$

d is the diameter of the shank of the stud, between 16 mm and 25 mm

f_u is the ultimate tensile strength of the stud $\leq 500 \, \text{N/mm}^2$ or $\leq 450 \, \text{N/mm}^2$ when equation 12.12b applies

f_{ck} is the cylinder characteristic compressive strength of the concrete

E_{cm} is the secant modulus of elasticity of the concrete, see table 1.1

A further reduction factor, k_1 or k_t is to be applied to P_{Rd} as specified in EC4, section 6.6.4 and its value depends on whether the ribs of the profiled sheeting are parallel or transverse to the supporting beam.

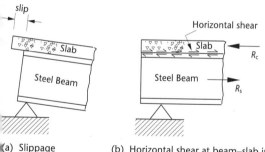

Figure 12.10
Slippage and horizontal shear

(a) Slippage

(b) Horizontal shear at beam–slab interface

(i) For ribs parallel to the supporting beam

$$k_1 = 0.6 \frac{b_0}{h_p} \left(\frac{h_{sc}}{h_p} - 1 \right) \le 1.0 \tag{12.12a}$$

(ii) For ribs transverse to the supporting beam

$$k_t = \frac{0.7}{\sqrt{n_r}} \frac{b_0}{h_p} \left(\frac{h_{sc}}{h_p} - 1 \right) \tag{12.12b}$$

n_r is the number of stud connectors in one rib at a beam connection, ≤ 2

h_{sc} is the overall nominal height of the stud

h_p is the overall depth of the profiled steel sheeting

b_0 is the mean width of the concrete rib

Also in this second case the ultimate tensile steel stress of the studs $f_u \le 450\,\text{N/mm}^2$. There is an upper limit $k_{t,\,max}$ for the reduction factor k_t which is given in EC4 section 6.6.4, table 6.2.

12.4.1 Full shear connection

The change in horizontal shear between zero and maximum moment is the lesser of the resistance R_s of the steel section and R_c the resistance of the concrete flange. Thus, to develop the full bending strength of the composite section the number of shear connectors n_f required over half the span is the lesser of

$$n_f = \frac{R_c}{k_t P_{Rd}} \quad \text{or} \quad n_f = \frac{R_s}{k_t P_{Rd}} \tag{12.13}$$

for a full shear connection, where

P_{Rd} is the effective strength of a shear stud

k_t is the reduction factor applied to the characteristic strength P_{Rd}

$R_c = 0.567 f_{ck} b_{eff} (h - h_p)$

$R_s = f_y A_a$

Full scale tests with uniformly distributed loading have shown that with plastic conditions during the ultimate limit state the shear studs can develop their full strength when spaced uniformly along the span of the beam.

12.4.2 Partial shear connection

In some cases it is not necessary to have a full shear connection in order to resist an ultimate design moment that is somewhat less than the full moment capacity of the composite section. Also, using fewer shear studs can often provide a simpler detail for the layout of the stud connectors. For partial shear connection the degree of shear connection η is defined as

$$\eta = \frac{n}{n_f} \tag{12.14}$$

where n_f is the number of shear connectors for full shear connection over a length of beam and n is the number of shear connectors provided in that length.

EC4 provides limits to the degree of shear connection η by two alternative equations according to the distance L_e for steel sections with equal flanges.

1. The nominal diameter d of the shank of the headed stud is within the range: $16\,\text{mm} \leq d \leq 25\,\text{mm}$ and the overall length of the stud after welding is $\geq 4d$:

$$L_e \leq 25 \qquad \eta \geq 1 - \left(\frac{355}{f_y}\right)(0.75 - 0.03L_e) \qquad \eta \geq 0.4 \qquad (12.15a)$$

2. The nominal diameter d of the shank of the headed stud is $d = 19\,\text{mm}$ and the overall length of the stud after welding $\geq 76\,\text{mm}$:

$$L_e \leq 25 \qquad \eta \geq 1 - \left(\frac{355}{f_y}\right)(1.00 - 0.04L_e) \qquad \eta \geq 0.4 \qquad (12.15b)$$

where L_e is the distance in sagging between the points of zero moment in metres. In both cases, where L_e is greater than 25 m, the factor should be greater than 1.

There are also a number of other conditions as listed in EC4 section 6.6.1.2. The ultimate moment resistance of the composite section with partial shear connection is derived from the analysis of the stress block systems shown in figure 12.11. In the analysis the depth of the concrete stress block s_q is

$$s_q = \frac{R_q}{0.567 f_{ck} b_{eff}}$$

where R_q is the shear resistance of the shear studs provided.

As previously shown in section 12.3 the depth of the section's neutral axis is obtained by considering the equilibrium of the material resistances R. The moment of resistance M_c is obtained by taking moments about a convenient axis such as the centreline of the

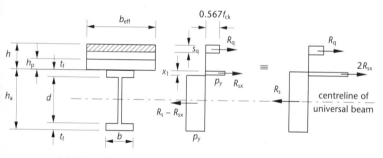

Figure 12.11
Stress blocks for partial shear connection

(a) Neutral axis in the steel flange $h < x < h + t_f : R_s > R_q > R_w$

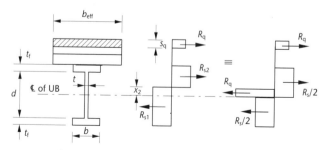

(b) Neutral axis in the steel web $x > h + t_f : R_c < R_w$

steel section, followed by some rearrangement of the equations. The diagrams for this analysis are shown in figure 12.11 for the two possible cases of:

(a) the neutral axis in the steel flange $R_q > R_w$

$$M_c = \frac{R_s h_a}{2} + R_q\left[h - \frac{R_q(h - h_p)}{2R_{cf}}\right] - \frac{(R_s - R_q)^2 t_f}{4R_{sf}}$$ (12.16a)

(b) the neutral axis in the steel web $R_q < R_w$

$$M_c = M_s + R_q\left[\frac{h_a}{2} + h - \frac{R_q(h - h_p)}{2R_{cf}}\right] - \frac{dR_q^2}{4R_v}$$ (12.16b)

Figure 12.12 shows the interaction diagram for the moment of resistance of the composite section against the degree of shear connection η where

$$\eta = \frac{n}{n_f}$$ (equation 12.14)

The curved interaction line (a) is based on the stress block equations of 12.16a and 12.16b which give the more precise results. The straight interaction line (b) represents a linear relation between the moment capacity and η which provides a simpler and safer but less economic solution.

12.4.3   Shear connection for concentrated loads

When the beam supports concentrated loads the slope dM/dx of the bending moment is greater and the shear is more intense. This means that the shear connectors have to be spaced closer together between the concentrated load and the adjacent support. The distribution of the shear connectors is then specified by the equation.

$$N_i = \frac{N_t(M_i - M_s)}{(M_c - M_s)}$$ (12.17)

where

N_i is the number of shear connectors between the concentrated load and the adjacent support

N_t is the total number of shear connectors required between the support and the point of maximum moment (M_{max})

M_i is the bending moment at the concentrated load

M_s is the moment capacity of the steel member

M_c is the moment capacity of the composite section

Figure 12.13 shows a beam supporting concentrated loads and the distribution of the shear connectors.

Figure 12.12
Interaction diagram for partial shear connection

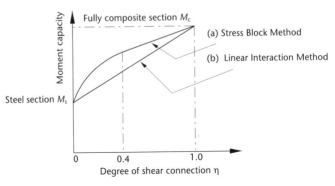

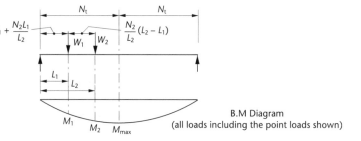

Figure 12.13
Distribution of shear
connectors with concentrated
loads

B.M Diagram
(all loads including the point loads shown)

12.5 Transverse reinforcement in the concrete flange

Transverse reinforcement is required to resist the longitudinal shear in the concrete flange. This shear acts on vertical planes either side of the shear connectors as shown in figure 12.14.

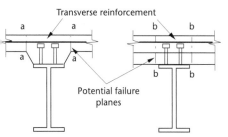

Ribs parallel to beam   Ribs perpendicular to beam

Figure 12.14
Transverse reinforcement in
the concrete flanges

The analysis and design for the transverse reinforcement to resist the longitudinal shear in a flanged beam follows the variable strut inclination method as required in EC2 and described in this book in sections 5.1.4 and 7.4, in conjunction with example 7.5 part (2).

EC2 specifies a minimum of transverse steel area equal to $(0.13h_f \times 1000/100)$ mm^2 per metre width.

The method of designing the transverse steel for a composite beam is shown in example 12.3 part (b) *Transverse reinforcement*.

EXAMPLE 12.3

Shear connectors and transverse reinforcement

The composite beam of example 12.2 and figure 12.9 spans 9.0 metres and is provided with 80 shear stud connectors in pairs at 225 mm centres. The studs are 19 mm diameter and of 100 mm height.

The plastic section of modulus of the steel section is $W_{pl,y} = 1653$ N/mm^2 and the design stress of the steel, $f_y = 355$ N/mm^2. The characteristic material strengths are $f_{ck} = 25$ N/mm^2 for the concrete and $f_{yk} = 500$ N/mm^2 for the reinforcing bars.

(a) Calculate the degree of shear resistance and the moment of resistance of the composite beam based upon the shear connectors provided.

(b) Design the transverse reinforcement required to resist the transverse shear in the concrete flange.

(a) Degree of shear connection and moment of resistance

The design shear resistance, P_{Rd}, of each shear stud is the lesser value obtained from equations 12.11a and 12.11b with f_u, the ultimate tensile strength of the steel, equal to $450 \, N/mm^2$.

Using these equations it is found from equation 12.11a that

$$P_{Rd} = \frac{0.8 f_u \pi d^2 / 4}{\gamma_v} = \frac{0.8 \times 450 \times \pi \times 19^2 / 4}{1.25} \times 10^{-3} = 81.7 \, kN$$

A reduction factor, k_t, is calculated from equation 12.12b with an upper limit taken from table 6.2 of EC4. From equation 12.12b, with reference to EC4, figure 6.13 and taking dimension $b_0 = 80 \, mm$ for the profiled steel sheeting

$$k_t = \frac{0.7}{\sqrt{n_r}} \frac{b_0}{h_p} \left(\frac{h_{sc}}{h_p} - 1 \right) = \frac{0.7}{\sqrt{2}} \times \frac{80}{50} \left(\frac{100}{50} - 1 \right) = 0.79$$

The upper limit of $k_t = 0.8$ from EC4 table 6.2.

Hence the design shear resistance, P_{Rd}, of a stud $= 0.79 \times 81.7 = 64.5 \, kN$.

For full shear connection the number of studs required over half the span is

$$n_f = \frac{R_s}{P_{Rd}} = \frac{3358}{64.5} = 52$$

($R_s = f_y A_a = 3358 \, kN$, is the resistance of the steel beam as obtained from example 12.2.)

Hence for full shear connection the total number of studs required over the whole span $= 104$.

The degree of shear connection, η, is

$$\eta = \frac{80}{104} = 0.77$$

The lower limit for η is calculated from equation 12.15a as

$$\eta \geq 1 - \left(\frac{355}{f_y} \right) (0.75 - 0.03 L_e) = 1 - \left(\frac{355}{355} \right) (0.75 - 0.03 \times 9) = 0.52 < 0.77 \quad OK$$

The moment of the resistance M_p of the composite beam based on the partial shear resistance can be obtained using the linear interaction method of figure 12.12. From the proportions of the straight line relationship

$$M_p = \eta (M_c - M_s) + M_s$$

where M_c is the moment capacity of the composite section with full shear connection from example 12.2. M_s is the moment capacity of the steel beam where

$$M_s = W_{pl, y} f_y = 1653 \times 355 \times 10^{-3} = 587 \, kN \, m$$

Therefore

$$M_p = \eta (M_c - M_s) + M_s$$
$$= 0.77 \times (1105 - 587) + 587 = 986 \, kN \, m$$

(b) Transverse reinforcement in the concrete flange

The design follows the procedures and equations set out in section 5.1.4 and example 7.5(2).

(i) Calculate the design longitudinal shear v_{Ed} at the web–flange interface

For a sagging moment the longitudinal shear stresses are the greatest over a distance of Δx measured from the point of zero moment and Δx is taken as half the distance to the maximum moment at mid-span, thus

$$\Delta x = 0.5 \times L/2$$
$$= 0.5 \times 9 \times 10^3/2 = 2250 \, \text{mm}$$

For a one span simply supported beam with a uniformly distributed load the change in moment, ΔM over distance $\Delta x = L/4$ from the zero moment at the support is

$$\Delta M = \frac{w_u \times L}{2} \times \frac{L}{4} - \frac{w_u \times L}{4} \times \frac{L}{8} = \frac{3w_u L^2}{32} = \frac{3}{4} \times \left(\frac{w_u L^2}{8} \right)$$

Therefore

$$\Delta M = 0.75 \times 986 = 740 \, \text{kN m}$$

The change in longitudinal force ΔF_d in the concrete flange at section b–b in figure 12.14 is

$$\Delta F_d = \frac{\Delta M}{z} \times \frac{0.5(b_{eff} - b)}{b_{eff}}$$

where the lever arm z is taken as the distance from the centre of the steel beam to the centre of the concrete flange, so that

$$z = h_a/2 + h - h_f/2$$
$$= 457/2 + 140 - 90/2 = 324 \, \text{mm}$$

Therefore

$$\Delta F_d = \frac{740 \times 10^3}{324} \times \frac{0.5(3000 - 191)}{3000} = 1069 \, \text{kN}$$

The longitudinal shear stress induced, v_{Ed}, is

$$v_{Ed} = \frac{\Delta F_d}{(h_f \times \Delta x)}$$
$$= \frac{1069 \times 10^3}{90 \times 2250} = 5.3 \, \text{N/mm}^2$$

(ii) Check the strength of the concrete strut

From equation 5.17, to prevent crushing of the concrete in the compressive strut in the flange

$$v_{Ed} \leq \frac{0.6(1 - f_{ck}/250)f_{ck}}{1.5(\cot \theta_f + \tan \theta_f)}$$

The moments are sagging so the flange is in compression and the limits for θ_f are

$$26.5° \leq \theta_f \leq 45°$$

With $\theta_f = $ the minimum value of $26.5°$

$$v_{Ed(max)} = \frac{0.6(1 - 25/250) \times 25}{(2.0 + 0.5)} = 5.4 \, \text{N/mm}^2 \quad (> 5.3 \, \text{N/mm}^2)$$

and the concrete strut has sufficient strength with $\theta_f = 26.5°$.

(iii) Design transverse steel reinforcement

Transverse shear reinforcement is required if $v_{Ed} \geq 0.27f_{ctk}$ where f_{ctk} is the characteristic axial tensile strength of concrete $= 1.8\,N/mm^2$ for class 25 concrete. Therefore

$$v_{Ed(max)} = 0.27f_{ck} = 0.27 \times 1.8 = 0.49\,N/mm^2 \quad (< 5.3\,N/mm^2)$$

and transverse shear reinforcement is required. The area required is given by:

$$\frac{A_{sf}}{s_f} = \frac{v_{Ed} \times h_f}{0.87f_{yk} \times \cot\theta_f} = \frac{5.3 \times 90}{0.87 \times 500 \times 2.0}$$
$$= 0.55$$

If bars are provided at (say) 175 mm centres then $A_{sf} = 0.55 \times 175 = 96\,mm^2$. Hence provide 12 mm bars ($A_s = 113\,mm^2$ for one bar).

This steel area satisfies the minimum requirement of $0.13\% = 117\,mm^2/m$.

Longitudinal reinforcement should also be provided in the flange.

12.6 Deflection checks at the serviceability limit state

At the serviceability limit state it is necessary to check the maximum deflections of the beam for the following conditions:

(a) During construction when the concrete flange has not hardened and the steel beam section alone has to carry all the loads due to the permanent and variable actions at that time.

(b) At service when the concrete has hardened and the composite steel and concrete section carries the additional permanent and variable loads.

12.6.1 Deflections during construction

The deflection δ at mid-span for a uniformly distributed load is

$$\delta = \frac{5wL^4}{384E_aI_a} \tag{12.18}$$

where

w is the serviceability load per metre at construction

L is the beam's span

E_a is the elastic modulus of the steel $= 210\,kN/mm^2$

I_a is the second moment of area of the steel section

The deflection due to the permanent action or dead load is locked into the beam as the concrete hardens.

12.6.2 Deflections at service during the working life of the structure

At this stage the concrete has hardened and forms a composite section together with the steel beam and the shear connectors.

The composite section is converted into a transformed section so that the area of concrete in compression is transformed into an equivalent steel area with a flange width as shown in figure 12.15, where

$$b_t = \frac{b_{eff}}{n}$$

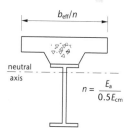

Figure 12.15
The transformed section at service

and $n = \dfrac{E_{steel}}{E_{c, eff}}$ is the modular ratio

For buildings, EC4 states that $E_{c, eff}$ may be taken as $E_{cm}/2$ where E_{cm} is the secant modulus of elasticity for concrete (see table 1.1).

It is also stated in EC4 that for calculating deflections at service the effects of partial shear connection can be ignored provided that the degree of shear connection, $\eta \geq 0.5$ and other practical requirements are satisfied.

The transformed composite section

For two areas A_1 and A_2 the position of their neutral axis may be found by taking area moments about the centroidal axis of A_1 such that

$$\bar{x} = \frac{A_2 s}{A_1 + A_2}$$

where

   $\bar{x}$ is the distance to the neutral axis from the centroid of A_1
   s is the distance between the centroids of A_1 and A_2.

The second moment of area of the total section about the neutral axis of A_1 and A_2 combined can be calculated from

$$I_T = I_1 + I_2 + \frac{A_1 A_2 s^2}{A_1 + A_2}$$
$$= I_1 + I_2 + A_1 s \bar{x}$$

where I_1 and I_2 are the second moments of areas of A_1 and A_2 respectively about their centroidal axes.

So with reference to figure 12.7 and taking A_2 as the steel area A_a, for the transformed composite section $s = (h_a + h + h_p)/2$ and the equations for $\bar{x}$ and I become

$$\bar{x} = \frac{A_a n (h_a + h + h_p)}{2\{A_a n + b_{eff}(h - h_p)\}} \tag{12.19}$$

$$I_{transf} = I_a + \frac{b_{eff}(h - h_p)^3}{12n} + \frac{b_{eff}(h - h_p)(h_a + h + h_p)\bar{x}}{2n} \tag{12.20}$$

where

   A_a is the area of the steel section and
   I_a is the second moment of area of the steel section.

The depth of the neutral axis, x_t from the top of the concrete flange is

$$x_t = \bar{x} + \frac{h - h_p}{2}$$

If $x < (h - h_p)$ then the neutral axis is within the flange and the concrete is cracked at service and these equations cannot apply.

Deflection at service due to the permanent and variable loads

The deflections are calculated for the unfactored actions. The second moment of area of the composite section is used in the calculations.

For the unpropped case the total deflection is:

$$\delta_{total} = \delta_{constr} + \delta_{composite}$$

where

δ_{constr} is the deflection of the steel beam due to the permanent load at construction

$\delta_{composite}$ is the deflection of the composite beam due to the quasi-permanent load which is the additional permanent load plus a proportion of the variable load depending on the type of structure. (See section 2.4 and table 2.4 in chapter 2.)

EXAMPLE 12.4

Serviceability checks for deflection

For the composite beam of the previous examples determine the deflections at service.

The relevant sectional properties for the $457 \times 191 \times 74 \, \text{kg/m}$ Universal Beam are:

Cross-sectional area $A_a = 94.6 \, \text{cm}^2$

Second moment of area $I_a = 33\,300 \, \text{cm}^4$

Assume the beam is part of a building floor system and is unpropped during construction.

The uniformly distributed characteristic actions are:

During construction – permanent load 9.8 kN/m,   variable load 2.25 kN/m

During service   – permanent load 11.0 kN/m,   variable load 18.0 kN/m
                                                  (Quasi-permanent component)

(a) At construction

The concrete has not hardened so that the steel beam supports the load of the wet concrete.

The deflection at mid-span of a beam with a uniformly distributed load is given by

$$\delta = \frac{5wL^4}{384E_aI_a}$$

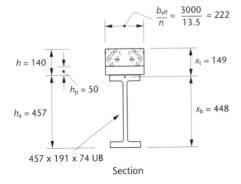

Figure 12.16
Transformed section example

$$\frac{b_{eff}}{n} = \frac{3000}{13.5} = 222$$

$h = 140$

$h_p = 50$

$h_a = 457$

$x_t = 149$

$x_b = 448$

457 x 191 x 74 UB

Section

Due to the permanent load, w_G and the variable load, w_Q the mid-span deflection for the steel beam is:

$$\delta = (w_G + w_Q) \times \frac{5L^4}{384 E_a I_a}$$

$$= (9.8 + 2.25) \times \frac{5 \times 9^4 \times 10^5}{384 \times 210 \times 33\,300}$$

$$= 12 + 3 = 15\,\text{mm} = \text{span}/600$$

The 12 mm deflection due to the permanent load at this stage is locked into the beam as the concrete hardens after construction.

(b) At service

The composite section is transformed into an equivalent steel section as shown in figure 12.16.

For a class C25/30 concrete the secant modulus of elasticity of the concrete, $E_{cm} = 31\,\text{kN/mm}^2$ (see table 1.1).

Take the modular ratio

$$n = \frac{E_a}{0.5 E_{cm}} = \frac{210}{0.5 \times 31} = 13.5$$

The position of the centroid of the transformed section is given by equation 12.19 as

$$\bar{x} = \frac{A_a n (h_a + h + h_p)}{2\{A_a n + b_{eff}(h - h_p)\}}$$

$$= \frac{94.6 \times 10^2 \times 13.5 \times (457 + 140 + 50)}{2 \times \{94.6 \times 10^2 \times 13.5 + 3000 \times (140 - 50)\}} = 104\,\text{mm}$$

$$x_t = \bar{x} + (h - h_p)/2 = 104 + (140 - 50)/2 = 149\,\text{mm} > h_f = 90\,\text{mm}$$

therefore the concrete is not cracked.

The second moment of area of the composite figure is given by equation 12.20 as

$$I_{transf} = I_a + \frac{b_{eff}(h - h_p)^3}{12n} + \frac{b_{eff}(h - h_p)(h_a + h + h_p)\bar{x}}{2n}$$

$$= 333 \times 10^6 + \frac{3000 \times (140 - 50)^3}{12 \times 13.5} +$$

$$\frac{3000 \times (140 - 50) \times (457 + 140 + 50)}{2 \times 13.5} \times 104$$

$$= (333 + 14 + 672) \times 10^6\,\text{mm}^4$$

$$= 1019 \times 10^6\,\text{mm}^4$$

Deflections at service

At service the additional permanent load $= 11.0 - 9.8 = 1.2\,\text{kN/m}$ and the quasi-permanent variable load $= 18.0\,\text{kN/m}$, thus deflection

$$\delta = (w_G + w_Q) \times \frac{5L^4}{384 E_a I_a}$$

$$= (1.2 + 18) \times \frac{5 \times 9^4 \times 10^3}{384 \times 210 \times 1019}$$

$$= 1 + 7 = 8\,\text{mm} = \text{span}/1125$$

Therefore the total final deflection including that at the construction stage is

$$\delta = 12 + 8 = 20 \, \text{mm} = \text{span}/450$$

At all stages the deflection is well within the normally acceptable limits of span/250. Deflections are seldom a problem for composite beams in buildings, but for long spans the beams could be pre-cambered for a proportion of the permanent load to avoid visible signs of the beam 'sagging'.

Appendix

Typical weights and live loads

$$1 \text{ kg} \qquad = 9.81 \text{ N force}$$
$$1 \text{ lb} = 0.454 \text{ kg} \qquad = 4.448 \text{ N force}$$
$$1 \text{ lb/ft}^2 = 4.88 \text{ kg/m}^2 \quad = 47.9 \text{ N/m}^2$$
$$1 \text{ lb/ft}^3 = 16.02 \text{ kg/m}^3 = 157 \text{ N/m}^3$$

Weights

	kN/m^3
Aluminium, cast	26
Asphalt paving	23
Bricks, common	19
Bricks, pressed	22
Clay, dry	19–22
Clay, wet	21–25
Concrete, reinforced	25
Glass, plate	27
Lead	112
Oak	9.5
Pine, white	5
Sand, dry	16–19
Sand, wet	18–21
Steel	77
Water	9.81

	kN/m^2
Brick wall, 115 mm thick	2.6
Gypsum plaster, 25 mm thick	0.5
Glazing, single	0.3

Floor and roof loads

	kN/m^2
Classrooms	3.0
Dance halls	5.0
Flats and houses	1.5
Garages, passenger cars	2.5
Gymnasiums	5.0
Hospital wards	2.0
Hotel bedrooms	2.0
Offices for general use	2.5
Flat roofs, with access	1.5
Flat roofs, no access	0.75

Bar areas and perimeters

Table A.1 Sectional areas of groups of bars (mm²)

Bar size (mm)	Number of bars									
	1	2	3	4	5	6	7	8	9	10
6	28.3	56.6	84.9	113	142	170	198	226	255	283
8	50.3	101	151	201	252	302	352	402	453	503
10	78.5	157	236	314	393	471	550	628	707	785
12	113	226	339	452	566	679	792	905	1020	1130
16	201	402	603	804	1010	1210	1410	1610	1810	2010
20	314	628	943	1260	1570	1890	2200	2510	2830	3140
25	491	982	1470	1960	2450	2950	3440	3930	4420	4910
32	804	1610	2410	3220	4020	4830	5630	6430	7240	8040
40	1260	2510	3770	5030	6280	7540	8800	10100	11300	12600

Table A.2 Perimeters and weights of bars

Bar size (mm)	6	8	10	12	16	20	25	32	40
Perimeter (mm)	18.85	25.1	31.4	37.7	50.2	62.8	78.5	100.5	125.6
Weight (kg/m)	0.222	0.395	0.616	0.888	1.579	2.466	3.854	6.313	9.864

Bar weights based on density of 7850 kg/m³.

Table A.3 Sectional areas per metre width for various bar spacings (mm²)

Bar size (mm)	Spacing of bars								
	50	75	100	125	150	175	200	250	300
6	566	377	283	226	189	162	142	113	94
8	1010	671	503	402	335	287	252	201	168
10	1570	1050	785	628	523	449	393	314	262
12	2260	1510	1130	905	754	646	566	452	377
16	4020	2680	2010	1610	1340	1150	1010	804	670
20	6280	4190	3140	2510	2090	1800	1570	1260	1050
25	9820	6550	4910	3930	3270	2810	2450	1960	1640
32	16100	10700	8040	6430	5360	4600	4020	3220	2680
40	25100	16800	12600	10100	8380	7180	6280	5030	4190

Shear reinforcement

Table A.4 A_{sw}/s for varying stirrup diameter and spacing

Stirrup diameter (mm)	Stirrup spacing (mm)										
	85	90	100	125	150	175	200	225	250	275	300
8	1.183	1.118	1.006	0.805	0.671	0.575	0.503	0.447	0.402	0.366	0.335
10	1.847	1.744	1.57	1.256	1.047	0.897	0.785	0.698	0.628	0.571	0.523
12	2.659	2.511	2.26	1.808	1.507	1.291	1.13	1.004	0.904	0.822	0.753
16	4.729	4.467	4.02	3.216	2.68	2.297	2.01	1.787	1.608	1.462	1.34

Note: A_{sw} is based on the cross-sectional area of two legs of the stirrup.

Wire fabric

Table A.5 Sectional areas for different fabric types

Fabric reference	Longitudinal wires			Cross wires		
	Wire size (mm)	Pitch (mm)	Area (mm²/m)	Wire size (mm)	Pitch (mm)	Area (mm²/m)
Square mesh						
A393	10	200	393	10	200	393
A252	8	200	252	8	200	252
A193	7	200	193	7	200	193
A142	6	200	142	6	200	142
A98	5	200	98	5	200	98
Structural mesh						
B1131	12	100	1131	8	200	252
B785	10	100	785	8	200	252
B503	8	100	503	8	200	252
B385	7	100	385	7	200	193
B283	6	100	283	7	200	193
B196	5	100	196	7	200	193
Long mesh						
C785	10	100	785	6	400	70.8
C636	9	100	636	6	400	70.8
C503	8	100	503	5	400	49
C385	7	100	385	5	400	49
C283	6	100	283	5	400	49
Wrapping mesh						
D98	5	200	98	5	200	98
D49	2.5	100	49	2.5	100	49

Anchorage and lap requirements

Table A.6 Anchorage and lap length coefficients (length $L = K_A \times$ bar size) for good bond conditions[1]

		K_A for concrete strength, f_{ck} (N/mm^2)						
		20	25	30	35	40	45	50
Straight bars								
Anchorage in tension and compression		47	40	36	32	29	27	25
Curved bars								
Anchorage in tension[3]		33	28	25	22	20	19	18
Anchorage in compression		47	40	36	32	29	27	25
	% of bars lapped at section							
Compression and tension laps[4]	<25%	47	40	36	32	29	27	25
	33%	54	46	42	37	33	31	29
	50%	66	56	51	45	41	38	35
	>50%	71	60	54	48	44	41	38

Notes:
1. For poor bond conditions (see figure 5.8) divide the coefficients by 0.7.
2. For bars greater than 32 mm divide the coefficients by $[(132 - \Phi)100]$ where Φ is the bar size.
3. For a curved bar in tension the anchorage length is generally that of a straight bar $\times$ 0.7 but also depends on the cover conditions – see table 5.2.
4. These figures apply for $\alpha_1 = \alpha_2 = \alpha_3 = \alpha_5 = 1$ (see table 5.2). Also see the additional requirements for minimum lap lengths and detailing specified in sections 5.2 and 5.3.

Maximum and minimum areas of reinforcement

Table A.7 Maximum areas of reinforcement

(a) For a slab or beam, tension or compression reinforcement

$100A_s/A_c \leq 4$ per cent other than at laps

(b) For a column

$100A_s/A_c \leq 4$ per cent other than at laps and 8 per cent at laps

(c) For a wall, vertical reinforcement

$100A_s/A_c \leq 4$ per cent

Table A.8   Minimum areas of reinforcement

Tension reinforcement in beams and slabs	Concrete class ($f_{yk} = 500\,N/mm^2$)			
	C25/30	C30/35	C40/50	C50/60
$\dfrac{A_{s,min}}{b_t d} > 0.26\dfrac{f_{ctm}}{f_{yk}}$ (> 0.0013)	0.0013	0.0015	0.0018	0.0021

Secondary reinforcement > 20% main reinforcement

Longitudinal reinforcement in columns
$A_{s,min} > 0.10 N_{sd}/0.87 f_{yk} > 0.002 A_c$ where N_{sd} is the axial compression force

Vertical reinforcement in walls
$A_{s,min} > 0.002 A_c$

Note: b_t is the mean width of the tension zone.

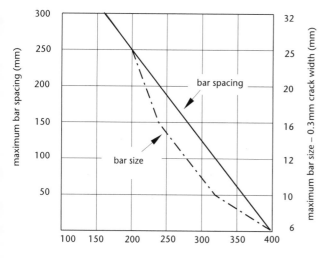

Stress in reinforcement under quasi-permanent load (N/mm²)
(see sections 6.1.3 and 6.1.7)

Figure A.1
Maximum bar size and spacing for crack control

Span–effective depth ratios

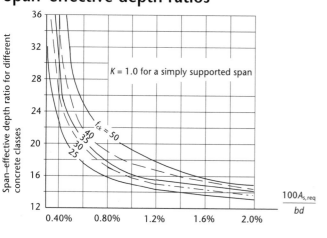

Figure A.2
Graph of basic span–effective depth ratios for different classes of concrete

Summary of basic design equations for the design of reinforced concrete

(a) Design for bending (see chapters 4 and 7)

For a singly reinforced section:

$$A_s = \frac{M}{0.87 f_{yk} z}$$

$$z = d\left\{0.5 + (0.25 - K/1.134)^{1/2}\right\}$$

$$K = M/bd^2 f_{ck}$$

For a doubly reinforced section ($K > K_{bal}$) – see figure A.3:

$$A'_s = \frac{(K - K_{bal}) f_{ck} bd^2}{0.87 f_{yk}(d - d')}$$

$$A_s = \frac{K_{bal} f_{ck} bd^2}{0.87 f_{yk} z_{bal}} + A'_s$$

When moment redistribution has been applied then the above equations must be modified – see table 4.2.

Table A.9 Limiting constant values

	Concrete class $\leq$ C50/60
Limiting x_{bal}/d	0.45
Maximum z_{bal}	0.82d
K_{bal} = limiting K	0.167
Limiting d'/d	0.171
Maximum percentage steel area $100A_{bal}/bd$	$23.4 f_{ck}/f_{yk}$

Figure A.3
Lever-arm curve

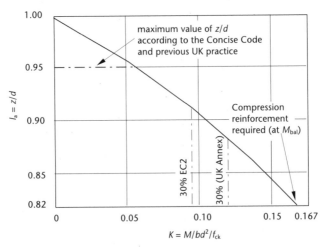

The percentage values on the K axis mark the limits for singly reinforced sections with moment redistribution applied (see section 4.7 and table 4.2)

(b) Design for shear (see chapters 5 and 7)

$$V_{Rd,max(22)} = 0.124b_w d(1 - f_{ck}/250)f_{ck}$$

$$V_{Rd,max(45)} = 0.18b_w d(1 - f_{ck}/250)f_{ck}$$

$$\theta = 0.5\sin^{-1}\left\{\frac{V_{Ed}}{0.18b_w df_{ck}(1 - f_{ck}/250)}\right\} \leq 45°$$

$$\frac{A_{sw}}{s} = \frac{V_{Ed}}{0.78df_{yk}\cot\theta}$$

$$\frac{A_{sw,min}}{s} = \frac{0.08f_{ck}^{0.5}b_w}{f_{yk}}$$

$$V_{min} = \frac{A_{sw}}{s} \times 0.78df_{yk}\cot\theta$$

$$\Delta F_{td} = 0.5V_{Ed}\cot\theta$$

(c) Design for torsion (see chapters 5 and 7)

$$t = \frac{\text{Area of the section}}{\text{Perimeter of the section}} = \frac{A}{u}$$

for a rectangular section $b \times h$

$$t = \frac{bh}{2(b+h)} \qquad A_k = (b-t)(h-t) \qquad u_k = 2(b+h-2t)$$

$$\frac{T_{Ed}}{T_{Rd,min}} + \frac{V_{Ed}}{V_{Rd,max}} \leq 1.0$$

$$T_{Rd,max} = \frac{1.33vf_{ck}A_k}{\cot\theta + \tan\theta}$$

$$\frac{A_{sw}}{s} = \frac{T_{Ed}}{2A_k 0.87f_{yk}\cot\theta}$$

$$A_{sl} = \frac{T_{Ed}u_k\cot\theta}{2A_k 0.87f_{ylk}}$$

(d) Design for punching shear in slabs (see chapter 8)

$$V_{Rd,max} = 0.5ud\left[0.6\left(1 - \frac{f_{ck}}{250}\right)\right]\frac{f_{ck}}{1.5}$$

$$u_1 = 2(a+b) + 4\pi d$$

$$A_{sw,min} = \frac{0.053\sqrt{f_{ck}}(s_r.s_t)}{f_{yk}}$$

$$A_{sw} \geq \frac{v_{Rd,cs} - 0.75v_{Rd,c}}{1.5\dfrac{f_{ywd,ef}}{s_r \times u_1}}$$

Proof of equation conversions used in Chapter 5 – section 5.1.2

To prove that $\sin\theta\cos\theta = \dfrac{1}{\tan\theta + \cot\theta}$:

Consider a right-angled triangle with sides length a, b and h, where h is the length of the hypotenuse and θ is the angle between sides of length a and b, and use the theorem of Pythagoras where $h^2 = a^2 + b^2$.

$$\sin\theta\cos\theta = \frac{a}{h} \times \frac{b}{h} = \frac{ab}{h^2} = \frac{ab}{a^2 + b^2} = \frac{1}{\dfrac{a^2}{ab} + \dfrac{b^2}{ab}} = \frac{1}{\tan\theta + \cot\theta}$$

A reason for using this type of conversion in the equations for the analysis for shear is that it facilitates the setting up of quadratic equations which can be more readily solved.

Typical design chart for rectangular columns

Figure A.4
Rectangular columns
($d'/h = 0.20$)

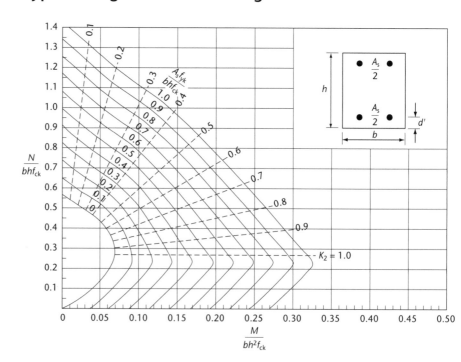

Reinforced concrete beam flowchart

EC2 Section

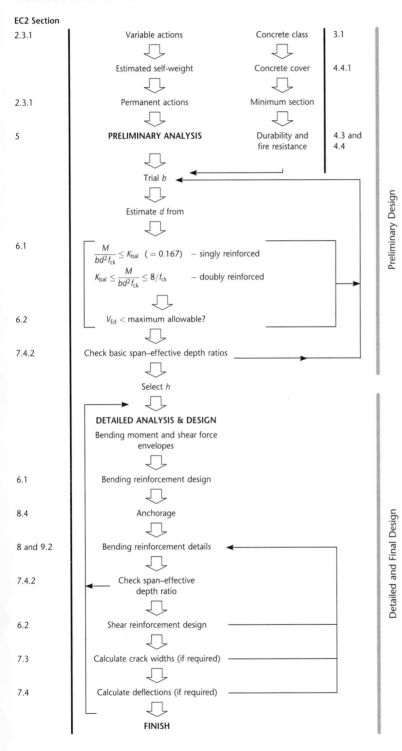

2.3.1   Variable actions   Concrete class   3.1

       Estimated self-weight   Concrete cover   4.4.1

2.3.1   Permanent actions   Minimum section

5   **PRELIMINARY ANALYSIS**   Durability and   4.3 and
                                        fire resistance   4.4

       Trial b

       Estimate d from

6.1 $\dfrac{M}{bd^2 f_{ck}} \le K_{bal}$ ($= 0.167$) – singly reinforced

         $K_{bal} \le \dfrac{M}{bd^2 f_{ck}} \le 8/f_{ck}$ – doubly reinforced

6.2 $V_{Ed} <$ maximum allowable?

7.4.2   Check basic span–effective depth ratios

       Select h

       DETAILED ANALYSIS & DESIGN
       Bending moment and shear force
       envelopes

6.1   Bending reinforcement design

8.4   Anchorage

8 and 9.2   Bending reinforcement details

7.4.2   Check span–effective
       depth ratio

6.2   Shear reinforcement design

7.3   Calculate crack widths (if required)

7.4   Calculate deflections (if required)

       FINISH

Preliminary Design

Detailed and Final Design

Prestressed concrete design flowchart

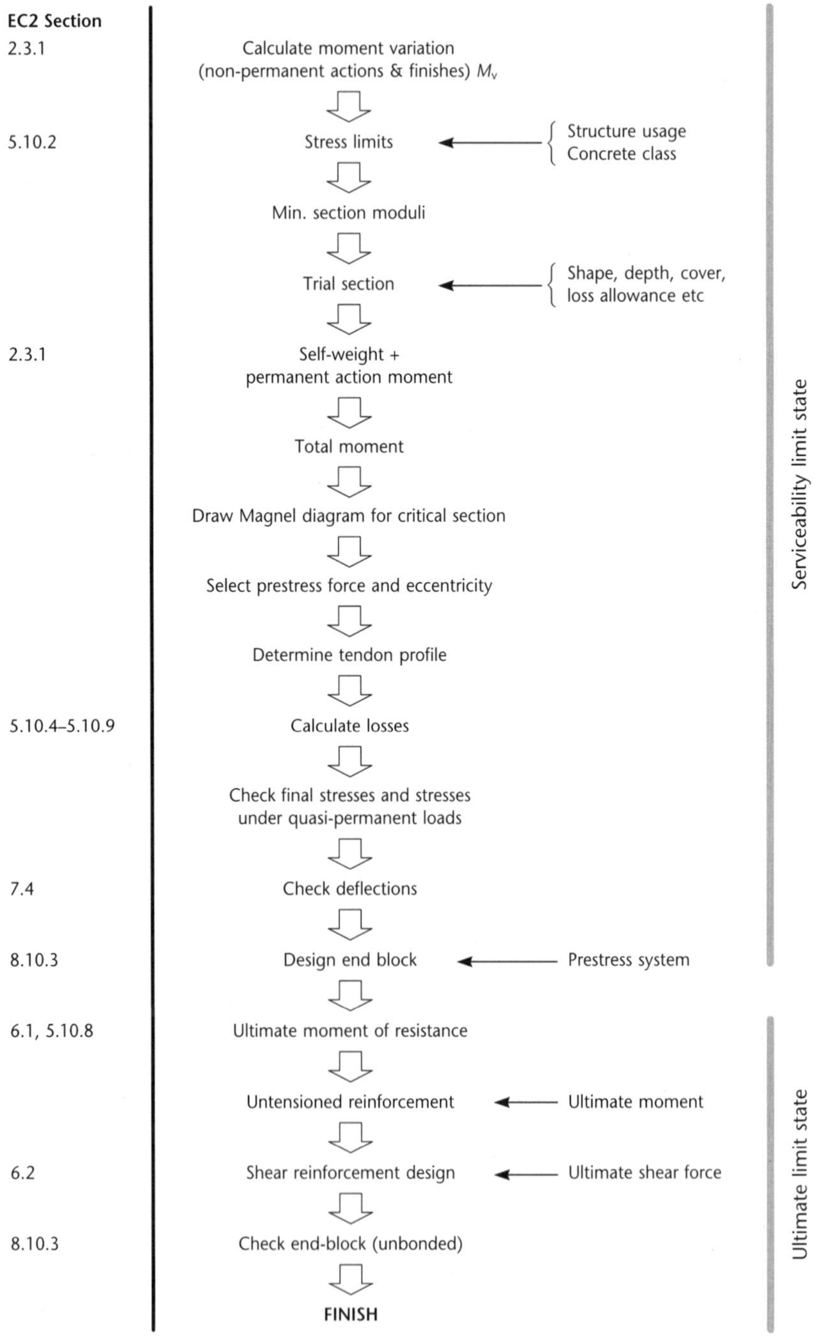

EC2 Section		
2.3.1	Calculate moment variation (non-permanent actions & finishes) M_v	
5.10.2	Stress limits ←	{ Structure usage Concrete class
	Min. section moduli	
	Trial section ←	{ Shape, depth, cover, loss allowance etc
2.3.1	Self-weight + permanent action moment	
	Total moment	
	Draw Magnel diagram for critical section	
	Select prestress force and eccentricity	
	Determine tendon profile	
5.10.4–5.10.9	Calculate losses	
	Check final stresses and stresses under quasi-permanent loads	
7.4	Check deflections	
8.10.3	Design end block ←	Prestress system
6.1, 5.10.8	Ultimate moment of resistance	
	Untensioned reinforcement ←	Ultimate moment
6.2	Shear reinforcement design ←	Ultimate shear force
8.10.3	Check end-block (unbonded)	
	FINISH	

Serviceability limit state

Ultimate limit state

Further reading

(a) Eurocodes and British Standards

1 EN 1990   *Eurocode 0*   *Basis of structural design*

2 EN 1991   *Eurocode 1*   *Actions on structures*

3 EN 1992   *Eurocode 2*   *Design of concrete structures*

4 EN1994   *Eurocode 4*   *Design of composite steel and concrete structures*

5 EN 1997   *Eurocode 7*   *Geotechnical design*

6 EN 1998   *Eurocode 8*   *Design of structures for earthquake resistance*

7 EN206-1   *Concrete specification, performance, production and conformity*

8 BS 8500-2:2002   *Complementary British Standard to BS EN 206-1. Specification for constituent materials and concrete*

9 EN 10080   *Steel for the reinforcement of concrete*

10 EN 10138   *Prestressing steels*

11 BS 4449 (2005)   *Steel for the reinforcement of concrete, weldable reinforcing steel, bar, coil and decoiled products*

12 BS 4482   *Cold reduced wire for the reinforcement of concrete*

13 BS 5057   *Concrete admixtures*

14 NA to EN1992   *UK National Annex to Eurocode 2*

15 BS PD 6687:2006   *Background paper to the UK National Annexes to BS EN 1992-1*

(b) Textbooks and other publications

16 J. Bungey, S. Millard, M. Grantham, *The Testing of Concrete in Structures*, 4th edn, Taylor & Francis, London, 2006

17 H. Gulvanessian, J.A Calgaro, M. Holicky, *Designers' Guide to EN 1990*, Thomas Telford, London 2005

18 M.K. Hurst, *Prestressed Concrete Design,* 2nd edn, Chapman & Hall, London, 1998

19 W. H. Mosley, J.H Bungey, R Hulse, *Reinforced Concrete Design,* 5th edn, Palgrave, London, 1999

20 R.S Narayanan, A Beeby, *Designers Guide to EN1992-1-1 and EN1992-1-2.* Thomas Telford, London, 2005

21 R.S Narayanan, C.H Goodchild, *The Concise Eurocode 2*, The Concrete Centre, Surrey, 2006

22 A.M. Neville, *Properties of Concrete,* 4th Edn, Pearson Education Ltd, Essex, 2000

23 *Manual for the Design of Concrete Building Structures to Eurocode 2,* The Institution of Structural Engineers, London, 2006

24 *Standard Method of Detailing Structural Concrete*, 3rd edn, The Institution of Structural Engineers, London, 2006

25 *Early-age Thermal Crack Control in Concrete*, Guide C660, CIRIA, London, 2007

(c) Websites

26 Eurocodes Expert – http://www.eurocodes.co.uk/

27 The Concrete Centre website – http://www.concretecentre.com

Index